AF352878

# Potential Health Benefits of Fruits and Vegetables II

# Potential Health Benefits of Fruits and Vegetables II

Editors

**Luca Mazzoni**
**Maria Teresa Ariza Fernández**
**Franco Capocasa**

Basel • Beijing • Wuhan • Barcelona • Belgrade • Novi Sad • Cluj • Manchester

*Editors*

Luca Mazzoni
Department of Agricultural,
Food and Environmental
Sciences
Università Politecnica
delle Marche
Ancona
Italy

Maria Teresa Ariza Fernández
Breeding and Biotechnology
Instituto Andaluz de
Investigación y Formación
Agraria y Pesquera (IFAPA)
Malaga
Spain

Franco Capocasa
Department of Agricultural,
Food and Environmental
Sciences
Università Politecnica
delle Marche
Ancona
Italy

*Editorial Office*
MDPI
St. Alban-Anlage 66
4052 Basel, Switzerland

This is a reprint of articles from the Special Issue published online in the open access journal *Applied Sciences* (ISSN 2076-3417) (available at: www.mdpi.com/journal/applsci/special_issues/ Benefits_Fruits_Vegetables_II).

For citation purposes, cite each article independently as indicated on the article page online and as indicated below:

Lastname, A.A.; Lastname, B.B. Article Title. *Journal Name* **Year**, *Volume Number*, Page Range.

**ISBN 978-3-0365-8517-8 (Hbk)**
**ISBN 978-3-0365-8516-1 (PDF)**
**doi.org/10.3390/books978-3-0365-8516-1**

© 2023 by the authors. Articles in this book are Open Access and distributed under the Creative Commons Attribution (CC BY) license. The book as a whole is distributed by MDPI under the terms and conditions of the Creative Commons Attribution-NonCommercial-NoDerivs (CC BY-NC-ND) license.

# Contents

*Editorial*

# Potential Health Benefits of Fruits and Vegetables II

Luca Mazzoni [1,*] , Franco Capocasa [1,*] and Maria Teresa Ariza Fernández [2,*]

1   Department of Agricultural, Food and Environmental Sciences, Università Politecnica delle Marche, Via Brecce Bianche 10, 60131 Ancona, Italy
2   Breeding and Biotechnology, Instituto Andaluz de Investigación y Formación Agraria y Pesquera (IFAPA), IFAPA-Centro de Churriana, Cortijo de la Cruz, s/n, 29140 Málaga, Spain
*   Correspondence: l.mazzoni@staff.univpm.it (L.M.); f.capocasa@staff.univpm.it (F.C.); mariat.ariza@juntadeandalucia.es (M.T.A.F.)

**Citation:** Mazzoni, L.; Capocasa, F.; Ariza Fernández, M.T. Potential Health Benefits of Fruits and Vegetables II. *Appl. Sci.* **2023**, *13*, 8524. https://doi.org/10.3390/app13148524

Received: 10 July 2023
Accepted: 18 July 2023
Published: 24 July 2023

**Copyright:** © 2023 by the authors. Licensee MDPI, Basel, Switzerland. This article is an open access article distributed under the terms and conditions of the Creative Commons Attribution (CC BY) license (https:// creativecommons.org/licenses/by/ 4.0/).

Consumer awareness regarding the significance of a well-balanced diet in preventing chronic diseases has increased significantly in recent years. Particularly, the consumption of plant-based foods, including vegetables and fruits, has been proven to play a crucial role in preventing various chronic diseases due to their abundance of bioactive compounds. Numerous researchers and scientists from diverse fields have dedicated substantial efforts to the study and characterization of the phytochemical profiles of numerous fruits and vegetables. They have also elucidated multiple mechanisms and metabolic pathways through which these plant-based foods exhibit their health-enhancing and disease-preventing properties.

The objective of this Special Issue was to gather the latest research on the phytochemical composition of fruits and vegetables, as well as on their health-promoting effects and mechanisms of action. This research encompassed various models, such as in vitro cellular models, animal trials, and human trials, to provide a comprehensive understanding of the applications and benefits of fruits and vegetables in promoting health.

In this Special Issue, 13 original contributions were received, comprising of 11 Research Articles (RA) and 2 Reviews (RV). The first level of research is usually the investigation of the antioxidant potential and the pattern of bioactive compounds in fruit and vegetables, which is achieved by analyzing their composition in different conditions (genotypes, environment, cultivation system, and post-harvest management). In this regard, most of the RAs received (5 out of 11) deal with the phytochemical composition of fruits (1) and vegetables (4). The only study reporting research on fruit was from Qaderi et al. (2023), which focused on strawberries, a perishable fruit rich in vitamins and phenolic compounds. The researchers examined the effects of different cold-storage temperatures ($-20\,°C$ and $-80\,°C$) on three strawberry cultivars ('Arianna', 'Francesca', and 'Silvia'), and their various treatments (whole and dried fruits), over seven months. The goal was to evaluate how storage conditions and duration influenced the stability of nutritional compounds such as vitamin C, phenolic acids, anthocyanins, and folate in strawberries. The results showed that storage temperature significantly affected the fruit's nutritional quality, with $-80\,°C$ storage preserving more nutritional compounds compared to $-20\,°C$. However, the storage time did not substantially impact the nutritional composition. Notably, oven drying had a detrimental effect on the vitamin C content, while folate levels increased during storage. The findings underscore the importance of considering storage conditions and duration for maintaining optimal nutritional quality in strawberries, thus informing future fruit storage strategies [1].

As mentioned above, this Special Issue attracted more RAs on vegetables, with five original contributions published. The study from Di Mola et al. (2023) aimed to assess the effects of two natural biostimulants on *Diplotaxis tenuifolia* L. plants grown under different salinity levels and harvested over six consecutive cropping cycles. The availability of quality irrigation water is declining due to soil salinization and aquifer deterioration, while at the same time, climate change requires intensified cropping systems for global food security.

The combination of factors had variable effects on the tested parameters, highlighting the significance of considering growing conditions and cropping periods when using biostimulants under salinity stress in *D. tenuifolia* plants. Salinity, biostimulant application, and harvesting time affected the antioxidant activity, bioactive compounds (e.g., total phenols, carotenoids, and total ascorbic acid), and mineral profile in different ways. Increasing the salinity led to a decrease in nitrate content, whereas biostimulant application resulted in higher nitrate accumulation compared to untreated plants. Although biostimulant application showed potential in mitigating salinity stress, the response of *D. tenuifolia* plants to saline conditions and biostimulant application also depended on growing conditions and successive crop cycles [2]. The study from Biel et al. (2023) aimed to evaluate the proximate composition, total polyphenolic compound content, and antioxidant activity of 27 plant materials collected in West Pomeranian, Poland. The samples were analyzed using established methods for chemical composition and antioxidant activity assessment. The dry matter content varied among the plant materials, with black chokeberry having the lowest concentrations and milk thistle and black cumin having the highest concentrations. The total polyphenolic compound content ranged from 291.832 to 7565.426 mg of chlorogenic acid equivalent per 100 g of dry matter. The antioxidant activity was measured using different methods and exhibited a wide range of values across the plant materials. Milk thistle fruit extract showed the lowest antioxidant activity and total polyphenolic compound content, whereas extracts from garlic, stinging nettle, and cleavers had the highest. The study suggests that certain plant parts with high antioxidant potential could be valuable sources of bioactive compounds, but further research is needed to identify any potentially harmful compounds [3]. On the topic of vegetables, the study by Mezzetti et al. (2022) examined the antioxidant compounds in two Italian broccoli cultivars ('Roya' and 'Santee') and black cabbage. Different plant portions and developmental stages were analyzed. Black cabbage seeds showed higher levels of antioxidants, phenols, and anthocyanins than the leaves. Similarly, broccoli heads had higher levels than the stems. The harvest date influenced the antioxidant capacity, with the second harvest of 'Roya' broccoli showing better results. These vegetables provide valuable antioxidants and potential health benefits [4]. Another study by Janiszewska-Turak et al. (2022) focused on beetroot and red bell pepper as rich sources of active compounds and their potential health benefits. To extend their shelf life and create a new product with coloring and probiotic potential, lactic fermentation was employed as a preservation method. The impact of fermentation on the content of active compounds in pickled juices and freeze-dried powders was evaluated. *Levilactobacillus brevis* and *Limosilactobacillus fermentum* were used for fermentation. The research showed no differences in the pigment content in fermented juices, but color coefficients varied in raw juices. Freeze-drying reduced the pigment content while simultaneously increasing dry matter and providing good storage conditions. Fermentation combined with marinade yielded higher pigments and lactic acid bacteria content. All powders were stable and can be used as a colorant source, while higher bacteria levels are required for probiotic properties [5].

Besides the compositional analyses, the second step in evaluating the positive effects of the bioactive compounds present in fruits and vegetables is to extend the analysis to other bioactive characteristics, such as antimicrobial, bactericidal, and anticancer effects. Regarding the collection of RAs, three of them deal with these issues. Witbooi et al. (2021) investigated potatoes, an important cultivation for global food security, and it has been reported that pigmented potato cultivars provide health benefits. However, there is limited information on their antioxidant, anticancer, and antimycobacterial activities. This study focused on the 'Salad Blue' (SB) pigmented cultivar and non-pigmented control (BP1) extracts. Chlorogenic acid was the prominent phenolic acid in both cultivars. The extracts showed no significant activity against *Mycobacterium smegmatis*. The antiproliferative activity against HepG2 liver cells varied, and the study provides valuable information for future oncology and nutritional research to enhance the health benefits of these cultivars [6]. Also, fruits of the *Bromelia* genus have compounds with health benefits and biotechnological applications.

*Bromelia karatas* fruits, for example, contain antioxidants and proteins with bactericidal activity. However, further studies are needed to explore the activity and potential benefits of these metabolites. In this study, the bactericidal activity of the methanolic extract and its fractions from ripe *B. karatas* fruit was evaluated against several bacterial strains. The methanolic extract showed minimum inhibitory concentration (MIC) and minimum bactericidal concentration (MBC) values against the tested bacteria. Gas chromatography mass spectrometry identified 131 compounds in the extract, some of which have known biological activities such as bactericidal, fungicide, anticancer, anti-inflammatory, enzyme inhibiting, and anti-allergic properties. The most abundant compounds in the extract were maleic anhydride, 5-hydroxymethylfurfural, and itaconic anhydride. This study highlights the potential health benefits of the metabolites found in *B. karatas* fruits [7]. Then, Phong et al. (2022) isolated eight known secondary metabolites, including two isocoumarins and six coumarins, from the stems and branches of *Acer mono* Maxim. Their structures were confirmed using nuclear magnetic resonance spectroscopy and comparing the results to published reports. For the first time, the inhibitory effects of these compounds on *Escherichia coli* β-glucuronidase were evaluated using in vitro assays. Compound 1 (3-(3,4-dihydroxyphenyl)-8-hydroxyisocoumarin) displayed significant inhibitory effects against β-glucuronidase (IC50 = 58.83 ± 1.36 μM). Kinetic studies indicated that compound 1 acts as a non-competitive inhibitor. Molecular docking studies revealed that compound 1 binds to the allosteric binding site of β-glucuronidase, which was consistent with the results of kinetic studies. Additionally, molecular dynamics simulations provided insights into the dynamic properties of the protein–ligand complex formed by compound 1. These findings suggest that compound 1 could serve as a lead metabolite for developing new β-glucuronidase inhibitors [8].

The last step in the evaluation of the potential health benefits of fruit and vegetables is to test them with in vivo models, which can comprise animal trials and human trials. In this collection, we published two RAs with in vivo studies in mice, and one RA conducted in vivo in humans. The study by Azevedo et al. (2022) aimed to explore the effects of an elderberry extract (EE) on mice over a period of 29 days and evaluate its safety as a natural colorant. Twenty-four female mice were divided into four groups: control, EE12 (12 mg/mL EE), EE24 (24 mg/mL EE), and EE48 (48 mg/mL EE). The main anthocyanins detected in the extract were cyanidin-3-O-sambubioside and cyanidin-3-O-glucoside. Food and drink intake was similar among the groups, except for EE48, which consumed significantly less. Histological analysis of the liver indicated no pathological significance. The EE, particularly at doses of 24 and 48 mg/mL, significantly reduced oxidative DNA damage compared to the non-supplemented group. The elderberry extract exhibited a favorable toxicological profile, suggesting its potential use in the food industry [9]. Alegiry et al. (2022) concentrate their research on MDD, a prevalent and serious health condition that remains a global challenge, despite numerous studies and available antidepressants. *Carthamus tinctorius* (safflower) is traditionally used in food and medicine. This study aimed to investigate the chemical composition of safflower and its antidepressant-like effects using a hot water extract in male mice. The mechanism of action was explored through the transcriptomic analysis of the hippocampus. GC-MS analysis revealed that the hot water extract contained a significant amount of oleamide, which is known for its activity. Neurobehavioral tests showed that safflower treatment significantly reduced immobility time in the TST and FST, and improved performance in the YMSAT compared to the control group. RNA-seq analysis identified differential gene expression in several genes related to MDD regulation. Overall, this study demonstrated the antidepressant-like effects of safflower hot water extract, attributed to its bioactive ingredient oleamide, as evidenced by behavioral changes and gene expression patterns [10]. The human in vivo trial was conducted by Siripun et al. (2022) and investigated dyslipidemia, which is a risk factor for cardiovascular disease and a leading cause of global mortality. Lipid-lowering drugs can have side effects; therefore, consuming vegetables and fruits with probiotics is a potential alternative to positively influence plasma lipid profiles. This study aimed to investigate the effects of consuming vegetable and fruit juice (VFJ) with and without probiotic *Lactobacillus paracasei* on various parameters in dyslipidemic patients over 30 days. The probiotic group showed significantly lower levels of total cholesterol, low-density lipoprotein cholesterol, triglycerides, and the TG/high-density lipoprotein cholesterol ratio compared to the placebo group. Additionally, the probiotic group had higher levels of high-density lipoprotein cholesterol. The probiotic group also demonstrated reduced levels of malondialdehyde (a marker of lipid peroxidation), increased levels of oxidative stress enzymes (catalase and glutathione peroxidase) in plasma, and increased bile acid levels in feces. These findings suggest that VFJ enriched with probiotic *L. paracasei* may serve as an alternative approach for preventing dyslipidemia in patients who have not yet started other medication, providing a primary intervention method [11].

The last two contributions to this collection are RVs, which provide a detailed overview of two different central topics for the evaluation of the potential health benefits of fruits and vegetables: the role of different vegetal-derived bioactive compounds in the modulation of colorectal cancer, and the main characteristics and health effects of berry fruit volatiles. According to Di Mola et al. (2023), colorectal cancer is a significant cause of illness and death, and drug resistance poses a major challenge in its treatment. Bioactive compounds derived from vegetables are being investigated as a potential strategy to enhance antitumor therapies by targeting key pathways involved in carcinogenesis and multidrug resistance. In both laboratory and animal studies, these compounds have shown the ability to reduce drug resistance and enhance therapeutic effectiveness when combined with cytotoxic drugs. This review aims to summarize the existing scientific literature on the antitumor and chemo-sensitizing properties of vegetable-derived biomolecules such as polyphenols, flavonoids, and terpenes. These compounds have the potential to offer promising prospects for improving the treatment of colorectal cancer [12]. In the second RV, Gu et al. (2022) state that volatile compounds give fruits their aroma, and berries are rich in these compounds, including esters, alcohols, terpenoids, and more. This review focuses on the volatile compounds in strawberries, blueberries, raspberries, blackberries, and cranberries. These compounds have various health benefits, such as anti-inflammatory, anti-cancer, anti-obesity, and anti-diabetic effects. Monoterpenes, like linalool, limonene, and geraniol, are particularly important in berry aromas and offer several health benefits. Further research is needed to explore the bioavailability and confirm the bioactivities of the volatile compounds from berries [13].

To summarize, this collection evidences an important step forward in the consolidation and verification of the potential health benefits of fruits and vegetables, starting from a compositional point of view and moving into in vivo studies, as well as reviewing crucial aspects of their effects. This work is in continuity with the previous Special Issue "Potential Health Benefits of Fruits and Vegetables I", to be followed by a similar collection (Potential Health Benefits of Fruits and Vegetables III), which will continue to underline the importance of fruit and vegetable consumption for human health.

**Funding:** Luca Mazzoni and María Teresa Ariza Fernández were supported by the European Union's Horizon 2020 research and innovation programme under grant agreement No. 101000747, and by PON "R&I" 2014–2020 with code ARS01_01224.

**Conflicts of Interest:** The authors declare no conflict of interest.

# References

1. Witbooi, H.; Bvenura, C.; Reid, A.; Lall, N.; Oguntibeju, O.; Kambizi, L. The Potential Effect of Elevated Root Zone Temperature on the Concentration of Chlorogenic, Caffeic, and Ferulic acids and the Biological Activity of Some Pigmented *Solanum tuberosum* L. Cultivar Extracts. *Appl. Sci.* **2021**, *11*, 6971. [CrossRef]
2. Siripun, P.; Chaiyasut, C.; Lailerd, N.; Makhamrueang, N.; Kaewarsar, E.; Sirilun, S. A Pilot Study of whether or Not Vegetable and Fruit Juice Containing *Lactobacillus paracasei* Lowers Blood Lipid Levels and Oxidative Stress Markers in Thai Patients with Dyslipidemia: A Randomized Controlled Clinical Trial. *Appl. Sci.* **2022**, *12*, 4913. [CrossRef]

3. Alegiry, M.; El Omri, A.; Bayoumi, A.; Alomar, M.; Rather, I.; Sabir, J. Antidepressant-Like Effect of Traditional Medicinal Plant Carthamus Tinctorius in Mice Model through Neuro-Behavioral Tests and Transcriptomic Approach. *Appl. Sci.* **2022**, *12*, 5594. [CrossRef]
4. Janiszewska-Turak, E.; Tracz, K.; Bielińska, P.; Rybak, K.; Pobiega, K.; Gniewosz, M.; Woźniak, Ł.; Gramza-Michałowska, A. The Impact of the Fermentation Method on the Pigment Content in Pickled Beetroot and Red Bell Pepper Juices and Freeze-Dried Powders. *Appl. Sci.* **2022**, *12*, 5766. [CrossRef]
5. Mezzetti, B.; Biondi, F.; Balducci, F.; Capocasa, F.; Mei, E.; Vagnoni, M.; Visciglio, M.; Mazzoni, L. Variation of Nutritional Quality Depending on Harvested Plant Portion of Broccoli and Black Cabbage. *Appl. Sci.* **2022**, *12*, 6668. [CrossRef]
6. Ayil-Gutiérrez, B.; Amaya-Guardia, K.; Alvarado-Segura, A.; Polanco-Hernández, G.; Uc-Chuc, M.; Acosta-Viana, K.; Guzmán-Marín, E.; Samaniego-Gámez, B.; Poot-Poot, W.; Lizama-Uc, G.; et al. Compound Identification from Bromelia karatas Fruit Juice Using Gas Chromatography-Mass Spectrometry and Evaluation of the Bactericidal Activity of the Extract. *Appl. Sci.* **2022**, *12*, 7275. [CrossRef]
7. Gu, I.; Howard, L.; Lee, S. Volatiles in Berries: Biosynthesis, Composition, Bioavailability, and Health Benefits. *Appl. Sci.* **2022**, *12*, 10238. [CrossRef]
8. Phong, N.; Min, B.; Yang, S.; Kim, J. Inhibitory Effect of Coumarins and Isocoumarins Isolated from the Stems and Branches of Acer mono Maxim. against Escherichia coli β-Glucuronidase. *Appl. Sci.* **2022**, *12*, 10685. [CrossRef]
9. Azevedo, T.; Ferreira, T.; Ferreira, J.; Teixeira, F.; Ferreira, D.; Silva-Reis, R.; Neuparth, M.; Pires, M.; Pinto, M.; Gil da Costa, R.; et al. Supplementation of an Anthocyanin-Rich Elderberry (*Sambucus nigra* L.) Extract in FVB/n Mice: A Healthier Alternative to Synthetic Colorants. *Appl. Sci.* **2022**, *12*, 11928. [CrossRef]
10. Qaderi, R.; Mezzetti, B.; Capocasa, F.; Mazzoni, L. Stability of Strawberry Fruit (*Fragaria* × *ananassa* Duch.) Nutritional Quality at Different Storage Conditions. *Appl. Sci.* **2023**, *13*, 313. [CrossRef]
11. Biel, W.; Pomietło, U.; Witkowicz, R.; Piątkowska, E.; Kopeć, A. Proximate Composition and Antioxidant Activity of Selected Morphological Parts of Herbs. *Appl. Sci.* **2023**, *13*, 1413. [CrossRef]
12. Di Mola, I.; Petropoulos, S.; Ottaiano, L.; Cozzolino, E.; El-Nakhel, C.; Rouphael, Y.; Mori, M. Bioactive Compounds, Antioxidant Activity, and Mineral Content of Wild Rocket (*Diplotaxis tenuifolia* L.) Leaves as Affected by Saline Stress and Biostimulant Application. *Appl. Sci.* **2023**, *13*, 1569. [CrossRef]
13. Quiñonero, F.; Mesas, C.; Peña, M.; Cabeza, L.; Perazzoli, G.; Melguizo, C.; Ortiz, R.; Prados, J. Vegetal-Derived Bioactive Compounds as Multidrug Resistance Modulators in Colorectal Cancer. *Appl. Sci.* **2023**, *13*, 2667. [CrossRef]

**Disclaimer/Publisher's Note:** The statements, opinions and data contained in all publications are solely those of the individual author(s) and contributor(s) and not of MDPI and/or the editor(s). MDPI and/or the editor(s) disclaim responsibility for any injury to people or property resulting from any ideas, methods, instructions or products referred to in the content.

*Review*

# Vegetal-Derived Bioactive Compounds as Multidrug Resistance Modulators in Colorectal Cancer

**Francisco Quiñonero** [1,2,3], **Cristina Mesas** [1,2,3], **Mercedes Peña** [1,2,3], **Laura Cabeza** [1,2,3], **Gloria Perazzoli** [1,2,3], **Consolación Melguizo** [1,2,3,*], **Raul Ortiz** [1,2,3,†] and **Jose Prados** [1,2,3,†]

[1]  Institute of Biopathology and Regenerative Medicine (IBIMER), Center of Biomedical Research (CIBM), University of Granada, 18100 Granada, Spain
[2]  Department of Anatomy and Embryology, Faculty of Medicine, University of Granada, 18071 Granada, Spain
[3]  Instituto de Investigación Biosanitaria de Granada (ibs.GRANADA), 18014 Granada, Spain
*   Correspondence: melguizo@ugr.es; Tel.: +34-958-243535
†   These authors contributed equally to this work.

**Abstract:** Colorectal cancer is one of the leading causes of morbidity and mortality today. Knowledge of its pathogenesis has made it possible to advance the development of different therapeutic strategies. However, the appearance of drug resistance constitutes one of the main causes of treatment failure. Bioactive compounds of vegetable origin are being studied as a new strategy to improve antitumor treatment, due to their ability to regulate the pathways involved in the development of carcinogenesis or processes that are decisive in its evolution, including multidrug resistance. In vitro and in vivo studies of these substances in combination with cytotoxic drugs have shown that they reduce resistance and increase therapeutic efficacy. The objective of this review is to summarize the knowledge that is described in the scientific literature on the antitumor and chemo-sensitizing capacity of vegetable-derived biomolecules such as polyphenols, flavonoids, and terpenes. These compounds may hold a promising future in improving the treatment of colorectal cancer.

**Keywords:** colorectal cancer; vegetables; multidrug resistance; biomolecules; resistance mechanisms

**Citation:** Quiñonero, F.; Mesas, C.; Peña, M.; Cabeza, L.; Perazzoli, G.; Melguizo, C.; Ortiz, R.; Prados, J. Vegetal-Derived Bioactive Compounds as Multidrug Resistance Modulators in Colorectal Cancer. *Appl. Sci.* **2023**, *13*, 2667. https://doi.org/10.3390/app13042667

Academic Editors: Luca Mazzoni, Maria Teresa Ariza Fernández and Franco Capocasa

Received: 13 December 2022
Revised: 15 February 2023
Accepted: 17 February 2023
Published: 19 February 2023

**Copyright:** © 2023 by the authors. Licensee MDPI, Basel, Switzerland. This article is an open access article distributed under the terms and conditions of the Creative Commons Attribution (CC BY) license (https://creativecommons.org/licenses/by/4.0/).

## 1. Introduction

Colorectal cancer (CRC) accounted for 9.39% of deaths by cancers in 2020 and is the third most commonly diagnosed cancer in the word. CRC incidence could double by 2035 due to an increase in the number of cases and their early diagnosis [1]. Current treatment is based on resection combined with chemotherapy and adjuvant radiotherapy in the early stages and larger resections that are associated with chemotherapy (oxaliplatin, irinotecan, 5-fluorouracil (5-FU) and capecitabine, among others) in metastatic stages [2]. However, despite the use of different therapeutic options, the average survival in metastatic patients is low (3 years) although the 5-year survival rates of patients that are diagnosed with localized tumors or with regional dissemination are 90% and 69%, respectively [3,4]. Acquired or intrinsic drug resistance in CRC is one the main causes of treatment failure. A clinical assay that was performed with patients from 13 European countries, Israel, and South Africa showed that more than 50% of CRC patients had resistance to the drug 5-FU [5,6]. In addition, the mechanisms that generate resistance to one drug produce resistance to others (multidrug resistance or MDR). The mechanism that is mediated by the ATP-dependent transporter family (ATP-binding cassette family, ABC) is prominent in CRC, including the p-glycoprotein (p-GP; ABCB1), a drug efflux pump which prevents cellular uptake of many structurally and functionally cytotoxic compounds [7,8]. Understanding resistance phenomena is essential for improving CRC patient prognosis.

In this context, it has been shown that some plant-derived natural products or their association with the classic cytotoxic drugs used in cancer treatment were capable of overcoming multidrug resistance and/or reducing the effective antitumor drug dose [9].

Despite the low proportion of plant species that have been explored for their antitumor activity [10], plant-derived natural products have played an important role in the treatment and prevention of cancer. Most of the compounds from vegetables and fruits with significant antitumor activity also showed clear advantages in relation to modern drugs that are used in chemotherapy [11]. Furthermore, this therapeutic strategy based on the use of natural compounds showed minimal side effects and lower cost than synthesized antitumor drugs [12]. However, biodistribution, biotransformation, and transport limitations of natural products may hinder their application in cancer patients. Furthermore, their effect on the immune system is not fully analyzed [13–15]. Concretely in CRC, the therapeutic potential of some natural compounds has been clearly demonstrated [16]. In addition, their combination with other conventional chemotherapeutic drugs has demonstrated a significant synergistic effect [17]. In fact, curcumin, resveratrol, and (-)-Epigallocatechin Gallate (EGCG) and terpenoids, secondary plant metabolites that are composed of isoprene units, enhanced the effect of cytotoxic drugs [18–20]. The antitumor effect exerted by these compounds is carried out through several pathways including cell cycle arrest and cell death by apoptosis [21]. On the other hand, polyphenols from plants also prevent multidrug resistance in several types of solid tumors, including CRC. In addition, these compounds inhibit cell proliferation, angiogenesis, and metastasis, as well as regulating the proinflammatory response [22]. Finally, quercetin, a flavonoid, not only modulates the drug resistance phenomena but is capable of increasing sensitivity to doxorubicin in CRC through the inhibition of a glutamine transporter (SLC1A5) [23,24].

The objective of this review is to study the different families of plant-derived compounds that modulate drug resistance in CRC, observing their ability to be used as a future therapy against this type of cancer.

## 2. Colorectal Cancer: Resistant Mechanisms

Despite the discovery of new drugs against CRC, the emergence of resistance to these agents is inevitable. Drug resistance can be innate (dysregulations of tumor cells before treatment) or acquired (resistance after treatment cycles) [25,26]. Drug efflux mediated by transmembrane transporters, specifically those of the ATP binding cassette superfamily (ABC), is one of most relevant mechanisms in CRC. These proteins are capable of expelling toxic substances from the inside of cells including different anticancer agents [27]. Specifically, p-GP, encoded by the ABCB1 gene, was overexpressed in different CRC cell lines, conferring resistance to treatment. In addition, resistant CRC cells overexpressed CD133, a protein that regulates p-GP expression through the AKT/NF-κB/MDR1 axis [28,29]. Other members of the ABC family, such as MRP1 and BCRP, are also overexpressed in some CRC cell lines, leading to multiple resistance to chemotherapeutic drugs as 5-FU, doxorubicin, irinotecan, vincristine, among others [30,31], while MRP2-mediated resistance to oxaliplatin and vincristine in CRC, being Nrf2, signaling is critical for its expression (Figure 1) [32]. Drug resistance in CRC can also arise when there are alterations in antitumor drug targets, such as mutations or changes in expression due to epigenetic variations [33]. Finally, an increase in the expression of repair protein-DNA, such as MGMT, was detected in some 5-FU-resistant CRC lines [34].

Apoptosis evasion promotes carcinogenesis and tumor progression, leading to the appearance of pharmacological resistance, especially to drugs that induce this pathway such as doxorubicin and cisplatin. Several apoptosis-resistant tumors were associated with the increased or decreased expression of antiapoptotic (BCL-2, MCL-1, and BCL-XL) and proapoptotic (p53, BAX, and BIM) genes, respectively [35]. In addition, modulation of DNA methylation, histones and chromatin remodeling can alter the expression of genes that are involved in the metabolism and activity of chemotherapeutic drugs, inducing resistance [33]. Moreover, tumor heterogeneity also plays a role in this phenomenon, as it makes treatment more difficult because of the presence of cancer stem cells which are more resistant to drugs. These cells have a self-healing and differentiation capacity and are associated with greater tumorigenicity. These cells are also capable of acquiring mesenchymal characteristics, which is related to the cell migration process and metastasis and a worse prognosis in patients [36,37]. On the other hand, it is known that the most resistant cells within the tumor sinus can transfer small miRNAs to their environment, inducing resistance in neighboring cells [38]. Finally, another important factor to highlight is the tumor microenvironment, including the extracellular matrix, blood vessels, fibroblast, and immune system cells. This microenvironment will be an additional layer of protection against drugs, making the entry of chemotherapeutics into the tumor sinus more limited [39].

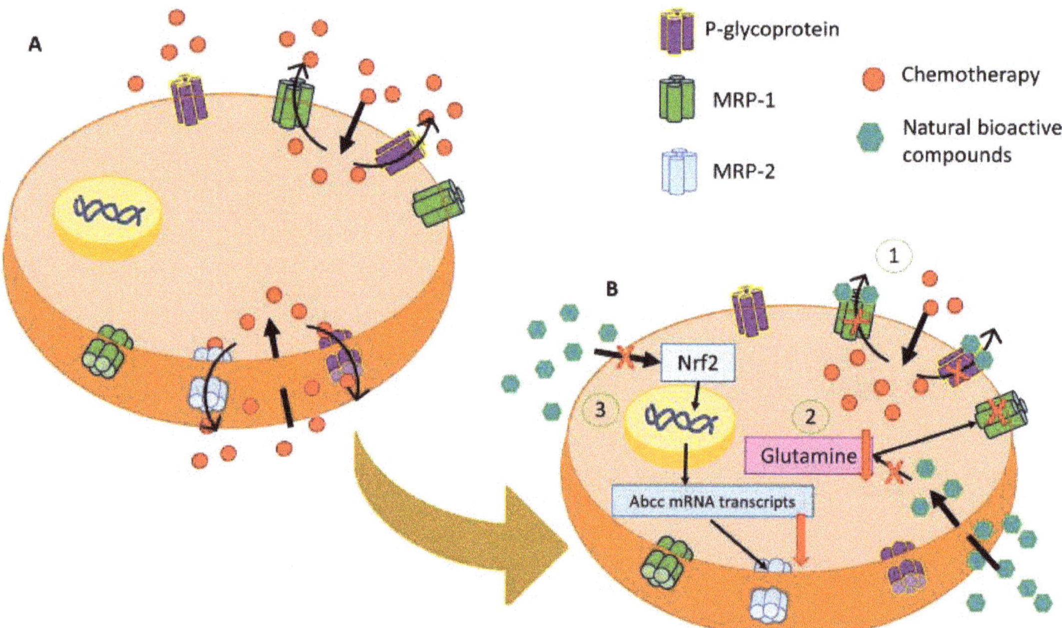


**Figure 1.** Representative image of modulation of the MDR mechanism of resistance by natural bioactive compounds. (**A**) Main membrane transporters that are involved in chemoresistance by expelling them into the extracellular medium. (**B**) Action of natural bioactive compounds in the regulation of MDR by (1) MRP-2 downregulation through the inhibition of the Nrf2/MRP2 pathway, (2) P-glycoprotein and MRP-1 blocking by natural bioactive compounds, and (3) inhibition of glutamine cell intake producing an MRP-1 disfunction.

### 3. Natural Products: A New Source of Molecules with Antitumor Activity

Many plant compounds are substances with known bioactive activity that can be used as chemotherapeutic agents in cancer. One of the best known, taxol, is a potent diterpene with antitumor activity that is derived from the bark of the yew tree (*Taxus brevifolia*). These compounds show advantages over traditional drugs, such as greater bioavailability, accessibility, cost-effectiveness, and lower systemic toxicity [40]. This last point is highly relevant since conventional drugs show many side effects [20]. In addition, these compounds show a synergy with conventional therapies and a benefit for the treatment of CRC by avoiding drug resistance [41–43]. Therefore, plant-derived natural compounds allow us to have more cancer treatment options. Polyphenols, flavonoids, and terpenes have been described as the main molecules that are able to modulate chemoresistance phenomena in CRC (Figure 2).

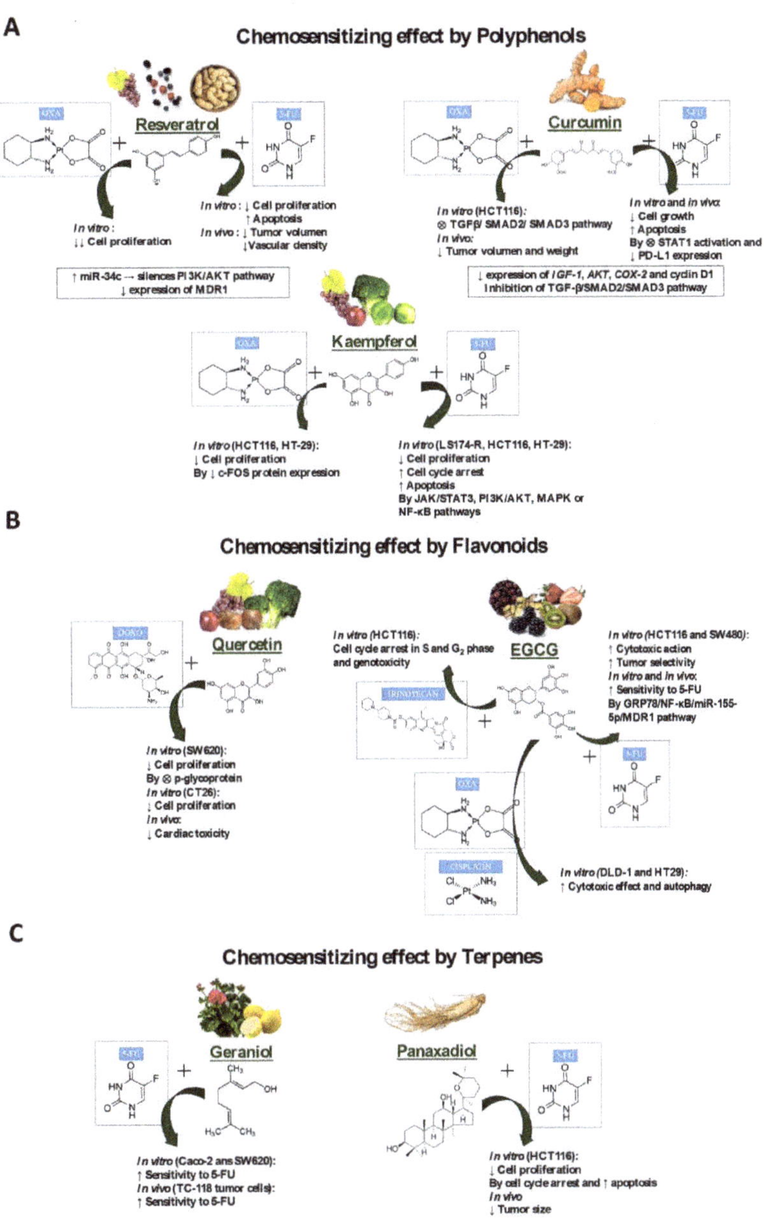

**Figure 2.** Sensitizing effect on chemotherapeutic drugs that are frequently used in CRC exerted by natural compounds of the following families (**A**) polyphenols, (**B**) flavonoids, and (**C**) terpenes, analyzed in this review. DOXO (doxorubicin); 5-FU (5-fluorouracil); OXA (oxaliplatin).

*3.1. Polyphenols*

Polyphenols (Figure 2A) comprise a heterogeneous group of phytochemicals that contain one or more phenolic rings in their chemical structure. They are divided into different groups, including stilbenes, phenolic acids, lignans, xanthones, tannins, and flavonoids [44]. These compounds are synthesized by different plants to carry out different functions, including protection against UV radiation and infection by microorganisms, as well as regulation of cell growth [45,46]. The three most studied polyphenols in CRC are resveratrol, curcumin, and kaempferol.

### 3.1.1. Resveratrol

Resveratrol is a polyphenol belonging to the stilbenes group. It is found in grapes, blueberries, and peanuts, intervening in the response to infection by microorganisms [47]. From its study it has been determined that it has antioxidant, anti-inflammatory, antitumor, cardioprotective, and neuroprotective properties. These effects are mainly based on its ability to suppress inflammation, regulate various enzymatic processes that are involved in the metabolism of numerous substances, eliminate reactive oxygen species, and inhibit cell proliferation through its action as a phytoestrogen. The therapeutic action of resveratrol comprises different mechanisms such as the induction of cell death through apoptosis and autophagy, the capacity to suppress tumor proliferation, or the inhibition of angiogenesis and metastasis [48]. The preventive and therapeutic effect has also been observed for CRC. In vitro studies that were carried out in different colon cancer cell lines have demonstrated the ability of this compound to inhibit cell growth and proliferation, promoting cell cycle stop and derived apoptosis activating apoptotic markers such as caspases 3 and 8. In the CRC line HT-29, this compound had an $IC_{50}$ of 10 mg/mL after 72 h of treatment [49]. Particularly, in HCT116 and CaCo-2 cell lines, it has been observed that resveratrol inhibits cell cycle initiation by the downregulation of the CCND1/CDK4 complex. In addition, resveratrol treatment decreases the synthesis of proangiogenic factors such as VEGF in CRC [50]. Furthermore, it has been observed that it is able to reverse the Warburg effect in tumor cells through modulation of the pyruvate dehydrogenase complex, influencing the process of glycolysis [51]. Its in vivo antitumor capacity has also been shown. Thus, in murine models of rats and mice that had the KRAS proto-oncogene overexpressed, dietary administration of 150 or 300 ppm resveratrol for 4 weeks was shown to prevent tumor growth through the deregulation of KRAS via miR-96 [52]. On the other hand, a model of colonic tumorigenesis in situ showed that resveratrol prevented the formation of precursor lesions (dysplasia or adenomas), in addition to reducing the volume of tumors if they had been generated. The downregulation of the pro-apoptotic gene BAX was observed in the treated tumor mucosa, as well as a reduction of COX-2 and NF-κB [47,53]. Human clinical trials showed that resveratrol intake was safe and well tolerated systemically, with side effects such as diarrhea, abdominal pain, and nausea being observed in patients that were treated with more than 1 g per day. Despite this, the bioavailability of resveratrol in the body is limited with 71–98% recovery of the compound or its derivatives in urine and feces after oral administration [47,54,55].

Another possibility that was studied was the combination of resveratrol together with traditional chemotherapy drugs, showing the sensitization of tumor cells to conventional drugs and the reduction of existing resistance to these drugs. Thus, its combination with 5-FU produced a decrease in cell proliferation and an increase in apoptosis in vitro, accompanied by a reduction in tumor volume and vascular density in vivo [56–58]. Synergistic effects have also been shown with oxaliplatin, observing that their combination produces a greater inhibition of cell proliferation compared to the chemotherapy alone. The effect of resveratrol is found to be exerted through increased expression of miR-34c, which silences the PI3K/AKT pathway and triggers reduced expression of the resistance gene MDR1 [59]. The death that is produced by this combination is through necrotic and apoptotic pathways [60].

Due to the low bioavailability of resveratrol, formulations have been designed to allow improved transport into the cell. Khayat et al. designed zein nanoformulations that had a high encapsulation of the plant compound and were more effective in inducing apoptosis and oxidative stress in cells compared to the free drug in HCT116, Caco-2, and HT-29 cell lines [61]. Another study designed gels that were capable of encapsulating resveratrol with high efficacy for efficient oral administration. These formulations were tested in

HCT116 cells, obtaining greater efficacy in cytotoxicity assays and significantly increasing the expression of pro-apoptotic genes such as Bcl-2 compared to free resveratrol [62]. Feng et al. synthesized lipid nanocapsules encapsulating resveratrol and showed 70% drug release at 48 h, increasing the cytotoxic effect of the free drug and significantly increasing the induced apoptosis in the HT-29 cell line [63]. Finally, other formulations encapsulating resveratrol in mesoporous silica nanoparticles and liposomes were developed. The former showed high drug loading capacity and improved efficacy against free drug in HT-29 and LS147T lines [64]. Meanwhile, liposomes showed a high drug release after 24 h and increased the cytotoxic effect of the free drug in the HT-29 line, increasing the delivery of this hydrophobic agent [65].

### 3.1.2. Curcumin

On the other hand, curcumin is one of the most studied bioactive compounds as an adjuvant in CRC. This phytochemical is derived from turmeric (*Curcuma longa*) and is used as a dietary supplement or food flavoring [66]. The mechanisms by which it influences CRC have been extensively studied. It is known to regulate the microbiome, protect the intestinal barrier and promote anti-inflammatory activity in the intestine through inactivation of the proinflammatory factor NF-κB and by increasing the expansion of CD4+ FOXP3+ regulatory T lymphocytes in the mucosa. In addition, it modulates tumor cell death via the autophagy mechanism by suppressing the PI3K/AKT/mTOR signaling pathway and it is able to trigger epigenetic modifications that increase its chemosensitizing power together with other drugs [67]. Pharmacokinetic and toxicity studies showed that its administration in humans is safe, although its absorption is low and it is a compound that is rapidly eliminated by the body, problems that currently prevent it from being tested as a therapeutic agent [68].

The chemopreventive action of curcumin in CRC has been demonstrated in numerous in vitro studies. In the HT116 cell line, it inhibits proliferation by blocking the WNT/β-catenin pathway that is mediated by c-MYC, producing cell cycle arrest in G2/M phase, and inducing apoptosis. In the HT-29 cell line, curcumin is also able to stop the cycle in the G1 phase and induce apoptosis by acting on the pAKT-AMPK-COX-2 signaling pathway at a dose of 20 μM [69]. In addition, curcumin has been shown to inhibit the migration and invasion of HT116 and LoVo cell lines [70]. In vivo studies that were conducted in murine models have shown that a dose of 300 mg/kg prevents progression to cancer from precancerous lesions, reducing the number of adenomas and tumor size [71,72]. Clinical studies that were conducted in humans observed that oral consumption of curcumin in CRC patients reduced serum TNF-α levels and increased apoptosis in the tumor tissue tested [73].

Curcumin has also been studied as an adjuvant and chemosensitizer in combination with drugs that were traditionally used in the treatment of CRC. Several in vitro and in vivo studies have demonstrated the ability of curcumin to decrease tumor cell resistance to drugs such as 5-FU and oxaliplatin. Particularly, in vitro studies indicate that this compound decreases the expression of several genes (IGF-1, AKT, COX-2, and cyclin D1) that are involved in resistance to these two drugs [73]. It has also been shown, in oxaliplatin-resistant HCT116 cell lines that were cultured with this chemotherapeutic and curcumin, that resistance is reversed through inhibition of the TGF-β/SMAD2/SMAD3 pathway. In mice that were transplanted with oxaliplatin-resistant HCT116 cells, the administration of curcumin together with oxaliplatin has also been observed to significantly reduce tumor volume and weight [74]. The combination of a slightly cytotoxic dose of curcumin (10 μM) with 5-FU (5 μM) produced a synergistic antitumor response, reducing cell growth, and inducing apoptosis as observed in in vitro and in vivo studies. This chemo-sensitizing effect is carried out by inhibiting STAT1 activation, which reduces PD-L1 expression [75]. Furthermore, this synergism could also lead to a decrease in CRC progression and metastasis, thanks to the ability of curcumin to decrease the expression of IGF-1 and the myc oncogene [76].

As curcumin is one of the most studied natural compounds, studies have attempted to encapsulate it to increase its bioavailability and absorption. Han et al. designed a nanoformulation based on fusion proteins that were capable of being efficiently introduced into cells with high expression of epidermal growth factor receptor (EGFR), being more effective in the HCT116 line (EGFR-positive) versus SW620 (EGFR-negative) [77]. Wang et al. encapsulated the compound in micelles and increased its bioavailability and cytotoxic activity in CRC cell lines. Moreover, after simulated in vitro digestions, 40% of the formulation was still effective against tumor cell lines, thus possessing potential for the prevention and treatment of this type of cancer [78]. Although this drug was encapsulated in other types of nanoformulations such as PLGA (poly lactic co-glycolic acid)-coated nanoparticles [79] or in β- cyclodextrins [80] showing efficacy, an interesting study that was carried out by Gupta et al. used poly(allylamine)/eudragit nanoparticles that were capable of encapsulating doxorubicin and curcumin with high efficiency. The combination of both drugs in the nanoformulation showed high cytotoxicity in CRC HCT116 cells and in Balb/C murine models, with retention of these nanoformulations in the colon region 24 h after treatment [81].

Therefore, the findings that were observed in the multiple studies employing the use of curcumin and resveratrol in CRC demonstrate their great capacity as chemopreventive and sensitizing agents in combination with other cytotoxic drugs. However, it is still essential to continue studying these compounds to be able to introduce them into clinical practice in the future.

### 3.1.3. Kaempferol

Kaempferol is a flavanol that is present in vegetables and plants such as broccoli, grapes, apples, brussels sprouts, or black tea. This bioactive compound possesses cardioprotective, neuroprotective, anti-inflammatory, antidiabetic, antioxidant, and antitumor properties [82]. The mechanisms of action underlying the anticancer effect of kaempferol are based on inhibition of the cell cycle in the G2/M phase to prevent tumor growth and proliferation, as well as induction of apoptosis by acting on different cell signaling pathways. Several in vitro studies have also observed its anti-angiogenic effect and anti-metastatic, inhibiting VEGF production and decreasing the expression of markers that are involved in epithelial-mesenchymal transition (EMT) at a dose of 0.1 μM in MDA cell lines [83,84]. Pharmacokinetic studies that were performed in rats determined that the bioavailable fraction of this compound after oral doses was 2% of the total administered [85]. To overcome this obstacle, the combination with quercetin has been tested, resulting in an increase in its bioavailability, in addition to improving efficacy [82]. In this way, this bioactive compound would act selectively on tumor cells, without affecting healthy ones. This compound exhibits cytotoxic effects on various CRC cell lines, such as HCT116, HT-29, HCT-15, LS174-R, and SW480, both when used alone and in combination with other chemotherapeutic agents. In HT-29 cell lines, the ability of this compound to inhibit the cell cycle has been studied, resulting in cell cycle arrest in G1 and G2/M phases through inhibition of CDK2, CDK4, and Cdc2 activity [86].

In vitro studies in the 5-FU-resistant LS174-R cell line and in other CRC lines such as HCT116 and HT-29 have shown that the antiproliferative activity that is exerted by the combination of kaempferol with this chemotherapeutic agent is through cell cycle arrest and induction of apoptosis, acting on signaling pathways such as JAK/STAT3, PI3K/AKT, MAPK, or NF-κB [87,88]. Other studies that were carried out in oxaliplatin-resistant HCT116, and HT-29 cell lines have shown that the combination of kaempferol with this drug generates a greater cytotoxic response, decreasing cell proliferation. The chemosensitization of these cells is mediated by the suppression of c-FOS protein expression [89].

Although not many nanoformulations have been synthesized to improve the bioavailability of kaemferol in the treatment of CRC, Meena et al. designed PEGylated gold nanoparticles that efficiently co-encapsulated DOX and the natural compound, improving the cytotoxic effect of both drugs separately and inducing apoptosis in the HT-29 line, without having high toxicity in the HTB-38 non-tumorous colorectal line. Finally, an in vivo experiment that was performed on mice showed a significant reduction in tumor volume without observing side effects [90].

Data on the use of kaempferol in vivo are still scarce, so it is imperative to continue its research to better understand its anticancer potential.

### 3.2. Flavonoids

Flavonoids represent a broad group of bioactive compounds from vegetables and plants. Although they have been catalogued as a subtype of polyphenols, as they are made up of a great variety of compounds, they are studied separately. The chemical structure of these substances is made up of a carbon skeleton with two aromatic rings joined by three carbons, with different modifications depending on the compound [43]. The subgroups that make up this family are chalcones, flavanols, flavones, flavanones, flavanols, isoflavones, and anthocyanins. Plants synthesize these extracts to provide color and aroma, for their protective properties against radiation or microbial infections, and their high detoxifying power [66]. There are numerous benefits of these compounds against various diseases, acting as antioxidants, anti-inflammatory, antibacterial, or antiviral agents, improving cognitive functions or preventing the onset of cancer and cardiovascular diseases [91]. Quercetin and EGCG (epigallocatechin gallate) are some of the most representative molecules of this group of compounds (Figure 2B).

### 3.2.1. Quercetin

Quercetin is the most representative bioactive compound of the flavanols group, and the one with the greatest presence in foods consumed daily. It can be found in onions, apples, grapes, broccoli, or tea [66]. This compound has been shown to be an excellent antioxidant and anti-inflammatory in vitro, also presenting an antitumor and antimicrobial effect [92]. Numerous in vivo and in vitro studies have analyzed the mechanisms of action by which quercetin exerts its antitumor effect. It mainly acts by regulating the cell cycle, inhibiting cell proliferation and growth through the modulation of different molecular pathways such as PI3K/AKT/mTOR and MAPK/ERK1/2. The cytotoxic effect of this compound has been shown in CRC lines HCT-15 and RKO at doses up to 300 $\mu$M [93] and with $IC_{50}$ of 50, 100, and 50 $\mu$M in CRC lines CT26, MC38, and HT29 after 24 h of treatment, respectively [94]. The combination of quercetin with other bioactive compounds, such as curcumin, has also been shown to enhance the antiproliferative effect in different types of cancer by modulating the Wnt/$\beta$-catenin signaling pathway as observed in in vitro studies [95]. This compound also induces cell death through autophagy and apoptosis of tumor cells by increasing the expression of pro-apoptotic proteins and reducing the expression of anti-apoptotic proteins. In addition, it possesses anti-angiogenic and anti-metastatic activity, inhibiting VEGF protein expression and the EMT process increasing E-cadherin and decreasing mesenchymal markers such as N-cadherin, vimentin, and Snail. Lastly, in CRC quercetin inhibits tumor invasion and migration by regulating the expression of matrix metalloproteinases (MMPs) [92]. In in vivo studies, it is involved in reducing tumor size, reducing the number of precancerous lesions, suppressing metastasis, and reducing resistance to chemotherapy drugs [96].

This compound has also been shown to synergize with doxorubicin. Thus, a study that was carried out in the SW620 cell line observed that quercetin improves the cytotoxic activity of doxorubicin acting at the p-GP level [24]. On the other hand, encapsulation of the compound and its use together with doxorubicin increases cell growth inhibition in the CT26 cell line, reducing cardiac toxicity in murine models [97].

Although quercetin has not been generally encased in formulations, Wen et al. encapsulated this compound in a film containing chitosan nanoparticles and showed it to be an effective delivery system in CRC. It produced an effective cytotoxic effect on the Caco-2 line, exerting its effect through cell cycle arrest in G0/G1 phase and inducing apoptosis [98].

### 3.2.2. Epigallocatechin Gallate

Epigallocatechin gallate (EGCG) belongs to the flavanol group and is the most studied compound of this group in CRC. It is found mainly in green tea but is also found in other foods such as pistachios, strawberries, kiwis, hazelnuts, blackberries, and cherries. This compound is characterized by its antitumoral properties, acting as a chemopreventive agent through the inhibition of carcinogenesis in numerous types of cancer. The mechanisms that are used by this catechin to carry out its antitumor effect are based on the regulation of various signaling pathways that are involved in proliferation, apoptosis, angiogenesis, and metastasis. Among them, EGCG inhibits colonic tumor cell proliferation and migration inducing apoptosis through the activation of caspase-3 and PARP, in addition to produce the downregulation of STAT3. This compound was tested on CRC lines SW480, SW620, and LS411N and $IC_{50}$ values of 74.6, 99.4, and 112.1 µg/mL were obtained after 24 h of treatment, respectively [99,100].

Its chemo-sensitizing activity has been studied in CRC cells that are resistant to chemotherapeutics. The combination of EGCG with 5-FU increases the efficacy of the latter, as well as its cytotoxic action, acting on tumor target cells. This effect has been observed in HCT116 and SW480 cell lines [101,102]. One of the mechanisms that is used by EGCG to enhance sensitivity to 5-FU is the inhibition of the GRP78/NF-κB/miR-155-5p/MDR1 pathway, having been demonstrated in in vitro and in vivo studies [103]. Furthermore, in vivo experiments demonstrated that administration of 30 mg/kg EGCG for two weeks decreased the number of hepatic metastatic lesions, reduced tumor growth, and increased apoptosis in tissues. In addition, less vascularization was found in treated tumors versus controls [104].

EGCG also has synergistic effects in combination with irinotecan in HCT116 cell lines. It has been observed that this drug produces S- and G2-phase arrest, inducing genotoxicity, an effect that was not observed in cells that were treated with EGCG alone [18]. Cisplatin and oxaliplatin are two chemotherapeutics that are used in the treatment of CRC. The combination of EGCG with these two agents in DLD-1 and HT-29 cell lines has shown an increased cytotoxic effect, causing cell proliferation inhibition and inducing autophagy [105]. These studies suggest that the combination of EGCG with different chemotherapeutics in CRC produce a synergism that increases antitumor activity.

To improve the bioavailability of this compound, Wang et al. synthesized gelatin and chitosan nanoparticles encapsulating 5-FU and ECGC, the latter compound inhibited tumor growth through its anti-angiogenic action and its ability to induce apoptosis. Encapsulation in this formulation allowed both compounds to increase blood circulation time in in vivo experiments, making them effective formulations against CRC [106].

A clinical trial was carried out with green tea extract, whose main bioactive compound was EGCG, trying to show whether the administration of 150 mg twice a day was able to reduce the risk of CRC, with no significant differences between the treatment group and the placebo [107]. Moreover, the chemo-sensitizing effect of EGCG is a breakthrough in overcoming tumor resistance. However, more evidence is needed to translate these results to clinical assays.

### 3.2.3. Other Flavonoids of Interest

In addition to the flavonoids that were analyzed above, which are the most studied as phytochemicals in CRC, there are other compounds that are derived from citrus fruits that show interest as active compounds against cancer. Among them, narangenin is a compound that is extracted from thyme (*Thymus vulgaris* L.), a shrub that grows in regions all over the world, especially in Mediterranean regions. Efficacy of this compound has been shown in CRC lines such as SW1116 and SW837, showing $IC_{50}$ at 24 h of 1 and 1.55 mM, respectively, while in the non-tumor line CRL1554 it did not reach an $IC_{50}$ dose at the 4 mM dose [108]. Another study showed that naringenin exerted its anti-tumor effect by deregulating cyclin D1 expression, leading to cell cycle arrest in HCT116 and SW480 lines [109]. This compound also showed an in vivo chemopreventive effect against the induction of precancerous lesions, reducing processes such as lipid peroxidation, ROS formation, and the activation of proinflammatory pathways in Wistar rat models [110]. Despite its interesting effects, this compound has low water solubility, cell permeability, and bioavailability. Therefore, Shabad et al. developed nanogels that were capable of increasing its dissolution capacity, obtaining a higher toxicity in in vitro models [111]. Another studied flavonoid is aromadendrin, derived from mandarin molasses. A study in the LoVo cell line and a DOX-resistant derivative (LoVo/Dx) showed that its effect is very limited in this type of cancer, producing a low cytotoxic effect [112]. Finally, another citrus-derived compound (tangeretin, present in the peel of several citrus fruits) showed synergistic activity with the chemotherapeutic drug 5-FU through early induction of oxidative stress and apoptosis in the HCT116 cell line. Bai et al. demonstrated that this synergistic effect occurs due to the natural compound that is able to decrease the expression of miR-21, whose expression increases after treatment with 5-FU, rescuing PTEN expression and inducing cellular autophagy [113].

### 3.3. Terpenes

Terpenes (Figure 2C) are another large group of bioactive compounds, many of which have CRC effects. These substances are part of the leaves, fruits, flowers, or roots of many plants, giving them a characteristic odor. When these compounds are oxidized, they become terpenoids, some of them being limonene, vitamin A, or $\beta$-carotene [42]. The study of these compounds as resistance modulators in CRC is not so recent or extensive, however, some evidence of the potential antitumor effect has been found in some of them.

### 3.3.1. Geraniol

Geraniol is a monoterpene that is found mainly in essential oils of aromatic plants and used in perfumes. It has been used as an active ingredient in many drugs since it exhibits analgesic and anti-inflammatory activity [114]. This compound presents preventive and therapeutic activity in many types of cancer, carrying out its effects by acting on the regulation of different signaling pathways such as PI3K/AKT/mTOR, MAPK/ERK1/2, or NF-κB, and modulating the expression of different molecules such as cyclins, CDKs, interleukins, and different growth factors [115].

In the Caco-2 cell line, it has been observed that this compound triggers the inhibition of cell proliferation by inducing an S-phase cycle arrest ($IC_{30}$ = 200 μM after 7 days). In addition, it was shown that it is able to induce apoptosis in in vivo models, showing a decrease in the expression of the anti-apoptotic protein BCL-2 in tumor tissue after treating mice with a dose of 250 mg/kg for 4 weeks. Its oral administration in in vivo models prevents the development of CRC, reducing the number and size of precursor lesions [116].

Furthermore, its chemosensitizing effect was observed in Caco-2 and SW620 cell lines, and in murine models originated with 5-FU-resistant TC-118 tumor cells, having shown that co-treatment sensitizes tumor cells to 5-FU [117,118]. However, these studies are scarce, which requires further analysis to understand the effects of geraniol on resistance phenomena.

### 3.3.2. Ginsenosides

Ginsenosides are a subgroup of bioactive chemical compounds that are found in the root of ginseng, an Asian plant. These substances are used in teas, in the preparation of creams, or capsules. Of all the compounds that make up this subgroup, panaxadiol is one of the most notable for its antitumor action in many types of cancer. In the CRC cell line HCT116, this terpene inhibits cell proliferation ($IC_{50}$ lower than 10 μM after 72 h of treatment) by suppressing PD-L1 expression; modulating different cellular pathways such as mTOR, MAPK/ERK, or JAK-STAT; and molecules such as HIF-1α. It has also been observed to decrease VEGF levels to exert antiangiogenic action. This has also been shown in studies that were performed on murine models treating mice at a dose of 30 mg/kg every 2 days for 3 weeks [119,120].

Another of the best known ginsenosides is Rg3. A study demonstrated that Rg3 was able to inhibit cell proliferation ($IC_{50}$ between 100–200 µM after 48 h of treatment) and promote apoptosis in the HT-29 tumor line, decreasing its pluripotency and decreasing its angiogenesis through deregulation of the AMPK molecular pathway [121]. Analysis of its in vivo activity showed that its treatment with 25 mg/kg of Rg3 for 12 consecutive days reduced tumor vascularization and increased the toxicity that was produced by the chemotherapeutics 5-FU and oxaliplatin, allowing evasion of chemoresistance to both drugs [122].

The study of panaxadiol in the HCT116 cell line resistant to 5-FU has demonstrated the ability to inhibit proliferation producing synergy in combination with this chemotherapeutic compound, stopping the cell cycle and inducing apoptosis. Its sensitizing effect has also been analyzed in mouse models combining 30 mg/kg of 5-FU and 15/30 mg/kg of panaxadiol, confirming the results that were obtained in vitro, where a significant reduction in tumor size has been observed [123]. Another in vitro study demonstrated that panaxadiol was synergistic with irinotecan in HCT116 and SW480 tumor lines [124]. On the other hand, Rg3 showed synergy with 5-FU both in vitro and in vivo in SW620 and LOVO lines, having shown suppression of proliferation and tumor development and metastasis generation through PI3K/Akt activation [125].

Among all the terpenes that were studied, Rg3 has been the most encapsulated in formulations to increase its biological availability. Qiu et al. synthesized nanoformulations of Rg3 encapsulated in pH-sensitive poly (ethylene glycol) that increased the blood circulation of the compound showing rapid release and enhanced cytotoxicity against free drug in SW480, SW620, and CL40 tumor cells. Meanwhile, these formulations did not possess toxicity in the non-tumor line CCD-18Co at the concentrations that were tested. Meanwhile, in vivo experiments showed a greater reduction of tumor volume versus the free compound, increasing apoptosis in the tumor sinus [126]. Finally, Sun et al. synthesized folato-targeted polyethylene glycol nanoparticles based on cyclodextrins that co-encapsulated Rg3 and quercetin. The microtumor environment modulating effect that was demonstrated by Rg3 together with the oxidative stress inducing ability of quercetin exhibited a high cytotoxic effect on CRC lines CT26 and HCT116. The in vitro effect was reflected in in vivo experiments, where increased survival of mice was shown in combination with an anti-PD-L1 drug [127].

Rg3 was tested in clinical trials as a possible hepatocellular carcinoma treatment (NCT02724358 and NCT01717066), although it was not tested for CRC therapy as there is no sufficient knowledge of this compound in this type of carcinoma.

## 4. Discussion

Chemotherapeutic treatment of colon cancer has been compromised mainly by the appearance of resistance, which reduces therapeutic efficacy and leads to a lower cure rate and worse prognosis. These can be produced by the existence of previous mutations in genes that are involved in resistance, by the activation of cellular pathways that are involved in cellular detoxification, and the existence of transmembrane transporters that expel the drugs to the exterior (such as p-glycoprotein) [128]. In addition to this protein, other members of the transmembrane protein family such as MRP1 and BCRP are also overexpressed in CRC, and a relationship has been observed between their expression and resistance phenomena against drugs that are frequently used in this type of tumor, such as 5-FU and doxorubicin [30,31]. It has long been shown that plant compounds can be used as a therapy for different types of cancer. Thus, a plant-derived compound such as taxol has been used for years as a chemotherapeutic agent and is currently used as a therapy in non-small cell lung cancer (NSCLC), breast, pancreatic, and cervical cancer [129]. These types of compounds can exert their actions at the molecular level through processes such as the regulation of oxidative stress or epigenetic modification in cells [130] (Figure 3).

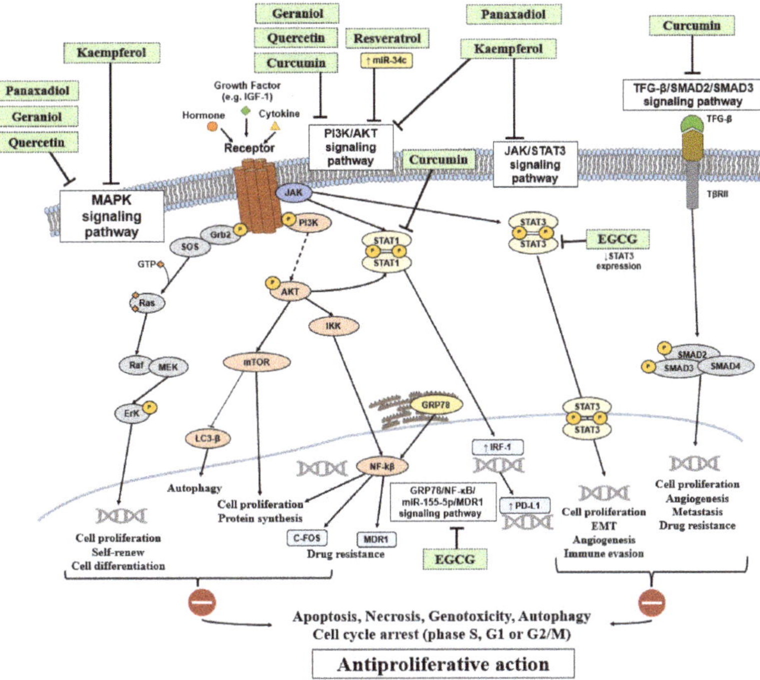


**Figure 3.** Main signaling pathways that were altered after treatment with several of the natural compounds that were analyzed in this review. Most pathways are linked to multidrug resistance processes and their inhibition leads to cell death processes that are induced by apoptosis, necrosis, autophagy, or mediated by genotoxicity.

It has been shown that the most studied compounds of plant origin in CRC are polyphenols, specifically resveratrol and curcumin (Table 1). Resveratrol belongs to the stilbenes group and has been shown to possess high antioxidant and antitumor activity, causing cell death through the induction of apoptosis and autophagy. Similar to resveratrol, the other polyphenols that were studied induce death through these pathways [48,67,73,83].

Polyphenols have been classified as chemopreventives and chemotherapeutics, although their sensitizing effect has also been observed in in vitro and in vivo models. These compounds can modulate drug resistance by increasing drug internalization into the cell, decreasing enzymes that are responsible for drug degradation (such as glutathione-S-transferases and cytochromes) and reducing the expression of transmembrane detoxifying proteins in the cell. In addition, they can induce apoptosis, oxidative stress damage, and inhibit metastasis-triggering processes such as EMT [131]. The alteration of all these resistance mechanisms implies that these polyphenols have been shown to be effective in combination with traditional drugs such as 5-FU or oxaliplatin [56,73,75,87–89]. This effect is also observed in foods with high polyphenol contents such as the strawberry tree honey, that chemosensitizes the drug 5-FU in colon cancer lines such as HCT116 and LoVo [132]. In addition to this, it has been observed that curcumin and resveratrol had the capacity to inhibit tumor proliferation in in vivo models and prevented the formation of tumor precursor lesions, with an increase in apoptosis of induced tumor tissues [47,53,76]. Pharmacokinetic studies have been carried out in humans, where low bioavailability has been observed [47,52,54,71,75]. Given the low bioavailability of these compounds, the use of nanotechnology for their encapsulation could help stabilize the compound and prevent its degradation in blood. In addition, the use of nanotechnology allows specific targeting of tumor cells through their functionalization with antibodies or peptides [14,15].

**Table 1.** Summary of the in vitro and in vivo effects that were exerted by the bioactive natural compounds that were analyzed.

| Fam. | Comp. | Sinergy | In Vitro | In Vivo | Clinical Trials | Refs. |
|---|---|---|---|---|---|---|
| Polyphenols | Resveratrol | DOX, 5-FU, OXA DOX, 5-FU, OXA DOX, 5-FU, OXA 5-FU, OXA | Apoptosis and decreased AC (VEGF inhibition). $IC_{50}$ of 10 mg/mL in HT-29 (72 h). | 150 or 300 ppm doses prevented cancerous lesions and induced apoptosis via BAX | Safe intake up to a 1g/day dose. Limited BAV (2–29%). | [20,47,49,50,52, 54,55,57,58] |
| | Curcumin | | G1-CCA and apoptosis at 20 µM. Reduced cell migration and invasion. | 300 mg/kg dose prevented precancerous lesions and decreased tumor size. | Low absorption and BAV. Increased tumor apoptosis. | [69–76,81] |
| | Kaempferol | | G1 and G2/M-CCA and apoptosis induction. Decreased AC at 0.1 µM in MDA cell line. | Decreased AC and MC. | Low toxicity and BAV (2%). | [82–90] |
| Flavonoids | Quercetin | DOX 5-FU, IRI, CPT, OXA | CCA, apoptosis and decreased AC and MC. CYT in up to 300 µM doses in HCT-15, RKO, CT26, MC38, and HT29 cells | 10 and 50 mg/kg reduced precancerous lesions and tumor size. Decreased MC and CHR. | Not performed. | [24,92–97] |
| | EGCG | | S and G2-CCA, apoptosis, CYT $IC_{50}$ between 74.6 and 112.1 in CRC cell lines SW480, SW620, and LS411N. Decreased MC. | 30 mg/kg for 2 weeks decreased MC, tumor growth and induced apoptosis | 150 mg twice a day of green tea extract did not show any effect in CRC development risk | [18,99–107] |
| Terpenes | Geraniol | 5-FU | S-CCA. CYT $IC_{30}$ of 200 µM for 7 days treatment in Caco-2 cell line. | 250 mg/kg for 4 weeks prevented CRC precancerous lesions. Reduced tumor growth and apoptosis induction. | Not performed. | [116–118] |
| | Panaxadiol | 5-FU, IRI | CYT $IC_{50}$ lower than 10 µM in HCT116 cell line (72 h). Decreased AC (VEGF inhibition) | 30 mg/kg for 3 weeks reduced tumor growth and AC. | Not performed. | [119,123–125] |
| | Rg3 | 5-FU | CYT $IC_{50}$ 100–200 µM in HT-29 cell line (48 h). Induction of apoptosis and AC via AMPK dysregulation. | 25 mg/kg for 12 days reduced tumor vascularization and decreased CHR to 5-FU and OXA. | Not performed. | [121,122,125] |

AC (antiangiogenic capacity); BAV (bioavailability); CCA (cell cycle arrest), Comp. (Compound); CPT (cisplatin); CHR (chemoresistance); CRC (colorectal cancer); CYT (cytotoxicity); DOX (doxorubicin); EGCG (epigallocatechin gallate); EMT (epithelial-mesenchymal transition); Fam. (Family); 5-FU (5-fluorouracil); IRI (irinotecan); MC (metastatic capacity); OXA (oxaliplatin); Refs. (references); VEGF (vascular endothelial growth factor).

On the other hand, it has been observed that flavonoid compounds also have anti-carcinogenic potential. Among them, the most investigated in CRC have been quercetin and EGCG. These compounds exerted their cytotoxic effect by producing cycle arrest, inhibiting key pathways in tumor development such as PI3K/AKT/mTOR and the MAPK pathway, and inhibiting processes that are linked to tumor progression such as cell migration [93,99,100]. In addition, they showed a great chemosensitizing capacity in combination with traditional drugs such as doxorubicin, irinotecan, 5-FU, cisplatin, and oxaliplatin under in vitro and in vivo conditions. A study that was conducted by Hassanein et al. [133] showed the chemopreventive effect of EGCG administration together with sulindac, a non-steroidal anti-inflammatory drug, showing that it was able to decrease the production of neoplastic lesions in in vivo models of CRC. As polyphenols, these compounds have a low bioavailability, which is a limitation for their use in humans [18,24,87–89,97,101–103,105]. In this context, the synthesis of gold NPs encapsulating EGCG has been shown to be an effective therapy against tumor cells while it has been shown that the co-encapsulation of EGCG and 5-FU in NPs allows an increase in the effect of both compounds separately, producing an anti-angiogenic and pro-apoptotic effect [106,134]. The encapsulation of this compound in nanoformulations would increase the half-life of the compound in serum, increasing its bioavailability and increasing its antitumor effect [135].

The bibliographic analysis showed that terpenes are the least studied bioactive compounds as chemosensitizers in this type of cancer. Among this family, geraniol and ginsenosides have been the most studied compounds with sensitizing properties in CRC [114,117,118]. Due to the small number of studies that have been conducted on these compounds, it is complicated for these results to be transferred to clinical studies at present. These compounds exert their antiproliferative activity by producing cell cycle arrest and inducing apoptotic pathways. In addition, it has been observed that they suppress pathways essential for tumor development such as PI3K/AKT/mTOR [115,119,120]. Results in in vivo models showed that geraniol sensitized tumors that were induced in mice from a 5-FU-resistant CRC line to the drug, while the major ginsenoside (Rg3) showed synergy with 5-FU in tumors that were generated from the SW620 and LoVo cell lines. However, it has been shown that these compounds have clear preventive and therapeutic properties in CRC [115,116,123].

Therefore, after reviewing the existing literature, it is necessary to continue investigating the antitumor properties and possible chemosensitizing actions of these compounds, trying to transfer these results to future clinical trials in humans. In addition, it is important to study their bioavailability, trying to limit their elimination in blood using these natural compounds encapsulated in nanoformulations.

## 5. Conclusions

The bioactive compounds of plant origin that were described in this review have been shown to have therapeutic and chemosensitizing action in vitro and in vivo. However, their low bioavailability in the human body presents a serious limitation for their application in therapy. Future larger studies including clinical trials and the development of future effective nanoformulations to increase their bioavailability will be necessary to determine its real utility in improving the treatment of CRC.

**Author Contributions:** Conceptualization, J.P. and C.M (Consolación Melguizo).; methodology, F.Q., C.M. (Cristina Mesas) and M.P; software, L.C.; validation, L.C. and G.P.; formal analysis, F.Q.; investigation, F.Q., C.M. (Cristina Mesas) and M.P.; data curation, G.P.; writing—original draft preparation, F.Q., C.M. (Cristina Mesas) and M.P.; writing—review and editing, L.C., G.P. and R.O.; visualization, C.M. (Consolación Melguizo) and R.O; supervision, C.M (Consolación Melguizo), R.O. and J.P.; funding acquisition, J.P. and C.M (Consolación Melguizo). All authors have read and agreed to the published version of the manuscript.

**Funding:** This work was partially supported by the Project P20_00540 (Proyectos I+D+i Junta de Andalucía 2020), PYC20 RE 035 and P18-TP-1420 (Junta de Andalucía), PI19/01478 (Instituto de Salud Carlos III, Innbio INB-009 (Granada University and ibs. GRANADA) and RTC2019-006870-1 (Spanish Minsistry of Science, Innovation and Universities). FQ and MP acknowledges the FPU (2019 and 2020) grant from Ministerio de Educación, Ciencia y Deporte y Competitividad (Spain).

**Institutional Review Board Statement:** Not applicable.

**Informed Consent Statement:** Not applicable.

**Data Availability Statement:** Not applicable.

**Acknowledgments:** We thank Instrumentation Scientific Center (CIC) from University of Granada for technical assistance.

**Conflicts of Interest:** The authors declare no conflict of interest.

## References


1.  Hossain, M.S.; Karuniawati, H.; Jairoun, A.A.; Urbi, Z.; Ooi, D.J.; John, A.; Lim, Y.C.; Kaderi Kibria, K.M.; Mohiuddin, A.K.M.; Ming, L.C.; et al. Colorectal Cancer: A Review of Carcinogenesis, Global Epidemiology, Current Challenges, Risk Factors, Preventive and Treatment Strategies. *Cancers* **2022**, *14*, 1732. [CrossRef]
2.  Cervantes, A.; Adam, R.; Roselló, S.; Arnold, D.; Normanno, N.; Taïeb, J.; Seligmann, J.; de Baere, T.; Osterlund, P.; Yoshino, T.; et al. Metastatic Colorectal Cancer: ESMO Clinical Practice Guideline for Diagnosis, Treatment and Follow-Up. *Ann. Oncol.* **2023**, *34*, 10–32. [CrossRef] [PubMed]
3.  Wu, C. Systemic Therapy for Colon Cancer. *Surg. Oncol. Clin. N. Am.* **2018**, *27*, 235–242. [CrossRef] [PubMed]
4.  Dekker, E.; Tanis, P.J.; Vleugels, J.L.A.; Kasi, P.M.; Wallace, M.B. Colorectal Cancer. *Lancet* **2019**, *394*, 1467–1480. [CrossRef] [PubMed]
5.  Van der Jeught, K.; Xu, H.C.; Li, Y.J.; Lu, X.B.; Ji, G. Drug Resistance and New Therapies in Colorectal Cancer. *World J. Gastroenterol.* **2018**, *24*, 3834–3848. [CrossRef]
6.  Douillard, J.Y.; Cunningham, D.; Roth, A.D.; Navarro, M.; James, R.D.; Karasek, P.; Jandik, P.; Iveson, T.; Carmichael, J.; Alakl, M.; et al. Irinotecan Combined with Fluorouracil Compared with Fluorouracil Alone. as First-Line Treatment for Metastatic Colorectal Cancer: A Multicentre Randomised Trial. *Lancet* **2000**, *355*, 1041–1047. [CrossRef]
7.  Karthika, C.; Sureshkumar, R.; Zehravi, M.; Akter, R.; Ali, F.; Ramproshad, S.; Mondal, B.; Kundu, M.K.; Dey, A.; Rahman, M.H.; et al. Multidrug Resistance in Cancer Cells: Focus on a Possible Strategy Plan to Address Colon Carcinoma Cells. *Life* **2022**, *12*, 811. [CrossRef]
8.  Elfadadny, A.; El-Husseiny, H.M.; Abugomaa, A.; Ragab, R.F.; Mady, E.A.; Aboubakr, M.; Samir, H.; Mandour, A.S.; El-Mleeh, A.; El-Far, A.H.; et al. Role of Multidrug Resistance-Associated Proteins in Cancer Therapeutics: Past, Present, and Future Perspectives. *Environ. Sci. Pollut. Res.* **2021**, *28*, 49447–49466. [CrossRef]
9.  Ng, C.X.; Affendi, M.M.; Chong, P.P.; Lee, S.H. The Potential of Plant-Derived Extracts and Compounds to Augment Anticancer Effects of Chemotherapeutic Drugs. *Nutr. Cancer* **2022**, *74*, 3058–3076. [CrossRef]
10. Veeresham, C. Natural Products Derived from Plants as a Source of Drugs. *J. Adv. Pharm. Technol. Res.* **2012**, *3*, 200–201. [CrossRef]
11. Najmi, A.; Javed, S.A.; al Bratty, M.; Alhazmi, H.A. Modern Approaches in the Discovery and Development of Plant-Based Natural Products and Their Analogues as Potential Therapeutic Agents. *Molecules* **2022**, *27*, 349. [CrossRef] [PubMed]
12. Islam, B.U.; Suhail, M.; Khan, M.K.; Zughaibi, T.A.; Alserihi, R.F.; Zaidi, S.K.; Tabrez, S. Polyphenols as Anticancer Agents: Toxicological Concern to Healthy Cells. *Phytother. Res.* **2021**, *35*, 6063–6079. [CrossRef] [PubMed]
13. Kubczak, M.; Szustka, A.; Rogalińska, M. Molecular Targets of Natural Compounds with Anti-Cancer Properties. *Int. J. Mol. Sci.* **2021**, *22*, 13659. [CrossRef] [PubMed]
14. Guo, Y.; Sun, Q.; Wu, F.-G.; Dai, Y.; Chen, X.; Guo, Y.; Sun, Q.; Wu, F.; Dai, Y.; Chen Yong Loo, X. Polyphenol-Containing Nanoparticles: Synthesis, Properties, and Therapeutic Delivery. *Adv. Mater.* **2021**, *33*, 2007356. [CrossRef]
15. Mitchell, M.J.; Billingsley, M.M.; Haley, R.M.; Wechsler, M.E.; Peppas, N.A.; Langer, R. Engineering Precision Nanoparticles for Drug Delivery. *Nat. Rev. Drug Discov.* **2020**, *20*, 101–124. [CrossRef]
16. Esmeeta, A.; Adhikary, S.; Dharshnaa, V.; Swarnamughi, P.; Ummul Maqsummiya, Z.; Banerjee, A.; Pathak, S.; Duttaroy, A.K. Plant-Derived Bioactive Compounds in Colon Cancer Treatment: An Updated Review. *Biomed. Pharmacother.* **2022**, *153*, 113384. [CrossRef]
17. Wei, Y.; Yang, P.; Cao, S.; Zhao, L. The Combination of Curcumin and 5-Fluorouracil in Cancer Therapy. *Arch. Pharmacal Res.* **2017**, *41*, 1–13. [CrossRef]
18. Wu, W.; Dong, J.; Gou, H.; Geng, R.; Yang, X.; Chen, D.; Xiang, B.; Zhang, Z.; Ren, S.; Chen, L.; et al. EGCG Synergizes the Therapeutic Effect of Irinotecan through Enhanced DNA Damage in Human Colorectal Cancer Cells. *J. Cell Mol. Med.* **2021**, *25*, 7913–7921. [CrossRef]
19. Lu, W.D.; Qin, Y.; Yang, C.; Li, L.; Fu, Z.X. Effect of Curcumin on Human Colon Cancer Multidrug Resistance in Vitro and in Vivo. *Clinics* **2013**, *68*, 694–701. [CrossRef]
20. Huang, L.; Zhang, S.; Zhou, J.; Li, X. Effect of Resveratrol on Drug Resistance in Colon Cancer Chemotherapy. *RSC Adv.* **2019**, *9*, 2572–2580. [CrossRef]
21. Kamran, S.; Sinniah, A.; Abdulghani, M.A.M.; Alshawsh, M.A. Therapeutic Potential of Certain Terpenoids as Anticancer Agents: A Scoping Review. *Cancers* **2022**, *14*, 1100. [CrossRef] [PubMed]
22. Costea, T.; Vlad, O.C.; Miclea, L.C.; Ganea, C.; Szöllősi, J.; Mocanu, M.M. Alleviation of Multidrug Resistance by Flavonoid and Non-Flavonoid Compounds in Breast, Lung, Colorectal and Prostate Cancer. *Int. J. Mol. Sci.* **2020**, *21*, 401. [CrossRef] [PubMed]
23. Liskova, A.; Samec, M.; Koklesova, L.; Brockmueller, A.; Zhai, K.; Abdellatif, B.; Siddiqui, M.; Biringer, K.; Kudela, E.; Pec, M.; et al. Flavonoids as an Effective Sensitizer for Anti-Cancer Therapy: Insights into Multi-Faceted Mechanisms and Applicability towards Individualized Patient Profiles. *EPMA J.* **2021**, *12*, 155–176. [CrossRef] [PubMed]

24. Zhou, Y.; Zhang, J.; Wang, K.; Han, W.; Wang, X.; Gao, M.; Wang, Z.; Sun, Y.; Yan, H.; Zhang, H.; et al. Quercetin Overcomes Colon Cancer Cells Resistance to Chemotherapy by Inhibiting Solute Carrier Family 1, Member 5 Transporter. *Eur. J. Pharmacol.* **2020**, *881*, 173185. [CrossRef] [PubMed]
25. Bukowski, K.; Kciuk, M.; Kontek, R. Mechanisms of Multidrug Resistance in Cancer Chemotherapy. *Int. J. Mol. Sci.* **2020**, *21*, 3233. [CrossRef] [PubMed]
26. Wang, X.; Zhang, H.; Chen, X. Drug Resistance and Combating Drug Resistance in Cancer. *Cancer Drug Resist.* **2019**, *2*, 141–160. [CrossRef]
27. Housman, G.; Byler, S.; Heerboth, S.; Lapinska, K.; Longacre, M.; Snyder, N.; Sarkar, S. Drug Resistance in Cancer: An Overview. *Cancers* **2014**, *6*, 1769–1792. [CrossRef]
28. Linn, S.C.; Giaccone, G. MDR1/P-Glycoprotein Expression in Colorectal Cancer. *Eur. J. Cancer* **1995**, *31A*, 1291–1294. [CrossRef]
29. Yuan, Z.; Liang, X.; Zhan, Y.; Wang, Z.; Xu, J.; Qiu, Y.; Wang, J.; Cao, Y.; Le, V.-M.; Ly, H.-T.; et al. Targeting CD133 Reverses Drug-Resistance via the AKT/NF-KB/MDR1 Pathway in Colorectal Cancer. *Br. J. Cancer* **2020**, *122*, 1342–1353. [CrossRef]
30. Cao, D.; Qin, S.; Mu, Y.; Zhong, M. The Role of MRP1 in the Multidrug Resistance of Colorectal Cancer. *Oncol. Lett.* **2017**, *13*, 2471–2476. [CrossRef]
31. Hu, T.; Li, Z.; Gao, C.-Y.; Cho, C.H. Mechanisms of Drug Resistance in Colon Cancer and Its Therapeutic Strategies. *World J. Gastroenterol.* **2016**, *22*, 6876–6889. [CrossRef] [PubMed]
32. Wang, Z.; Sun, X.; Feng, Y.; Wang, Y.; Zhang, L.; Wang, Y.; Fang, Z.; Azami, N.L.B.; Sun, M.; Li, Q. Dihydromyricetin Reverses MRP2-Induced Multidrug Resistance by Preventing NF-KB-Nrf2 Signaling in Colorectal Cancer Cell. *Phytomedicine* **2021**, *82*, 153414. [CrossRef] [PubMed]
33. Asano, T. Drug Resistance in Cancer Therapy and the Role of Epigenetics. *J. Nippon. Med. Sch.* **2020**, *87*, 244–251. [CrossRef] [PubMed]
34. Das, P.K.; Islam, F.; Lam, A.K. The Roles of Cancer Stem Cells and Therapy Resistance in Colorectal Carcinoma. *Cells* **2020**, *9*, 1392. [CrossRef]
35. Neophytou, C.M.; Trougakos, I.P.; Erin, N.; Papageorgis, P. Apoptosis Deregulation and the Development of Cancer Multi-Drug Resistance. *Cancers* **2021**, *13*, 4363. [CrossRef]
36. Prasetyanti, P.R.; Medema, J.P. Intra-Tumor Heterogeneity from a Cancer Stem Cell Perspective. *Mol. Cancer* **2017**, *16*, 41. [CrossRef]
37. Phi, L.T.H.; Sari, I.N.; Yang, Y.G.; Lee, S.H.; Jun, N.; Kim, K.S.; Lee, Y.K.; Kwon, H.Y. Cancer Stem Cells (CSCs) in Drug Resistance and Their Therapeutic Implications in Cancer Treatment. *Stem. Cells Int.* **2018**, *2018*, 5416923. [CrossRef]
38. Burrell, R.A.; Swanton, C. Tumour Heterogeneity and the Evolution of Polyclonal Drug Resistance. *Mol. Oncol.* **2014**, *8*, 1095–1111. [CrossRef]
39. Junttila, M.R.; de Sauvage, F.J. Influence of Tumour Micro-Environment Heterogeneity on Therapeutic Response. *Nature* **2013**, *501*, 346–354. [CrossRef]
40. Dutta, S.; Mahalanobish, S.; Saha, S.; Ghosh, S.; Sil, P.C. Natural Products: An Upcoming Therapeutic Approach to Cancer. *Food Chem. Toxicol.* **2019**, *128*, 240–255. [CrossRef]
41. Castañeda, A.M.; Meléndez, C.M.; Uribe, D.; Pedroza-Díaz, J. Synergistic Effects of Natural Compounds and Conventional Chemotherapeutic Agents: Recent Insights for the Development of Cancer Treatment Strategies. *Heliyon* **2022**, *8*, e09519. [CrossRef]
42. Rejhová, A.; Opattová, A.; Čumová, A.; Slíva, D.; Vodička, P. Natural Compounds and Combination Therapy in Colorectal Cancer Treatment. *Eur. J. Med. Chem.* **2018**, *144*, 582–594. [CrossRef]
43. Redondo-Blanco, S.; Fernández, J.; Gutiérrez-del-Río, I.; Villar, C.J.; Lombó, F. New Insights toward Colorectal Cancer Chemotherapy Using Natural Bioactive Compounds. *Front. Pharmacol.* **2017**, *8*, 109. [CrossRef]
44. Durazzo, A.; Lucarini, M.; Souto, E.B.; Cicala, C.; Caiazzo, E.; Izzo, A.A.; Novellino, E.; Santini, A. Polyphenols: A Concise Overview on the Chemistry, Occurrence, and Human Health. *Phytother. Res.* **2019**, *33*, 2221–2243. [CrossRef]
45. Mileo, A.M.; Nisticò, P.; Miccadei, S. Polyphenols: Immunomodulatory and Therapeutic Implication in Colorectal Cancer. *Front. Immunol.* **2019**, *10*, 729. [CrossRef]
46. Bracci, L.; Fabbri, A.; del Cornò, M.; Conti, L. Dietary Polyphenols: Promising Adjuvants for Colorectal Cancer Therapies. *Cancers* **2021**, *13*, 4499. [CrossRef]
47. Ko, J.H.; Sethi, G.; Um, J.Y.; Shanmugam, M.K.; Arfuso, F.; Kumar, A.P.; Bishayee, A.; Ahn, K.S. The Role of Resveratrol in Cancer Therapy. *Int. J. Mol. Sci.* **2017**, *18*, 2589. [CrossRef]
48. Elshaer, M.; Chen, Y.; Wang, X.J.; Tang, X. Resveratrol: An Overview of Its Anti-Cancer Mechanisms. *Life Sci.* **2018**, *207*, 340–349. [CrossRef] [PubMed]
49. Fuggetta, M.P.; Lanzilli, G.; Tricarico, M.; Cottarelli, A.; Falchetti, R.; Ravagnan, G.; Bonmassar, E. Effect of Resveratrol on Proliferation and Telomerase Activity of Human Colon Cancer Cells in Vitro. *J. Exp. Clin. Cancer Res.* **2006**, *25*, 189–193. [PubMed]
50. Rotelli, M.T.; Bocale, D.; de Fazio, M.; Ancona, P.; Scalera, I.; Memeo, R.; Travaglio, E.; Zbar, A.P.; Altomare, D.F. IN-VITRO Evidence for the Protective Properties of the Main Components of the Mediterranean Diet against Colorectal Cancer: A Systematic Review. *Surg Oncol.* **2015**, *24*, 145–152. [CrossRef] [PubMed]

51. Saunier, E.; Antonio, S.; Regazzetti, A.; Auzeil, N.; Laprévote, O.; Shay, J.W.; Coumoul, X.; Barouki, R.; Benelli, C.; Huc, L.; et al. Resveratrol Reverses the Warburg Effect by Targeting the Pyruvate Dehydrogenase Complex in Colon Cancer Cells. *Sci. Rep.* **2017**, *7*, 6945. [CrossRef]
52. Saud, S.M.; Li, W.; Morris, N.L.; Matter, M.S.; Colburn, N.H.; Kim, Y.S.; Young, M.R. Resveratrol Prevents Tumorigenesis in Mouse Model of Kras Activated Sporadic Colorectal Cancer by Suppressing Oncogenic Kras Expression. *Carcinogenesis* **2014**, *35*, 2778–2786. [CrossRef] [PubMed]
53. Carter, L.G.; D'Orazio, J.A.; Pearson, K.J. Resveratrol and Cancer: Focus on in Vivo Evidence. *Endocr. Relat. Cancer* **2014**, *21*, R209–R225. [CrossRef] [PubMed]
54. Aires, V.; Limagne, E.; Cotte, A.K.; Latruffe, N.; Ghiringhelli, F.; Delmas, D. Resveratrol Metabolites Inhibit Human Metastatic Colon Cancer Cells Progression and Synergize with Chemotherapeutic Drugs to Induce Cell Death. *Mol. Nutr. Food Res.* **2013**, *57*, 1170–1181. [CrossRef] [PubMed]
55. Cottart, C.H.; Nivet-Antoine, V.; Laguillier-Morizot, C.; Beaudeux, J.L. Resveratrol Bioavailability and Toxicity in Humans. *Mol. Nutr. Food Res.* **2010**, *54*, 7–16. [CrossRef]
56. Huang, X.M.; Yang, Z.J.; Xie, Q.; Zhang, Z.K.; Zhang, H.; Ma, J.Y. Natural Products for Treating Colorectal Cancer: A Mechanistic Review. *Biomed Pharm.* **2019**, *117*, 109142. [CrossRef] [PubMed]
57. Abdel Latif, Y.; El-Bana, M.; Hussein, J.; El-Khayat, Z.; Farrag, A.R. Effects of Resveratrol in Combination with 5-Fluorouracil on N-Methylnitrosourea-Induced Colon Cancer in Rats. *Comp. Clin. Path* **2019**, *28*, 1351–1362. [CrossRef]
58. Buhrmann, C.; Yazdi, M.; Popper, B.; Shayan, P.; Goel, A.; Aggarwal, B.B.; Shakibaei, M. Resveratrol Chemosensitizes TNF-β-Induced Survival of 5-FU-Treated Colorectal Cancer Cells. *Nutrients* **2018**, *10*, 888. [CrossRef] [PubMed]
59. Yang, S.; Li, W.; Sun, H.; Wu, B.; Ji, F.; Sun, T.; Chang, H.; Shen, P.; Wang, Y.; Zhou, D. Resveratrol Elicits Anti-Colorectal Cancer Effect by Activating MiR-34c-KITLG in Vitro and in Vivo. *BMC Cancer* **2015**, *15*, 969. [CrossRef] [PubMed]
60. Kaminski, B.M.; Weigert, A.; Scherzberg, M.C.; Ley, S.; Gilbert, B.; Brecht, K.; Brüne, B.; Steinhilber, D.; Stein, J.; Ulrich-Rückert, S. Resveratrol-Induced Potentiation of the Antitumor Effects of Oxaliplatin Is Accompanied by an Altered Cytokine Profile of Human Monocyte-Derived Macrophages. *Apoptosis* **2014**, *19*, 1136–1147. [CrossRef]
61. Khayat, M.T.; Zarka, M.A.; El-Telbany, D.; Farag, A.; El-Halawany, A.M.; Kutbi, H.I.; Elkhatib, W.F.; Noreddin, A.M.; Khayyat, A.N.; El-Telbany, R.F.A.; et al. Intensification of Resveratrol Cytotoxicity, pro-Apoptosis, Oxidant Potentials in Human Colorectal Carcinoma HCT-116 Cells Using Zein Nanoparticles. *Sci. Rep.* **2022**, *12*, 15235. [CrossRef] [PubMed]
62. Md, S.; Abdullah, S.; Alhakamy, N.A.; Alharbi, W.S.; Ahmad, J.; Shaik, R.A.; Ansari, M.J.; Ibrahim, I.M.; Ali, J. Development, Optimization, and In Vitro Evaluation of Novel Oral Long-Acting Resveratrol Nanocomposite In-Situ Gelling Film in the Treatment of Colorectal Cancer. *Gels* **2021**, *7*, 276. [CrossRef] [PubMed]
63. Feng, M.; Zhong, L.-X.; Zhan, Z.-Y.; Huang, Z.-H.; Xiong, J.-P. Enhanced Antitumor Efficacy of Resveratrol-Loaded Nanocapsules in Colon Cancer Cells: Physicochemical and Biological Characterization. *Eur. Rev. Med. Pharmacol. Sci.* **2017**, *21*, 375–382. [PubMed]
64. Summerlin, N.; Qu, Z.; Pujara, N.; Sheng, Y.; Jambhrunkar, S.; McGuckin, M.; Popat, A. Colloidal Mesoporous Silica Nanoparticles Enhance the Biological Activity of Resveratrol. *Colloids Surf. B Biointerfaces* **2016**, *144*, 1–7. [CrossRef]
65. Soo, E.; Thakur, S.; Qu, Z.; Jambhrunkar, S.; Parekh, H.S.; Popat, A. Enhancing Delivery and Cytotoxicity of Resveratrol through a Dual Nanoencapsulation Approach. *J. Colloid Interface Sci.* **2016**, *462*, 368–374. [CrossRef]
66. Montané, X.; Kowalczyk, O.; Reig-Vano, B.; Bajek, A.; Roszkowski, K.; Tomczyk, R.; Pawliszak, W.; Giamberini, M.; Mocek-Płóciniak, A.; Tylkowski, B. Current Perspectives of the Applications of Polyphenols and Flavonoids in Cancer Therapy. *Molecules* **2020**, *25*, 3342. [CrossRef]
67. Weng, W.; Goel, A. Curcumin and Colorectal Cancer: An Update and Current Perspective on This Natural Medicine. *Semin. Cancer Biol.* **2022**, *80*, 73–86. [CrossRef]
68. Shehzad, A.; Wahid, F.; Lee, Y.S. Curcumin in Cancer Chemoprevention: Molecular Targets, Pharmacokinetics, Bioavailability, and Clinical Trials. *Arch. Pharm.* **2010**, *343*, 489–499. [CrossRef]
69. Jaiswal, A.S.; Marlow, B.P.; Gupta, N.; Narayan, S. Beta-Catenin-Mediated Transactivation and Cell-Cell Adhesion Pathways Are Important in Curcumin (Diferuylmethane)-Induced Growth Arrest and Apoptosis in Colon Cancer Cells. *Oncogene* **2002**, *21*, 8414–8427. [CrossRef]
70. Basbinar, Y.; Calibasi-Kocal, G.; Pakdemirli, A.; Bayrak, S.; Ozupek, N.M.; Sever, T.; Ellidokuz, H.; Yigitbasi, T. Curcumin Effects on Cell Proliferation, Angiogenesis and Metastasis in Colorectal Cancer. *JBUON* **2019**, *24*, 1482–1487.
71. Pricci, M.; Girardi, B.; Giorgio, F.; Losurdo, G.; Ierardi, E.; di Leo, A. Curcumin and Colorectal Cancer: From Basic to Clinical Evidences. *Int. J. Mol. Sci.* **2020**, *21*, 2364. [CrossRef] [PubMed]
72. Tunstall, R.G.; Sharma, R.A.; Perkins, S.; Sale, S.; Singh, R.; Farmer, P.B.; Steward, W.P.; Gescher, A.J. Cyclooxygenase-2 Expression and Oxidative DNA Adducts in Murine Intestinal Adenomas: Modification by Dietary Curcumin and Implications for Clinical Trials. *Eur. J. Cancer* **2006**, *42*, 415–421. [CrossRef] [PubMed]
73. Yin, T.F.; Wang, M.; Qing, Y.; Lin, Y.M.; Wu, D. Research Progress on Chemopreventive Effects of Phytochemicals on Colorectal Cancer and Their Mechanisms. *World J. Gastroenterol.* **2016**, *22*, 7068. [CrossRef] [PubMed]
74. Yin, J.; Wang, L.; Wang, Y.; Shen, H.; Wang, X.; Wu, L. Curcumin Reverses Oxaliplatin Resistance in Human Colorectal Cancer via Regulation of TGF-β/Smad2/3 Signaling Pathway. *Onco. Targets Ther.* **2019**, *12*, 3893–3903. [CrossRef]

75. Zheng, X.; Yang, X.; Lin, J.; Song, F.; Shao, Y. Low Curcumin Concentration Enhances the Anticancer Effect of 5-Fluorouracil against Colorectal Cancer. *Phytomedicine* **2021**, *85*, 153547. [CrossRef]
76. Hosseini, S.A.; Zand, H.; Cheraghpour, M. The Influence of Curcumin on the Downregulation of MYC, Insulin and IGF-1 Receptors: A Possible Mechanism Underlying the Anti-Growth and Anti-Migration in Chemoresistant Colorectal Cancer Cells. *Medicina* **2019**, *55*, 90. [CrossRef]
77. Han, Z.; Song, B.; Yang, J.; Wang, B.; Ma, Z.; Yu, L.; Li, Y.; Xu, H.; Qiao, M. Curcumin-Encapsulated Fusion Protein-Based Nanocarrier Demonstrated Highly Efficient Epidermal Growth Factor Receptor-Targeted Treatment of Colorectal Cancer. *J. Agric. Food Chem.* **2022**, *70*, 15464–15473. [CrossRef]
78. Wang, A.; Jain, S.; Dia, V.; Lenaghan, S.C.; Zhong, Q. Shellac Micelles Loaded with Curcumin Using a PH Cycle to Improve Dispersibility, Bioaccessibility, and Potential for Colon Delivery. *J. Agric. Food Chem.* **2022**, *70*, 15166–15177. [CrossRef]
79. Elbassiouni, F.E.; El-Kholy, W.M.; Elhabibi, E.-S.M.; Albogami, S.; Fayad, E. Comparative Study between Curcumin and Nanocurcumin Loaded PLGA on Colon Carcinogenesis Induced Mice. *Nanomaterials* **2022**, *12*, 324. [CrossRef]
80. Low, Z.X.; Teo, M.Y.M.; Nordin, F.J.; Dewi, F.R.P.; Palanirajan, V.K.; In, L.L.A. Biophysical Evaluation of Water-Soluble Curcumin Encapsulated in β-Cyclodextrins on Colorectal Cancer Cells. *Int. J. Mol. Sci.* **2022**, *23*, 12866. [CrossRef]
81. Gupta, A.; Sood, A.; Dhiman, A.; Shrimali, N.; Singhmar, R.; Guchhait, P.; Agrawal, G. Redox Responsive Poly(Allylamine)/Eudragit S-100 Nanoparticles for Dual Drug Delivery in Colorectal Cancer. *Biomater. Adv.* **2022**, *143*, 213184. [CrossRef]
82. Imran, M.; Salehi, B.; Sharifi-Rad, J.; Gondal, T.A.; Saeed, F.; Imran, A.; Shahbaz, M.; Fokou, P.V.T.; Arshad, M.U.; Khan, H.; et al. Kaempferol: A Key Emphasis to Its Anticancer Potential. *Molecules* **2019**, *24*, 2277. [CrossRef] [PubMed]
83. Chen, A.Y.; Chen, Y.C. A Review of the Dietary Flavonoid, Kaempferol on Human Health and Cancer Chemoprevention. *Food Chem.* **2013**, *138*, 2099–2107. [CrossRef]
84. Schindler, R.; Mentlein, R. Flavonoids and Vitamin E Reduce the Release of the Angiogenic Peptide Vascular Endothelial Growth Factor from Human Tumor Cells. *J. Nutr.* **2006**, *136*, 1477–1482. [CrossRef]
85. Barve, A.; Chen, C.; Hebbar, V.; Desiderio, J.; Saw, C.L.-L.; Kong, A.-N. Metabolism, Oral Bioavailability and Pharmacokinetics of Chemopreventive Kaempferol in Rats. *Biopharm. Drug Dispos.* **2009**, *30*, 356–365. [CrossRef] [PubMed]
86. Cho, H.J.; Park, J.H.Y. Kaempferol Induces Cell Cycle Arrest in HT-29 Human Colon Cancer Cells. *J. Cancer Prev.* **2013**, *18*, 257–263. [CrossRef] [PubMed]
87. Riahi-Chebbi, I.; Souid, S.; Othman, H.; Haoues, M.; Karoui, H.; Morel, A.; Srairi-Abid, N.; Essafi, M.; Essafi-Benkhadir, K. The Phenolic Compound Kaempferol Overcomes 5-Fluorouracil Resistance in Human Resistant LS174 Colon Cancer Cells. *Sci. Rep.* **2019**, *9*, 195. [CrossRef] [PubMed]
88. Li, Q.; Wei, L.; Lin, S.; Chen, Y.; Lin, J.; Peng, J. Synergistic Effect of Kaempferol and 5-fluorouracil on the Growth of Colorectal Cancer Cells by Regulating the PI3K/Akt Signaling Pathway. *Mol. Med. Rep.* **2019**, *20*, 728–734. [CrossRef] [PubMed]
89. Park, J.; Lee, G.E.; An, H.J.; Lee, C.J.; Cho, E.S.; Kang, H.C.; Lee, J.Y.; Lee, H.S.; Choi, J.S.; Kim, D.J.; et al. Kaempferol Sensitizes Cell Proliferation Inhibition in Oxaliplatin-Resistant Colon Cancer Cells. *Arch. Pharm. Res.* **2021**, *44*, 1091–1108. [CrossRef] [PubMed]
90. Meena, D.; Vimala, K.; Kannan, S. Combined Delivery of DOX and Kaempferol Using PEGylated Gold Nanoparticles to Target Colon Cancer. *J. Clust. Sci.* **2022**, *33*, 173–187. [CrossRef]
91. Kumar, S.; Pandey, A.K. Chemistry and Biological Activities of Flavonoids: An Overview. *Sci. World J.* **2013**, *2013*, 162750. [CrossRef]
92. Tang, S.M.; Deng, X.T.; Zhou, J.; Li, Q.P.; Ge, X.X.; Miao, L. Pharmacological Basis and New Insights of Quercetin Action in Respect to Its Anti-Cancer Effects. *Biomed Pharm.* **2020**, *121*, 109604. [CrossRef] [PubMed]
93. Reyes-Farias, M.; Carrasco-Pozo, C. The Anti-Cancer Effect of Quercetin: Molecular Implications in Cancer Metabolism. *Int. J. Mol. Sci.* **2019**, *20*, 3177. [CrossRef]
94. Kee, J.Y.; Han, Y.H.; Kim, D.S.; Mun, J.G.; Park, J.; Jeong, M.Y.; Um, J.Y.; Hong, S.H. Inhibitory Effect of Quercetin on Colorectal Lung Metastasis through Inducing Apoptosis, and Suppression of Metastatic Ability. *Phytomedicine* **2016**, *23*, 1680–1690. [CrossRef] [PubMed]
95. Srivastava, N.S.; Srivastava, R.A.K. Curcumin and Quercetin Synergistically Inhibit Cancer Cell Proliferation in Multiple Cancer Cells and Modulate Wnt/β-Catenin Signaling and Apoptotic Pathways in A375 Cells. *Phytomedicine* **2019**, *52*, 117–128. [CrossRef]
96. Neamtu, A.A.; Maghiar, T.A.; Alaya, A.; Olah, N.K.; Turcus, V.; Pelea, D.; Totolici, B.D.; Neamtu, C.; Maghiar, A.M.; Mathe, E. A Comprehensive View on the Quercetin Impact on Colorectal Cancer. *Molecules* **2022**, *27*, 1973. [CrossRef] [PubMed]
97. Chang, C.E.; Hsieh, C.M.; Huang, S.C.; Su, C.Y.; Sheu, M.T.; Ho, H.O. Lecithin-Stabilized Polymeric Micelles (LsbPMs) for Delivering Quercetin: Pharmacokinetic Studies and Therapeutic Effects of Quercetin Alone and in Combination with Doxorubicin. *Sci. Rep.* **2018**, *8*, 17640. [CrossRef]
98. Wen, P.; Hu, T.-G.; Li, L.; Zong, M.-H.; Wu, H. A Colon-Specific Delivery System for Quercetin with Enhanced Cancer Prevention Based on Co-Axial Electrospinning. *Food Funct.* **2018**, *9*, 5999–6009. [CrossRef]
99. Almatrood, S.A.; Almatroudi, A.; Khan, A.A.; Alhumaydh, F.A.; Alsahl, M.A.; Rahmani, A.H. Potential Therapeutic Targets of Epigallocatechin Gallate (EGCG), the Most Abundant Catechin in Green Tea, and Its Role in the Therapy of Various Types of Cancer. *Molecules* **2020**, *25*, 3146. [CrossRef] [PubMed]
100. Luo, K.W.; Xia, J.; Cheng, B.H.; Gao, H.C.; Fu, L.W.; Luo, X. le Tea Polyphenol EGCG Inhibited Colorectal-Cancer-Cell Proliferation and Migration via Downregulation of STAT3. *Gastroenterol. Rep.* **2020**, *9*, 59–70. [CrossRef] [PubMed]

101. Wubetu, G.Y.; Shimada, M.; Morine, Y.; Ikemoto, T.; Ishikawa, D.; Iwahashi, S.; Yamada, S.; Saito, Y.; Arakawa, Y.; Imura, S. Epigallocatechin Gallate Hinders Human Hepatoma and Colon Cancer Sphere Formation. *J. Gastroenterol. Hepatol.* **2016**, *31*, 256–264. [CrossRef] [PubMed]

102. Toden, S.; Tran, H.M.; Tovar-Camargo, O.A.; Okugawa, Y.; Goel, A. Epigallocatechin-3-Gallate Targets Cancer Stem-like Cells and Enhances 5-Fluorouracil Chemosensitivity in Colorectal Cancer. *Oncotarget* **2016**, *7*, 16158–16171. [CrossRef] [PubMed]

103. La, X.; Zhang, L.; Li, Z.; Li, H.; Yang, Y. (-)-Epigallocatechin Gallate (EGCG) Enhances the Sensitivity of Colorectal Cancer Cells to 5-FU by Inhibiting GRP78/NF-KB/MiR-155-5p/MDR1 Pathway. *J. Agric. Food Chem.* **2019**, *67*, 2510–2518. [CrossRef]

104. Maruyama, T.; Murata, S.; Nakayama, K.; Sano, N.; Ogawa, K.; Nowatari, T.; Tamura, T.; Nozaki, R.; Fukunaga, K.; Ohkohchi, N. (-)-Epigallocatechin-3-Gallate Suppresses Liver Metastasis of Human Colorectal Cancer. *Oncol. Rep.* **2014**, *31*, 625–633. [CrossRef]

105. Hu, F.; Wei, F.; Wang, Y.; Wu, B.; Fang, Y.; Xiong, B. EGCG Synergizes the Therapeutic Effect of Cisplatin and Oxaliplatin through Autophagic Pathway in Human Colorectal Cancer Cells. *J. Pharmacol. Sci.* **2015**, *128*, 27–34. [CrossRef] [PubMed]

106. Wang, R.; Huang, J.; Chen, J.; Yang, M.; Wang, H.; Qiao, H.; Chen, Z.; Hu, L.; Di, L.; Li, J. Enhanced Anti-Colon Cancer Efficacy of 5-Fluorouracil by Epigallocatechin-3- Gallate Co-Loaded in Wheat Germ Agglutinin-Conjugated Nanoparticles. *Nanomedicine* **2019**, *21*, 102068. [CrossRef]

107. Seufferlein, T.; Ettrich, T.J.; Menzler, S.; Messmann, H.; Kleber, G.; Zipprich, A.; Frank-Gleich, S.; Algül, H.; Metter, K.; Odemar, F.; et al. Green Tea Extract to Prevent Colorectal Adenomas, Results of a Randomized, Placebo-Controlled Clinical Trial. *Am. J. Gastroenterol.* **2022**, *117*, 884–894. [CrossRef]

108. Abaza, M.S.I.; Orabi, K.Y.; Al-Quattan, E.; Al-Attiyah, R.J. Growth Inhibitory and Chemo-Sensitization Effects of Naringenin, a Natural Flavanone Purified from Thymus Vulgaris, on Human Breast and Colorectal Cancer. *Cancer Cell Int.* **2015**, *15*, 46. [CrossRef]

109. Song, H.M.; Park, G.H.; Eo, H.J.; Lee, J.W.; Kim, M.K.; Lee, J.R.; Lee, M.H.; Koo, J.S.; Jeong, J.B. Anti-Proliferative Effect of Naringenin through P38-Dependent Downregulation of Cyclin D1 in Human Colorectal Cancer Cells. *Biomol. Ther.* **2015**, *23*, 339–344. [CrossRef] [PubMed]

110. Rehman, M.U.; Rahman Mir, M.U.; Farooq, A.; Rashid, S.M.; Ahmad, B.; Bilal Ahmad, S.; Ali, R.; Hussain, I.; Masoodi, M.; Muzamil, S.; et al. Naringenin (4,5,7-Trihydroxyflavanone) Suppresses the Development of Precancerous Lesions via Controlling Hyperproliferation and Inflammation in the Colon of Wistar Rats. *Environ. Toxicol.* **2018**, *33*, 422–435. [CrossRef]

111. Md, S.; Abdullah, S.; Awan, Z.A.; Alhakamy, N.A. Smart Oral PH-Responsive Dual Layer Nano-Hydrogel for Dissolution Enhancement and Targeted Delivery of Naringenin Using Protein-Polysaccharides Complexation Against Colorectal Cancer. *J. Pharm. Sci.* **2022**, *111*, 3155–3164. [CrossRef] [PubMed]

112. Wesołowska, O.; Wiśniewski, J.; Środa-Pomianek, K.; Bielawska-Pohl, A.; Paprocka, M.; Duś, D.; Duarte, N.; Ferreira, M.J.U.; Michalak, K. Multidrug Resistance Reversal and Apoptosis Induction in Human Colon Cancer Cells by Some Flavonoids Present in Citrus Plants. *J. Nat. Prod.* **2012**, *75*, 1896–1902. [CrossRef] [PubMed]

113. Bai, Y.; Xiong, Y.; Zhang, Y.Y.; Cheng, L.; Liu, H.; Xu, K.; Wu, Y.Y.; Field, J.; Wang, X.D.; Zhou, L.M. Tangeretin Synergizes with 5-Fluorouracil to Induce Autophagy through MicroRNA-21 in Colorectal Cancer Cells. *Am. J. Chin. Med.* **2022**, *50*, 1681–1701. [CrossRef]

114. De Cássia Da Silveira, E.; Sá, R.; Andrade, L.N.; de Sousa, D.P. A Review on Anti-Inflammatory Activity of Monoterpenes. *Molecules* **2013**, *18*, 1227–1254. [CrossRef] [PubMed]

115. Cho, M.; So, I.; Chun, J.N.; Jeon, J.H. The Antitumor Effects of Geraniol: Modulation of Cancer Hallmark Pathways (Review). *Int. J. Oncol.* **2016**, *48*, 1772–1782. [CrossRef]

116. Vieira, A.; Heidor, R.; Cardozo, M.T.; Scolastici, C.; Purgatto, E.; Shiga, T.M.; Barbisan, L.F.; Ong, T.P.; Moreno, F.S. Efficacy of Geraniol but Not of β-Ionone or Their Combination for the Chemoprevention of Rat Colon Carcinogenesis. *Braz. J. Med. Biol. Res.* **2011**, *44*, 538–545. [CrossRef]

117. Carnesecchi, S.; Bras-Gonçalves, R.; Bradaia, A.; Zeisel, M.; Gossé, F.; Poupon, M.F.; Raul, F. Geraniol, a Component of Plant Essential Oils, Modulates DNA Synthesis and Potentiates 5-Fluorouracil Efficacy on Human Colon Tumor Xenografts. *Cancer Lett.* **2004**, *215*, 53–59. [CrossRef]

118. Carnesecchi, S.; Langley, K.; Exinger, F.; Gosse, F.; Raul, F. Geraniol, a Component of Plant Essential Oils, Sensitizes Human Colonic Cancer Cells to 5-Fluorouracil Treatment. *J. Pharmacol. Exp. Ther.* **2002**, *301*, 625–630. [CrossRef]

119. Gao, J.L.; Lv, G.Y.; He, B.C.; Zhang, B.Q.; Zhang, H.; Wang, N.; Wang, C.Z.; Du, W.; Yuan, C.S.; He, T.C. Ginseng Saponin Metabolite 20(S)-Protopanaxadiol Inhibits Tumor Growth by Targeting Multiple Cancer Signaling Pathways. *Oncol. Rep.* **2013**, *30*, 292–298. [CrossRef]

120. Wang, Z.; Li, M.Y.; Zhang, Z.H.; Zuo, H.X.; Wang, J.Y.; Xing, Y.; Ri, M.H.; Jin, H.L.; Jin, C.H.; Xu, G.H.; et al. Panaxadiol Inhibits Programmed Cell Death-Ligand 1 Expression and Tumour Proliferation via Hypoxia-Inducible Factor (HIF)-1$\alpha$ and STAT3 in Human Colon Cancer Cells. *Pharmacol. Res.* **2020**, *155*, 104727. [CrossRef]

121. Yuan, H.D.; Quan, H.Y.; Zhang, Y.; Kim, S.H.; Chung, S.H. 20(S)-Ginsenoside Rg3-Induced Apoptosis in HT-29 Colon Cancer Cells Is Associated with AMPK Signaling Pathway. *Mol. Med. Rep.* **2010**, *3*, 825–831. [CrossRef] [PubMed]

122. Tang, Y.C.; Zhang, Y.; Zhou, J.; Zhi, Q.; Wu, M.Y.; Gong, F.R.; Shen, M.; Liu, L.; Tao, M.; Shen, B.; et al. Ginsenoside Rg3 Targets Cancer Stem Cells and Tumor Angiogenesis to Inhibit Colorectal Cancer Progression in Vivo. *Int. J. Oncol.* **2018**, *52*, 127–138. [CrossRef]

123. Wang, C.Z.; Zhang, Z.; Wan, J.Y.; Zhang, C.F.; Anderson, S.; He, X.; Yu, C.; He, T.C.; Qi, L.W.; Yuan, C.S. Protopanaxadiol, an Active Ginseng Metabolite, Significantly Enhances the Effects of Fluorouracil on Colon Cancer. *Nutrients* **2015**, *7*, 799–814. [CrossRef] [PubMed]
124. Du, G.-J.; Wang, C.-Z.; Zhang, Z.-Y.; Wen, X.-D.; Somogyi, J.; Calway, T.; He, T.-C.; Du, W.; Yuan, C.-S. Caspase-Mediated pro-Apoptotic Interaction of Panaxadiol and Irinotecan in Human Colorectal Cancer Cells. *J. Pharm. Pharmacol.* **2012**, *64*, 727–734. [CrossRef] [PubMed]
125. Hong, S.; Cai, W.; Huang, Z.; Wang, Y.; Mi, X.; Huang, Y.; Lin, Z.; Chen, X. Ginsenoside Rg3 Enhances the Anticancer Effect of 5-FU in Colon Cancer Cells via the PI3K/AKTpathway. *Oncol. Rep.* **2020**, *44*, 1333–1342. [CrossRef]
126. Qiu, R.; Qian, F.; Wang, X.; Li, H.; Wang, L. Targeted Delivery of 20(S)-Ginsenoside Rg3-Based Polypeptide Nanoparticles to Treat Colon Cancer. *Biomed Microdevices* **2019**, *21*, 18. [CrossRef]
127. Sun, D.; Zou, Y.; Song, L.; Han, S.; Yang, H.; Chu, D.; Dai, Y.; Ma, J.; O'Driscoll, C.M.; Yu, Z.; et al. A Cyclodextrin-Based Nanoformulation Achieves Co-Delivery of Ginsenoside Rg3 and Quercetin for Chemo-Immunotherapy in Colorectal Cancer. *Acta Pharm. Sin. B* **2022**, *12*, 378–393. [CrossRef]
128. Ortíz, R.; Quiñonero, F.; García-Pinel, B.; Fuel, M.; Mesas, C.; Cabeza, L.; Melguizo, C.; Prados, J. Nanomedicine to Overcome Multidrug Resistance Mechanisms in Colon and Pancreatic Cancer: Recent Progress. *Cancers* **2021**, *13*, 2058. [CrossRef]
129. Gallego-Jara, J.; Lozano-Terol, G.; Sola-Martínez, R.A.; Cánovas-Díaz, M.; Puente, T.d.D. A Compressive Review about Taxol®: History and Future Challenges. *Molecules* **2020**, *25*, 5986. [CrossRef]
130. Yuan, M.; Zhang, G.; Bai, W.; Han, X.; Li, C.; Bian, S. The Role of Bioactive Compounds in Natural Products Extracted from Plants in Cancer Treatment and Their Mechanisms Related to Anticancer Effects. *Oxid. Med. Cell Longev.* **2022**, *2022*, 1429869. [CrossRef]
131. Maleki Dana, P.; Sadoughi, F.; Asemi, Z.; Yousefi, B. The Role of Polyphenols in Overcoming Cancer Drug Resistance: A Comprehensive Review. *Cell Mol. Biol. Lett.* **2022**, *27*, 1. [CrossRef]
132. Afrin, S.; Giampieri, F.; Cianciosi, D.; Alvarez-Suarez, J.M.; Bullon, B.; Amici, A.; Quiles, J.L.; Forbes-Hernández, T.Y.; Battino, M. Strawberry Tree Honey in Combination with 5-Fluorouracil Enhances Chemosensitivity in Human Colon Adenocarcinoma Cells. *Food Chem. Toxicol.* **2021**, *156*, 112484. [CrossRef]
133. Hassanein, N.M.A.; Hassan, E.S.G.; Hegab, A.M.; Elahl, H.M.S. Chemopreventive Effect of Sulindac in Combination with Epigallocatechin Gallate or Kaempferol against 1,2-Dimethyl Hydrazine-Induced Preneoplastic Lesions in Rats: A Comparative Study. *J. Biochem. Mol. Toxicol.* **2018**, *32*, e22198. [CrossRef] [PubMed]
134. Chavva, S.R.; Deshmukh, S.K.; Kanchanapally, R.; Tyagi, N.; Coym, J.W.; Singh, A.P.; Singh, S. Epigallocatechin Gallate-Gold Nanoparticles Exhibit Superior Antitumor Activity Compared to Conventional Gold Nanoparticles: Potential Synergistic Interactions. *Nanomaterials* **2019**, *9*, 396. [CrossRef] [PubMed]
135. Li, Z.; Jiang, H.; Xu, C.; Gu, L. A Review: Using Nanoparticles to Enhance Absorption and Bioavailability of Phenolic Phytochemicals. *Food Hydrocoll* **2015**, *43*, 153–164. [CrossRef]

**Disclaimer/Publisher's Note:** The statements, opinions and data contained in all publications are solely those of the individual author(s) and contributor(s) and not of MDPI and/or the editor(s). MDPI and/or the editor(s) disclaim responsibility for any injury to people or property resulting from any ideas, methods, instructions or products referred to in the content.

*Article*

# Bioactive Compounds, Antioxidant Activity, and Mineral Content of Wild Rocket (*Diplotaxis tenuifolia* L.) Leaves as Affected by Saline Stress and Biostimulant Application

Ida Di Mola [1,†], Spyridon A. Petropoulos [2,*,†], Lucia Ottaiano [1], Eugenio Cozzolino [3], Christophe El-Nakhel [1], Youssef Rouphael [1] and Mauro Mori [1]

[1] Department of Agricultural Sciences, University of Naples Federico II, 80055 Portici, Italy
[2] Department of Agriculture Crop Production and Rural Environment, University of Thessaly, Fytokou Street, 38446 Volos, Greece
[3] Council for Agricultural Research and Economics (CREA)—Research Center for Cereal and Industrial Crops, 81100 Caserta, Italy
* Correspondence: spetropoulos@uth.gr
† These authors have contributed equally to this work.

**Featured Application: The results of the present work provide useful insights into the response of *Diplotaxis tenuifolia* L. plants to salinity stress and biostimulant application under successive crop cycles. Considering the complex effects of biostimulants on vegetable crops, providing new information regarding the positive effects of this agronomic practice under salinity stress and variable growing conditions will be useful for farmers and crop production stakeholders in their efforts to address the increasing soil salinization and degradation of the quality of irrigation water.**

**Abstract:** The availability of irrigation water of good quality is decreasing due to soil salinization and the deterioration of aquifers. Moreover, ongoing climate change severely affects crop production and necessitates the intensification of cropping systems in order to ensure food security at a global scale. For this purpose, the aim of the present study was to evaluate the mitigating effects of two natural biostimulants on *Diplotaxis tenuifolia* L. plants cultivated at different salinity levels (EC of 0 dS m$^{-1}$, 2 dS m$^{-1}$, 4 dS m$^{-1}$, and 6 dS m$^{-1}$) and harvested at six consecutive cropping cycles. The tested factors showed a varied combinatorial effect on the tested parameters. These findings indicate the importance of considering growing conditions and cropping periods when applying biostimulants in *D. tenuifolia* plants under salinity stress. Antioxidant activity and bioactive compounds, such as total phenols, carotenoids, and total ascorbic acid, were variably affected by salinity, biostimulant application, and harvesting time, while mineral profile was also affected by the tested factors depending on the combination of factors. Finally, nitrate content showed decreasing trends with increasing salinity, while biostimulant application resulted in the higher accumulation of nitrates compared to the untreated plants. Although biostimulant application seems to alleviate the negative effects of salinity stress, the effect of growing conditions, as indicated by successive crop cycles, is also important for the response of *D. tenuifolia* plants to saline conditions and biostimulant application.

**Keywords:** antioxidant activity; chlorophylls; carotenoids; tropical plants extracts; protein hydrolysates; total phenols

**Citation:** Di Mola, I.; Petropoulos, S.A.; Ottaiano, L.; Cozzolino, E.; El-Nakhel, C.; Rouphael, Y.; Mori, M. Bioactive Compounds, Antioxidant Activity, and Mineral Content of Wild Rocket (*Diplotaxis tenuifolia* L.) Leaves as Affected by Saline Stress and Biostimulant Application. *Appl. Sci.* **2023**, *13*, 1569. https://doi.org/10.3390/app13031569

Academic Editor: Luca Mazzoni

Received: 31 December 2022
Revised: 21 January 2023
Accepted: 24 January 2023
Published: 26 January 2023

**Copyright:** © 2023 by the authors. Licensee MDPI, Basel, Switzerland. This article is an open access article distributed under the terms and conditions of the Creative Commons Attribution (CC BY) license (https://creativecommons.org/licenses/by/4.0/).

## 1. Introduction

Modern crop production has to cope with increasing soil degradation and the low availability of irrigation water due to salinization, with major consequences on total and marketable crop yield as well as on the quality of the final products [1–3]. Moreover, approximately 20% of available agricultural land is affected by salinity [3], while it is expected that the current soil salinization trends will result in 50% of arable land being salt-affected by 2050 [4], mainly due to poor agricultural practices and the impact of climate change [5]. In this context, vegetable crops are highly affected by the changing conditions, especially the leafy species that are considered to be more prone to salinity than other crops [4,5]. Therefore, cropping systems and agronomic practices have to be reconsidered in order to address the current scenario and ensure food security in the mid- and long-term, especially in the arid and semi-arid regions of the world that are most affected by salinization and irrigation water shortages [6–8]. Moreover, halophyte species, which include important bioactive compounds with health-beneficial properties, could be integrated into farming systems and contribute to food security as well as land reclamation [9–11]. Among the several cropping strategies for mitigating the adverse effects of salinity, biostimulant application is proposed as an innovative and ecofriendly tool that allows cultivation under unfavorable conditions due to abiotic and biotic stressors [12–15].

Several studies have reported the beneficial effects of biostimulants on various crops grown under saline conditions. For example, El-Nakhel et al. [16] recently suggested that the application of legume-derived protein hydrolysates significantly ameliorated the negative impacts of high salinity (EC levels up to 6 dS m$^{-1}$) in pot-grown spinach plants. Moreover, Lucini et al. [17] reported similar beneficial effects on the crop performance and metabolite profile of pot-grown lettuce plants subjected to high salinity. Rouphael et al. [18,19] also suggested that the biostimulants obtained from seaweeds or vegetal proteins may increase tolerance to high salinity in lettuce plants through alterations in metabolic processes that induce the biosynthesis of stress-related compounds. Van Oosten et al. [20] suggested that the key mechanisms of action that are related to the beneficial effects of protein hydrolysates and plant extracts rich in amino acids as well as peptides on crops are due to osmoprotection and the scavenging of radicals at the shoot level, in addition to metal chelation and improved nutrient availability at the rhizosphere level.

Irrigation with saline water may induce metabolic changes that affect the chemical composition of vegetable products, while it may also affect the visual appearance and marketability of leafy products [21]. Apart from negative effects, moderate salinity levels may improve the quality of vegetable products through the enhancement of nutraceutical compound content (e.g., phenolic compounds, carotenoids, tocopherols, vitamins, and other antioxidant compounds) or by contributing to the taste and aroma attributes [22]. Therefore, the combinatory effects of salinity and biostimulant application have to be extensively studied in order to identify those salinity thresholds and biostimulant products that allow the production of high-quality end-products without compromising marketable yield. Moreover, a thorough evaluation of germplasm variability has to be considered, since a differential response to salinity is expected depending on the genotype [23].

Successive harvesting is a common practice in various leafy greens and, according to the literature, may significantly increase the overall yield when compared to single harvesting practices [24]. Moreover, aside from increased yields this technique is cost- as well as labor-efficient and may allow for more growth cycles throughout the growing period [25]. The number of harvests is dependent on the genotype and the growing conditions, since not all of the cultivars and growing periods are suitable for this technique due to susceptibility to inflorescence formation under specific temperature and photoperiod conditions [26]. Considering that successive harvesting may occur under variable growing conditions, this practice may also affect the chemical composition and quality of leafy vegetables, such as rocket [27].

Rocket (*Eruca sativa* Miller) and wild rocket (*Diplotaxis tenuifolia* L.) are two leafy vegetables of the Brassicaceae family with increasing commercial interest during recent years due to their special characteristics in terms of nutritional value, aroma, and taste [28]; however, in addition to the high nutritional value of its edible leaves there is great concern regarding the tendency of nitrate accumulation, which is considered to be an anti-nutritional factor and may pose significant threats to human health [29–31]. Therefore, special attention should be given to those agronomic practices that may contribute to the reduction in leaves' nitrate content, such as the cropping season and time of harvesting, nitrogen fertilization rates, or nitrogen form, among others [29,32–34]. In this context, the use of biostimulants has been associated with an increased content of nitrate in leafy vegetables, although contrasting results have been reported in the literature [35–37]. These findings indicate that biostimulant application should be carefully considered in order to avoid any undesirable effects on the quality of the final product, since it seems that the impact on nitrate content depends on the species, the biostimulant product, and the cropping system [30].

In this work, a pot experiment was conducted in order to evaluate the effect of two biostimulatory products on the mineral contents and chemical profiles of *Diplotaxis tenuifolia* plants grown under saline conditions. Moreover, the agronomic practice of successive harvesting was implemented, aiming to determine any effects of growth stage on plants' responses to salinity and biostimulant application.

## 2. Materials and Methods

### 2.1. Experimental Settings and Design, Crop Management, and Soil Sampling

The experiment was carried out at the experimental field of the Department of Agricultural Science in Portici (Naples, 40° 48.870′ N; 14° 20.821′ E; 70 m a.s.l.) under a plastic greenhouse (2020–2021). The species chosen for the test was wild rocket (*Diplotaxis tenuifolia* L.) cv. "Reset", marketed by Maraldi Sementi Srl (Cesena—FC, Italy), with the following characteristics: green leaves with medium-sized lobes, high potential yield, and a noticeable flexibility in terms of growing conditions that makes it useful for cultivation in any season.

For the test, pots with a 0.38 $m^2$ surface and a 0.70 m height were used; they were filled with a sandy soil, whose physical and chemical properties are reported in Table 1.

**Table 1.** Physical and chemical properties of the test soil.

| Parameters | Unit | Mean Value |
|---|---|---|
| Sand | % | 91.0 |
| Silt | % | 4.5 |
| Clay | % | 4.5 |
| N—total (Kjeldahl method) | % | 0.101 |
| $P_2O_5$ (Olsen method) | mg kg$^{-1}$ | 253.0 |
| $K_2O$ (tetraphenylborate method) | mg kg$^{-1}$ | 490.0 |
| Organic matter | % | 2.5 |
| Electrical conductivity | dS m$^{-1}$ | |
| pH | | 7.4 |

Three seedling groups per pot (about 15–20 seedlings each) were transplanted on 8 October 2020, and harvested 6 consecutive times from 25 November 2020 until 20 May 2021, hereafter referred to as I, II, III, IV, V, and VI. Regarding fertilization, only nitrogen in the form of ammonium nitrate (26%) was applied, provided at a rate corresponding to 18 kg ha$^{-1}$ per each cycle. For the first cycle, it was added 18 days after transplant (DAT), while in the successive cycles this was carried out at approximately 4 days after harvest (DAH) of the previous growing cycle, ranging between 2 and 9 DAH depending on the conditions. No pesticide treatments were carried out.

The experimental design was a split-plot design, with saline irrigation as the main experimental factor and biostimulant application as the subfactor. The saline irrigation treatments included the following: EC0: irrigation with tap water; EC2: irrigation with water that had an EC of 2.0 dS m$^{-1}$; EC4: irrigation with water that had an EC of 4.0 dS m$^{-1}$; and EC6: irrigation with water that had an EC of 6.0 dS m$^{-1}$. There were three biostimulant application treatments: 1. untreated—control; 2. treated with Auxym®, a tropical plant extract hereafter referred to BA; and 3. treated with Trainer®, a protein hydrolysate derived from legumes, hereafter referred to BT. Both biostimulants are marketed by Hello-Nature Italia Srl (Rivoli Veronese, Italy), and they were applied as a foliar spray three times per each cycle (except for the last cycle, which was shorter; therefore, only two applications were performed) at doses of 2 mL L$^{-1}$ and 3 mL L$^{-1}$ for Auxym®and Trainer®, respectively.

Starting with harvest II, the formation of new leaves was taken into account to decide when to perform the first biostimulant application. All of the treatments were replicated 3 times for a total of 36 pots.

The water losses were calculated by the Hargreaves method [38], and were fully restored by 26 irrigations over the six cycles. The desired EC for saline treatments was obtained by adding NaCl to tap water, as reported by Di Mola et al. [39]. A total of 28.5 L per pot was applied with 34.6, 69.1, and 103.7 g of NaCl per pot in the EC2, EC4, and EC6 treatments, respectively. At the first cycle, the first three irrigations were made with tap water for all of the treatments in order to promote the rooting and establishment of seedlings. To monitor the trend in the soil electrical conductivity, at each harvest the soil was sampled at a 0–20 cm depth and the soil water solution extraction method (1:5 dilution) was used to measure soil EC ($EC_{1:5}$; Basic 30 CRISON electrical conductivity meter; Crison Hach, Barcelona, Spain).

The air temperature under the plastic greenhouse was monitored by a Vantage Pro2 (Davis Instruments, Hayward, CA, USA) weather station on an hourly basis. Data are reported as the daily minimum and maximum temperatures (Supplementary Figure S1).

### 2.2. ABTS and Hydrophilic Antioxidant Activity, Total Phenols, and Total Ascorbic Acid Analysis

At each harvest, fresh leaves for each treatment and replicate were collected and stored in a freezer at −80 °C, after which they were lyophilized for the determination of ABTS and hydrophilic antioxidant activity (ABTS AA and HAA, respectively), total phenols, and total ascorbic acid (TAA).

The ABTS AA was determined on 200 mg of freeze-dried sample extracted with methanol. The measurement of the absorbance was assessed spectrophotometrically (Hach DR 2000, Hach Co., Loveland, CO, USA) at 734 nm according to the methods of Re et al. [40]. The ABTS AA was expressed as mmol of Trolox per 100 g of dry weight (dw).

The HAA was assessed, after extraction with distilled water, by the N, N-dimethyl-p-phenylenediamine (DMPD) method [41]. It was measured spectrophotometrically at 505 nm, and the values were expressed as mmol of ascorbic acid per 100 g of dry weight (dw).

The TAA was also spectrophotometrically measured at 525 nm according to the procedure of Kampfenkel et al. [42].

The total phenolic content was determined spectrophotometrically at 765 nm according to the Singleton et al. method [43] and expressed as mg of gallic acid per 100 $g^{-1}$ of dw.

### 2.3. Chlorophylls and Carotenoids Analysis

Chlorophylls (chlorophylls a and b) and carotenoids were assessed spectrophotometrically on 1 g of fresh leaves, after extraction with ammoniacal acetone, according to the method described by Wellburn [44] at 662 and 647 nm for chlorophylls a and b, respectively, as well as at 470 nm for carotenoids. They were expressed as mg $g^{-1}$ fresh weight (fw).

### 2.4. Mineral Content Analysis

The determination of nitrate, Na, K, Ca, Mg, Cl, S, and P was carried out on 250 mg of dried and finely ground leaf tissues, suspended in 50 ml of ultrapure water (Milli-Q, Merck Millipore, Darmstadt, Germany), followed by a shaking water bath (ShakeTemp SW22, Julabo, Seelbach, Germany) at 80 °C for 10 min. The supernatant was filtered and processed on an ion chromatograph (ICS3000, Thermo Scientific™ Dionex™, Sunnyvale, CA, USA) as described previously by Rouphael et al. [45]. Mineral concentrations were expressed as mg $g^{-1}$ dw, while nitrate concentration was converted into mg $kg^{-1}$ fw.

### 2.5. Statistical Analysis

All of the results were subjected to a three-way analysis of variance (ANOVA), considering salinity levels (S), biostimulant application (B), and harvesting time (H) as factors. The analysis was performed with the SPSS software package (version 22, Chicago, IL, USA).

When significant effects were detected, the means were separated using Tukey's honestly significant difference (HSD) test at $p < 0.05$.

## 3. Results and Discussion

### 3.1. Soil Electrical Conductivity

The soil electrical conductivity linearly increased in all of the saline treatments, but with different trends; at the last harvest the increase over the first harvest was more than double in EC2, and about three times in EC4 as well as EC6 (Figure 1). Moreover, the EC values were significantly higher in EC6 compared to the rest of the salinity levels, while the EC0 treatment presented the lowest overall values in all of the harvestings, being significantly lower than the rest of the treatments at the last four harvestings (II–IV). Finally, all of the salinity treatments differed among each other in the last three harvestings. Similarly to our findings, Mori et al. [46] also suggested an increase in soil EC values with the increasing salinity of the nutrient solution, which is attributed to the gradual buildup of Na and Cl in soil solutions due to inhibited water uptake from plants. Moreover, Schiattone et al. [47] also suggested a gradual increase in soil EC with increasing salinity and multiple harvestings in wild rocket plants, although the increase was less profound between different harvesting times of the same salinity level compared to the increase between the different salinity levels. According to Feng et al. [48], the application of saline irrigation water resulted in an increase in EC levels in soil solutions which evolved with the progression of the growing cycle, while similar findings were recorded for the use of brackish water in cotton plant irrigation [49].

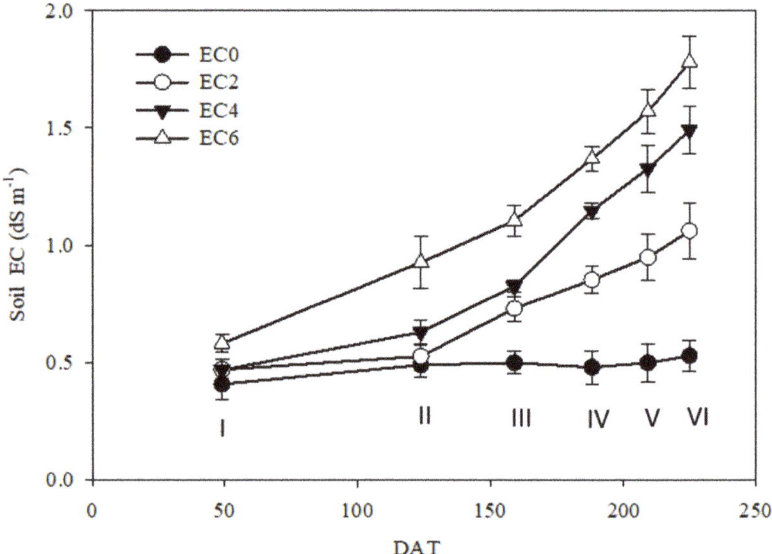


**Figure 1.** Trends in soil electrical conductivity (dS m$^{-1}$) over the six harvesting times (I–VI). Vertical bars above each mean indicate standard deviation (SD; n = 3).

### 3.2. Antioxidant Activity and Compounds

The analysis of variance of the data about antioxidant activity (hydrophilic antioxidant activity (HAA) and ABTS assay), total phenol content, total ascorbic acid content (TAA), chlorophylls (a, b, and total chlorophyll), and total carotenoids is presented in the Supplementary Material (Table S1). The analysis did not show any significant interaction of the three tested factors, namely salinity (S), biostimulants (B), and harvesting time (H). On the other hand, significant interactions were recorded between B × H, S × H, and S × B for specific parameters, e.g., total phenols for the B × H interaction; HAA, ABTS, and TAA for the S × H interaction; and HAA, ABTS, total phenols, and TAA for the S × B interaction.

Supplementary Table S3 presents the effect of biostimulant application and harvesting time on HAA, the ABTS assay, total phenols, and TAA content in *Diplotaxis tenuifolia* leaves, regardless of salinity level. As indicated in Supplementary Table S1, the only significant interaction was recorded in the case of total phenol content. The highest total phenol content (Figure 2) was recorded in *D. tenuifolia* leaves collected from plants treated with the Auxym®biostimulant and harvested at the fourth harvesting time, without being significantly different from the untreated plants harvested at the same harvesting time. On the other hand, the lowest overall value was recorded for the Auxym®biostimulant and the third harvesting time, without being significantly different from untreated plants of the first harvesting time. Moreover, for each biostimulant an increasing trend in total phenols was recorded until the fourth harvest, followed by a significant decline for the last two harvestings, especially in plants treated with biostimulants. In contrast to our study, Schiattone et al. [50] did not report a significant effect of biostimulants on total phenol content, while they suggested a significant effect of two biostimulants (azoxystrobin and extracts of yeast as well as brown algae) in wild rocket plants grown under nitrogen deprivation conditions. Moreover, Candido et al. [51] suggested a differential response of wild rocket plants to azoxystrobin and nitrogen deprivation depending on the cropping cycle, thus indicating that growing conditions are pivotal for plants' responses to abiotic stressors and/or biostimulant application. Similarly, Giordano et al. [52] did not record a significant difference in two successive harvests in terms of the total phenol content of perennial wall rocket plants treated with two biostimulants, whereas TAA content significantly increased for both biostimulants over the control treatment. Moreover, Carillo et al. [53] reported a decrease in total phenols in lettuce plants where two successive harvests were implemented, while TAA content showed the opposite trend. On the other hand, Corrado et al. [54] did not observe any significant differences in the total phenol content of basil plants in two successive harvests, while Caruso et al. [35] indicated an increase in total phenol content in lettuce plants grown in two different seasons (winter and winter–spring) and subjected to two biostimulants (tropical plant extracts and legume-derived protein hydrolysates). The contrasting results in literature reports could be associated with differences in the genotypes tested, since, according to Ciriello et al. [55], a significant difference in the total phenol content of different basil genotypes was recorded in two successive harvestings.

Table 2 presents the effect of biostimulant application and salinity level on HAA, the ABTS assay, total phenols, and TAA content in *D. tenuifolia* leaves, regardless of the harvesting time. A varied response was recorded depending on the antioxidant activity assay. In particular, the non-salinized untreated plants (EC0 × BC treatment) recorded the highest and lowest values for HAA and ABTS AA assays, respectively, whereas an opposite trend was observed for plants that were grown under mid- to high-salinity levels and did not receive any biostimulant application (EC4 × BC). This result is in agreement with the findings of Di Mola et al. [56], who also suggested that non-salinized lettuce plants had the highest HAA, while El-Nakhel et al. [16] also reported a contrasting response of spinach plants to salinity stress and biostimulant application in regard to HAA and ABTS AA assays. Regarding total phenols and TAA content, no specific trends were recorded since no significant differences were recorded between most of the treatments, although the lowest overall content was recorded for the highest salinity level and the plants treated with the Trainer®biostimulant; however, El-Nakhel et al. [16] suggested a significant decrease in total phenols and TAA content for spinach plants treated with a legume-derived protein hydrolysate, whereas increasing salinity resulted in an increase in total phenol content and had no effect on TAA content. On the other hand, Rouphael et al. [36] noted a beneficial effect of biostimulant application on the total phenol content of spinach plants. According to Carillo et al. [53], no significant effects of salinity on the total phenol content, as well as the lipophilic and hydrophilic antioxidant activity, of lettuce plants grown under salinity stress were noted, while the same authors suggested fluctuating trends for TAA content.

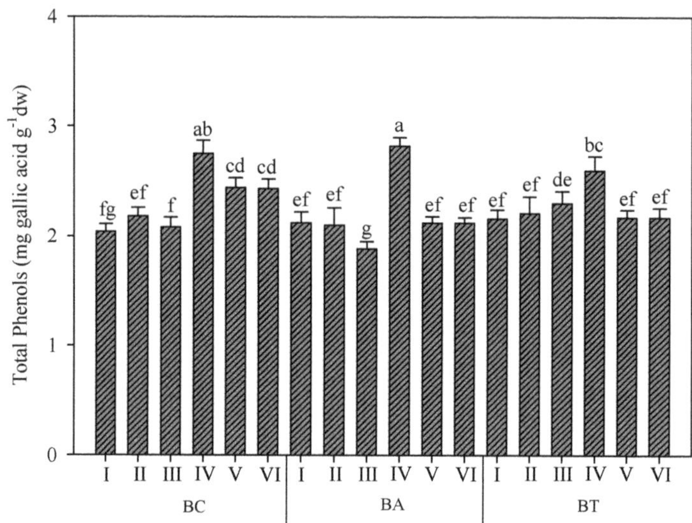

**Figure 2.** Effect of biostimulant and harvesting time on total phenol content in *Diplotaxis tenuifolia* leaves, regardless of salinity level. Different letters indicate significant differences according to Tukey's honestly significant difference (HSD) test at $p = 0.05$; Roman numbers (I–VI) indicate the successive harvests; BC: no biostimulants added; BA: plants treated with Auxym®; and BT: plants treated with Trainer®. All data are expressed as mean $\pm$ SE (standard error), n = 3.

**Table 2.** Effect of salinity and biostimulant on hydrophylic antioxidant activity (HAA), ABTS antioxidant activity, total phenols, and total ascorbic acid (TAA) content in *Diplotaxis tenuifolia* leaves, regardless of harvesting time.

| Treatment | | HAA mmol Ascorbic Acid equ. 100 g$^{-1}$ dw | ABTS AA mmol Trolox equ. 100 g$^{-1}$ dw | Total Phenols mg Aallic Acid g$^{-1}$ dw | TAA Total Ascorbic Acid mg 100 g$^{-1}$ fw |
|---|---|---|---|---|---|
| EC0 | BC | $10.17 \pm 0.44$ a | $10.09 \pm 0.62$ d | $2.29 \pm 0.08$ ab | $32.88 \pm 7.05$ a |
| | BA | $9.75 \pm 0.30$ a–c | $11.49 \pm 0.35$ b–d | $2.19 \pm 0.08$ a–c | $30.27 \pm 8.00$ ab |
| | BT | $9.70 \pm 0.45$ a–c | $12.34 \pm 0.57$ a–c | $2.37 \pm 0.09$ a | $27.29 \pm 5.42$ a–c |
| EC2 | BC | $8.52 \pm 0.32$ d–f | $11.61 \pm 0.70$ b–d | $2.31 \pm 0.12$ a | $26.31 \pm 5.40$ a–c |
| | BA | $9.89 \pm 0.32$ ab | $10.86 \pm 0.56$ cd | $2.32 \pm 0.09$ a | $31.88 \pm 5.75$ a |
| | BT | $9.11 \pm 0.25$ b–d | $12.32 \pm 0.51$ a–c | $2.40 \pm 0.07$ a | $32.37 \pm 8.23$ a |
| EC4 | BC | $7.95 \pm 0.40$ f | $13.88 \pm 0.66$ a | $2.41 \pm 0.11$ a | $22.85 \pm 4.75$ bc |
| | BA | $8.93 \pm 0.31$ c–e | $12.69 \pm 0.36$ ab | $2.07 \pm 0.11$ bc | $26.74 \pm 4.53$ a–c |
| | BT | $8.59 \pm 0.28$ d–f | $12.14 \pm 0.41$ a–c | $2.27 \pm 0.10$ ab | $30.63 \pm 4.60$ ab |
| EC6 | BC | $8.10 \pm 0.36$ ef | $11.73 \pm 0.67$ b–d | $2.28 \pm 0.05$ ab | $23.52 \pm 2.52$ bc |
| | BA | $8.62 \pm 0.38$ d–f | $13.19 \pm 0.77$ ab | $2.21 \pm 0.12$ a–c | $22.57 \pm 4.47$ bc |
| | BT | $8.21 \pm 0.35$ ef | $12.55 \pm 0.43$ a–c | $2.03 \pm 0.08$ c | $19.75 \pm 2.82$ c |

Different letters within each column indicate significant differences according to Tukey's honestly significant difference (HSD) test at $p = 0.05$. EC0: tap water (0 ds m$^{-1}$); EC2: 2.0 dS m$^{-1}$; EC4: 4.0 dS m$^{-1}$; EC6: 6.0 dS m$^{-1}$; BC: no biostimulants added; BA: plants treated with Auxym®; and BT: plants treated with Trainer®. All data are expressed as mean $\pm$ SE (standard error), n = 3.

Table 3 presents the effect of salinity level and harvesting time on HAA, the ABTS assay, and total ascorbic acid content in *D. tenuifolia* leaves, regardless of biostimulant application. HAA varied depending on the salinity level and the harvesting time, with the highest overall value being recorded for the non-salinized plants and the fourth harvesting time, whereas the highest salinity level and late harvesting (EC6 × VI) resulted in the lowest values for this particular assay. On the other hand, the EC6 × II and EC0 × VI treatments resulted in the highest and lowest values for the ABTS assay. Total phenol content was not affected by salinity levels and harvesting time, while for TAA content a decreasing trend with harvesting time was recorded, with the highest values being measured at the first harvesting time for all of the tested salinity levels. The lack of effect of harvesting time on the total phenol content (data not shown) of perennial wall rocket plants has been previously suggested by Giordano et al. [52], while Carillo et al. [53] did not record a significant effect of salinity on the antioxidant activity and total phenol content of lettuce plants.

**Table 3.** Effect of salinity and biostimulant on hydrophylic antioxidant activity (HAA), ABTS antioxidant activity, and total ascorbic acid (TAA) content in *Diplotaxis tenuifolia* leaves, regardless of biostimulant application.

| Treatment | | HAA<br>mmol Ascorbic Acid<br>equ. 100 g$^{-1}$ dw | ABTS AA<br>mmol Trolox<br>equ. 100 g$^{-1}$ dw | TAA<br>Total Ascorbic Acid<br>mg 100 g$^{-1}$ fw |
|---|---|---|---|---|
| EC0 | I | 8.00 ± 0.34 ik | 13.33 ± 0.56 a–c | 86.76 ± 7.68 a |
| | II | 9.54 ± 0.37 d–f | 12.78 ± 0.95 b–e | 33.41 ± 2.24 d |
| | III | 9.90 ± 0.22 c–e | 11.38 ± 0.58 e–h | 19.64 ± 1.44 e |
| | IV | 12.61 ± 0.57 a | 11.00 ± 0.58 g–i | 18.83 ± 4.01 e–g |
| | V | 9.74 ± 0.17 c–e | 9.93 ± 0.64 ij | 11.37 ± 1.95 hi |
| | VI | 9.44 ± 0.17 ef | 9.43 ± 0.64 j | 10.87 ± 1.93 i |
| EC2 | I | 9.26 ± 0.40 e–g | 9.83 ± 1.24 ij | 85.80 ± 6.49 a |
| | II | 10.78 ± 0.39 b | 12.30 ± 0.66 b–g | 30.89 ± 3.01 d |
| | III | 8.97 ± 0.42 f–h | 12.94 ± 0.40 a–c | 18.76 ± 1.85 e–g |
| | IV | 8.51 ± 0.40 h–j | 12.85 ± 0.69 b–d | 17.00 ± 3.00 e–i |
| | V | 8.93 ± 0.38 f–h | 11.09 ± 0.75 f–i | 14.58 ± 1.82 e–i |
| | VI | 8.63 ± 0.38 g–i | 10.59 ± 0.75 h–j | 14.08 ± 1.84 e–i |
| EC4 | I | 8.02 ± 0.22 i–k | 12.31 ± 0.62 b–g | 60.86 ± 6.95 b |
| | II | 10.18 ± 0.18 b–d | 13.01 ± 0.82 a–c | 26.93 ± 2.76 d |
| | III | 8.09 ± 0.37 i–k | 13.65 ± 1.02 ab | 18.12 ± 1.08 e–g |
| | IV | 8.51 ± 0.64 h–j | 13.08 ± 0.64 a–c | 17.65 ± 4.00 e–h |
| | V | 8.22 ± 0.47 ij | 12.94 ± 0.65 a–c | 18.69 ± 3.56 e–g |
| | VI | 7.92 ± 0.47 jk | 12.44 ± 0.65 b–f | 18.19 ± 3.58 e–g |
| EC6 | I | 8.13 ± 0.55 i–k | 12.18 ± 0.79 c–g | 45.37 ± 5.32 c |
| | II | 10.30 ± 0.31 bc | 14.28 ± 1.16 a | 27.01 ± 1.68 d |
| | III | 8.53 ± 0.42 h–j | 12.53 ± 0.98 b–e | 19.63 ± 2.49 ef |
| | IV | 8.30 ± 0.55 h–j | 12.47 ± 0.49 b–f | 12.90 ± 1.73 g–i |
| | V | 7.45 ± 0.17 kl | 12.00 ± 0.94 c–g | 13.65 ± 1.79 e–i |
| | VI | 7.15 ± 0.17 l | 11.50 ± 0.94 d–h | 13.15 ± 1.77 f–i |

Different letters within each column indicate significant differences according to Tukey's honestly significant difference (HSD) test at $p = 0.05$. EC0: tap water (0 ds m$^{-1}$); EC2: 2.0 dS m$^{-1}$; EC4: 4.0 dS m$^{-1}$; and EC6: 6.0 dS m$^{-1}$. Roman numbers (I–VI) indicate the successive harvests. All data are expressed as mean ± SE (standard error), n = 3.

These contradictory results indicate that growing conditions are pivotal for the responses of *D. tenuifolia* plants to salinity stress and biostimulant application. This could be due to the fact that there is a combinatory effect on the plant protective mechanisms that induces the biosynthesis of antioxidant compounds as well as their activity. As reported in the literature and suggested by the results of the present study, when several factors are tested at the same time (e.g., biostimulant applications, harvesting time, salinity stress) the response of plants to studied parameters may vary depending on the experimental conditions. This response may also differ compared to the response to single factors. For example, in our study total phenol content was affected by salinity and harvesting time or salinity and biostimulant application, but no effect was recorded when salinity and harvesting time were considered. In the study by Bulgari et al. [57], a significant increase in total phenol content was observed for rocket plants grown in a floating system and subjected to EC levels of 3.5 dS m$^{-1}$. Hamilton and Fonseca [58] reported a variable effect of salinity levels (up to 9.6 dS m$^{-1}$) on rocket plants (*Eruca sativa* and *Diplotaxis tenuifolia*) depending on the growing period (March to April and May to June). Similarly, El-Nakhel et al. [16]

recorded an increase in total phenols in spinach plants with increasing salinity. The same authors suggested a negative effect of a legume-derived protein hydrolysate compared to the control plants [16]. On the other hand, Corrado et al. [59] did not record a significant effect of salinity on total phenol content in two lettuce varieties, while they suggested a significant effect of harvesting time and genotype. According to Franzoni et al. [60], the benefits from biostimulant application on plants under stress could be associated with an increase in the expression of transcription factors involved in plant responses to stress, such as the induction of cuticular waxes, phospholipid and brassinosteroid biosynthesis, the regulation of sugar metabolism, and intracellular transport. Therefore, it seems that plant responses to external factors related to stress (e.g., salinity) or growth promotion (e.g., biostimulant) are highly affected by environmental and growing conditions. For this reason, further research is needed in order to define those conditions (e.g., the combinations of harvesting time, salinity level, and biostimulant product) that are favorable to the induction of secondary metabolism, allowing plants to better cope with salinity stress.

### 3.3. Chlorophyll and Carotenoid Content

Table 4 presents the result of chlorophyll and carotenoid content in relation to salinity level, biostimulant application, and harvesting time. Considering that no significant interactions between the tested factors were detected, only the main effects of each factor are presented. Chlorophylls a and b as well as total chlorophyll content was significantly affected by biostimulant application, with Trainer®showing the highest content compared to the rest of the treatments. On the other hand, salinity level affected chlorophyll b and total chlorophyll content, with increasing salinity resulting in a significant decrease, especially at the highest salinity level tested. The effect of harvesting time on chlorophyll content did not show specific trends except for chlorophyll b, where a decrease was recorded with increasing salinity, whereas chlorophyll a and total chlorophyll content fluctuated over the growing period. Moreover, total carotenoid content showed a significant increase at low salinity (EC2), followed by a reduction with increasing salinity to levels similar to the control treatment. Finally, harvesting time seems to have a varied effect, with a slight increase at the second harvest followed by a decrease at subsequent harvesting, especially the late ones (harvests V and VI), where the lowest values were recorded.

The findings of our study are in agreement with reports in the literature, where it is suggested that salinity stress induces the disruption of the photosynthetic apparatus through the damage of chloroplasts and the decrease in chlorophyll content, especially in older leaves, which tend to accumulate more ions than younger ones [3]. Moreover, the application of biostimulants may mitigate the negative effects of abiotic stressors on chlorophyll content in leafy vegetables such as lettuce [61]. In harmony, El-Nakhel et al. [16] indicated a negative effect of increasing salinity on the chlorophyll content of spinach plants, while suggesting a positive effect from the application of a biostimulant that contained protein hydrolysates. The same authors recorded an increase in total carotenoids with increasing salinity, whereas biostimulant application had no significant effect on this parameter. Lucini et al. [17] noted a significant decrease in SPAD index values and chlorophyll fluorescence in lettuce plants grown under saline conditions, while biostimulant application (of plant-derived protein hydrolysates) only mitigated the negative effects on chlorophyll fluorescence and not those on the SPAD index. A similar finding with our study was reported by Caruso et al. [27], who indicated a positive effect of biostimulants on chlorophyll b content and no effects on carotenoid content, while they also mentioned that cropping season did not affect the abovementioned parameters. The single effect of biostimulant application (borage leaf or flower extracts) did not affect chlorophyll and carotenoid content in wild rocket plants, indicating the lack of effect in unstressed plants [62]. This argument was confirmed by Franzoni et al. [60], who suggested that borage extracts mitigated the negative effects of salinity on chlorophyll a fluorescence but had no effect on unstressed plants. It seems that growing conditions may affect the responses of plants to salinity and biostimulant application, since, according to Giordano et al. [52], a variable effect of protein hydrolysates and plant extracts on the chlorophyll content of perennial wall rocket leaves was recorded, depending on harvesting time.

**Table 4.** Effect of salinity, biostimulant, and harvest on chlorophyll a, chlorophyll b, total chlorophylls, and carotenoid averages in *Diplotaxis tenuifolia*.

| Treatment | Chlorophyll a mg g$^{-1}$ fw | Chlorophyll b mg g$^{-1}$ fw | Total Chlorophyll mg g$^{-1}$ fw | Carotenoids mg g$^{-1}$ fw |
|---|---|---|---|---|
| EC0 | 1.03 ± 0.01 | 0.57 ± 0.02 a | 1.61 ± 0.03 a | 0.323 ± 0.012 b |
| EC2 | 1.01 ± 0.01 | 0.52 ± 0.01 ab | 1.52 ± 0.02 b | 0.348 ± 0.006 a |
| EC4 | 1.02 ± 0.01 | 0.53 ± 0.01 ab | 1.58 ± 0.04 ab | 0.329 ± 0.009 ab |
| EC6 | 1.00 ± 0.01 | 0.49 ± 0.02 b | 1.50 ± 0.03 b | 0.328 ± 0.005 ab |
| BC | 1.00 ± 0.01 b | 0.50 ± 0.01 b | 1.52 ± 0.03 b | 0.326 ± 0.010 |
| BA | 1.01 ± 0.01 b | 0.51 ± 0.01 b | 1.52 ± 0.02 b | 0.340 ± 0.006 |
| BT | 1.05 ± 0.01 a | 0.57 ± 0.02 a | 1.62 ± 0.02 a | 0.329 ± 0.005 |
| I | 1.04 ± 0.02 a | 0.57 ± 0.02 a | 1.60 ± 0.03 ab | 0.348 ± 0.005 b |
| II | 1.00 ± 0.01 ab | 0.57 ± 0.02 a | 1.54 ± 0.03 a–c | 0.397 ± 0.005 a |
| III | 1.04 ± 0.01 a | 0.55 ± 0.02 ab | 1.59 ± 0.03 ab | 0.342 ± 0.003 b |
| IV | 0.98 ± 0.01 b | 0.52 ± 0.02 a–c | 1.47 ± 0.03 c | 0.334 ± 0.004 b |
| V | 1.03 ± 0.01 ab | 0.50 ± 0.01 bc | 1.63 ± 0.05 a | 0.296 ± 0.010 c |
| VI | 1.02 ± 0.01 ab | 0.47 ± 0.01 c | 1.49 ± 0.02 bc | 0.276 ± 0.010 c |

Different letters within each column and for the same factor indicate significant differences according to Tukey's honestly significant difference (HSD) test at $p = 0.05$. EC0: tap water (0 ds m$^{-1}$); EC2: 2.0 dS m$^{-1}$; EC4: 4.0 dS m$^{-1}$; EC6: 6.0 dS m$^{-1}$; BC: no biostimulants added; BA: plants treated with Auxym®; and BT: plants treated with Trainer®. The Roman numbers (I–VI) indicate the successive harvests. All data are expressed as mean ± SE (standard error), n = 3.

### 3.4. Leaf Nutrient Composition

The cation (Na, K, Ca, and Mg) concentrations were significantly affected by the second-degree interactions of salinity × biostimulant and salinity × harvest, except for K, which was also affected by the interaction of biostimulant × harvest (Supplementary Material, Table S2). In regard to the anions, Cl content was significantly affected by all three of the second-degree interactions; S content only by the S × H interaction and P content by the S × B interaction (Supplementary Material, Table S2). As for the nitrate content, it was affected by all of the second-degree interactions.

At lower salinity levels (EC0 and EC4) no significant differences were recorded between the biostimulant treatments, while at EC4 Auxym®application resulted in a significant decrease in Na content, which was not the case at the highest salinity level (EC6). At this salinity level, the highest overall Na content was recorded for this particular biostimulant without being significantly different from the other biostimulant treatment (Table 5). In regard to K concentration, it reached the highest value in plants that were not treated with biostimulants and which were irrigated with tap water, not being different from EC0 × BA, all of the treatments of EC2, and BA × EC4 (Table 5). For Mg concentration, only in EC2 was the treatment with Trainer®different from BC, while significant differences were also recorded from EC4 × BA and EC6 (both BC and BT treatments) (Table 5). In the plants irrigated with tap water or water with low salinity (EC2), BT elicited a higher value of Ca concentration in the leaves, but in EC2 it was not different from BA; in higher salinity levels no differences were recorded between plants treated and untreated with biostimulants (Table 5). Biostimulant application reduced Cl concentration in leaves up to moderate salinity levels (EC4 treatment), while at the highest salinity tested (EC6) no significant differences were recorded between BA, BT, and BC (Table 5). Sulfur content was not affected by the factors tested (data not shown). In regard to P concentration, at low (EC2) and moderate salinity stress (EC4) the plants treated with Trainer®showed higher values, contrary to what occurred in EC0 and EC6, where the treatment with BA elicited higher values, although no significant differences were recorded between the two biostimulant products (Table 5). In accordance with our study, Lucini et al. [17] reported that increasing salinity in a nutrient solution resulted in increased Na content in lettuce leaves, whereas biostimulant application showed no significant effect. Similarly, K, P, Ca, and Mg content decreased with increasing salinity, while biostimulant application only affected P content [17]. Moreover, increased salinity has been associated with the disruption of nutrient and osmotic balance due to increased ratios of Na/K, Na/Ca, and Na/Mg [63].

**Table 5.** Effect of salinity and biostimulant application on minerals (Na, K, Ca, Mg, Cl, and P) and nitrate content in *Diplotaxis tenuifolia* leaves, regardless of harvesting time.

| Treatment | | Mineral Composition (g kg$^{-1}$ dw) | | | | | | Nitrate (mg kg$^{-1}$ fw) |
|---|---|---|---|---|---|---|---|---|
| | | Na | K | Ca | Mg | Cl | P | |
| EC0 | BC | 5.14 ± 0.31 e | 60.06 ± 2.31 a | 22.50 ± 0.98 bc | 5.09 ± 0.13 a | 25.75 ± 1.57 de | 2.66 ± 0.11 ab | 3791.5 ± 391.0 de |
| | BA | 4.81 ± 0.18 e | 57.16 ± 1.50 ab | 22.00 ± 0.69 c–e | 4.79 ± 0.12 ab | 20.47 ± 1.00 ef | 2.81 ± 0.08 a | 5403.9 ± 320.2 a |
| | BT | 4.72 ± 0.23 e | 53.66 ± 1.44 bc | 24.50 ± 0.96 a | 5.12 ± 0.16 a | 19.20 ± 0.63 f | 2.68 ± 0.09 ab | 5799.0 ± 365.8 a |
| EC2 | BC | 10.39 ± 0.64 d | 55.41 ± 1.58 a–c | 22.24 ± 0.70 b–d | 4.63 ± 0.10 b | 37.57 ± 2.41 c | 2.60 ± 0.08 ac | 3286.7 ± 279.5 f |
| | BA | 8.45 ± 0.55 d | 54.35 ± 2.63 a–c | 24.07 ± 0.51 ab | 4.86 ± 0.12 ab | 28.23 ± 1.52 d | 2.49 ± 0.07 b–d | 4569.8 ± 245.8 bc |
| | BT | 9.43 ± 0.54 d | 55.62 ± 1.32 a–c | 22.90 ± 0.56 a–c | 5.09 ± 0.08 a | 29.83 ± 1.69 d | 2.69 ± 0.09 ab | 4810.2 ± 210.0 b |
| EC4 | BC | 14.83 ± 1.26 ab | 52.00 ± 1.07 bc | 21.08 ± 0.99 c–f | 4.86 ± 0.09 ab | 50.81 ± 3.85 a | 2.60 ± 0.07 a–c | 2746.0 ± 396.1 g |
| | BA | 10.88 ± 0.82 cd | 55.46 ± 1.38 a–c | 19.91 ± 0.65 f | 4.62 ± 0.09 b | 37.05 ± 2.03 c | 2.69 ± 0.08 ab | 4230.9 ± 228.1 cd |
| | BT | 14.17 ± 1.11 ab | 53.89 ± 2.00 bc | 20.53 ± 0.56 d–f | 4.85 ± 0.10 ab | 44.84 ± 2.33 b | 2.81 ± 0.10 a | 3962.6 ± 227.6 d |
| EC6 | BC | 13.86 ± 1.02 bc | 51.22 ± 1.40 bc | 19.88 ± 0.86 f | 4.66 ± 0.11 b | 49.91 ± 2.91 ab | 2.19 ± 0.08 d | 2422.4 ± 453.5 g |
| | BA | 17.06 ± 1.29 a | 52.07 ± 1.82 bc | 19.99 ± 0.77 f | 4.76 ± 0.09 ab | 50.34 ± 3.16 ab | 2.42 ± 0.08 b–d | 3471.0 ± 285.5 ef |
| | BT | 15.95 ± 1.27 ab | 50.55 ± 1.38 c | 20.18 ± 0.78 ef | 4.59 ± 0.11 b | 49.14 ± 2.98 ab | 2.32 ± 0.07 cd | 3740.4 ± 254.5 d–f |

Different letters within each column indicate significant differences according to Tukey's honestly significant difference (HSD) test at $p = 0.05$. EC0: tap water (0 ds m$^{-1}$); EC2: 2.0 dS m$^{-1}$; EC4: 4.0 dS m$^{-1}$; EC6: 6.0 dS m$^{-1}$; BC: no biostimulants added; BA: plants treated with Auxym®; and BT: plants treated with Trainer®. All data are expressed as mean ± SE (standard error), n = 3.

Nitrate content was significantly increased by biostimulant application for all of the tested salinity levels compared to the untreated plants (no biostimulants added), while a decreasing trend with increasing salinity was recorded for all of the biostimulant treatments (with or no biostimulants added) (Table 5). Similar results were recorded by El-Nakhel et al. [16], who also suggested an increase in nitrate content in spinach plants treated with a legume-derived protein hydrolysate, while increasing salinity also resulted in a significant decrease in nitrate. It seems that the high availability of Cl in the nutrient solution may impair nitrate uptake and decrease its content in plant tissue, while biostimulants may serve as nitrogen pools and contribute to the accumulation of nitrate in leaves without an increased uptake of exogenous nitrogen being observed [64,65]. In contrast, Bulgari et al. [62] reported a decrease in nitrate content in wild rocket plants treated with borage extracts, which indicates that biostimulant composition and the presence of nitrogenous compounds could be responsible for the increase in nitrate observed in other studies. Moreover, Bonasia et al. [57] recorded a variable effect of increasing salinity on nitrate content in wild rocket plants, depending on the cropping system and the genotype tested. Similarly, Giordano et al. [52] did not observe a significant effect of biostimulant application on the nitrate content of perennial wall rocket leaves despite the presence of nitrogen in one of the biostimulant products, which highlights the importance of harvesting time for the combined responses of plants.

As expected, the concentrations of Na and Cl increased when the salinity levels in the nutrient solution increased; in fact, the mean values of Na and Cl concentration were 4.89, 9.42, 13.29, and 15.62 g kg$^{-1}$ as well as 21.81, 31.87, 44.23, and 49.79 g kg$^{-1}$ for EC0, EC2, EC4, and EC6, respectively (Table 6). In addition, both elements also increased over the harvest periods, with values that were about double at harvest VI compared to I: 12.40 vs. 5.71 for Na and 43.44 vs. 22.40 for Cl, respectively. In contrast, Ca and Mg decreased with increasing salinity levels from EC0 (23.00 and 5.00 g kg$^{-1}$, respectively) to EC6 (20.02 and 4.67 g kg$^{-1}$, respectively), while both elements showed the highest value at harvest II (Table 6). In regard to the other two anions, S increased with the increase in salinity stress, but only EC2 was different from EC0; over the harvest periods, S concentration decreased until harvest IV and then increased, reaching, at harvest VI, a value that was not significantly different from that of harvest I (Table 6). Instead, P concentration decreased when the salt concentration incremented, but only EC6 was different from all of the other saline treatments; moreover, its concentration significantly decreased over the harvest periods (Table 6). Nitrate content recorded fluctuating trends, with a varied effect being observed at the various salinity levels and harvesting times, although a decreasing trend at the late harvesting times (harvests IV to VI) was noticed.

**Table 6.** Effect of salinity and harvesting time on minerals (Na, Ca, Mg, Cl, and S) and nitrate content in *Diplotaxis tenuifolia* leaves, regardless of biostimulant application.

| Treatment | | Mineral Composition (g kg$^{-1}$ dw) | | | | | Nitrate (mg kg$^{-1}$ fw) |
|---|---|---|---|---|---|---|---|
| | | Na | Ca | Mg | Cl | S | |
| EC0 | I | 3.65 ± 0.36 m | 20.42 ± 0.86 fgh | 4.29 ± 0.13 ij | 18.61 ± 1.16 h | 7.52 ± 0.34 b–e | 4386.2 ± 338.4 ef |
| | II | 5.23 ± 0.21 lm | 28.97 ± 0.67 a | 5.73 ± 0.16 a | 18.86 ± 1.17 h | 7.44 ± 0.42 b–f | 5159.8 ± 238.9 d |
| | III | 5.44 ± 0.31 lm | 23.89 ± 1.23 cd | 5.46 ± 0.11 a | 22.19 ± 1.78 f–h | 5.18 ± 0.55 i | 6729.8 ± 487.5 a |
| | IV | 4.50 ± 0.20 lm | 22.58 ± 0.99 d–f | 4.81 ± 0.14 c–h | 21.09 ± 1.59 gh | 3.73 ± 0.46 j | 4780.7 ± 534.5 cd |
| | V | 4.87 ± 0.37 lm | 21.22 ± 0.47 fg | 4.89 ± 0.13 b–g | 24.28 ± 1.90 fg | 5.14 ± 0.94 i | 4738.7 ± 164.4 d |
| | VI | 5.63 ± 0.16 l | 20.93 ± 0.81 fg | 4.80 ± 0.07 c–h | 25.82 ± 2.27 f | 7.00 ± 0.96 c–f | 4139.5 ± 390.1 fg |
| EC2 | I | 5.25 ± 0.23 lm | 24.89 ± 0.71 bc | 4.63 ± 0.15 gh | 19.59 ± 1.52 h | 7.69 ± 0.31 b–d | 5136.5 ± 3510.5 b |
| | II | 8.61 ± 0.38 hi | 26.14 ± 0.56 b | 5.02 ± 0.15 b–d | 25.53 ± 1.41 f | 6.39 ± 0.45 f–h | 4017.4 ± 144.5 gh |
| | III | 11.34 ± 0.43 g | 21.86 ± 0.74 ef | 5.14 ± 0.16 b | 36.05 ± 2.38 e | 5.32 ± 0.43 hi | 5077.9 ± 235.6 bc |
| | IV | 10.68 ± 0.36 g | 21.85 ± 0.81 ef | 4.97 ± 0.14 b–e | 36.20 ± 2.27 e | 6.83 ± 0.62 d–f | 3915.4 ± 313.6 g–i |
| | V | 10.17 ± 0.80 gh | 21.66 ± 0.59 ef | 4.72 ± 0.13 d–h | 36.01 ± 2.11 e | 7.77 ± 0.70 b–d | 3825.5 ± 276.2 h–j |
| | VI | 10.49 ± 0.68 gh | 22.02 ± 0.52 ef | 4.70 ± 0.15 e–h | 37.87 ± 2.12 e | 9.31 ± 0.26 a | 3360.5 ± 149.1 k–m |
| EC4 | I | 6.14 ± 0.28 jk | 21.92 ± 0.79 ef | 4.57 ± 0.14 hi | 25.55 ± 1.23 f | 8.01 ± 0.24 bc | 4526.7 ± 224.6 de |
| | II | 11.69 ± 0.41 fg | 25.06 ± 0.51 bc | 5.04 ± 0.11 bc | 37.95 ± 2.04 e | 5.62 ± 0.24 g–i | 3277.7 ± 214.5 mn |
| | III | 13.32 ± 0.46 ef | 18.59 ± 0.63 ij | 4.83 ± 0.08 b–h | 47.24 ± 2.88 d | 4.51 ± 0.48 ij | 3904.6 ± 334.0 g–i |
| | IV | 16.70 ± 1.45 ac | 18.81 ± 0.98 hi | 4.93 ± 0.20 b–f | 50.66 ± 3.46 cd | 5.31 ± 0.80 hi | 3642.0 ± 296.3 ij |
| | V | 16.31 ± 1.35 bc | 18.90 ± 0.83 hi | 4.69 ± 0.09 e–h | 51.23 ± 4.31 cd | 6.76 ± 0.69 d–g | 3568.3 ± 287.3 j–l |
| | VI | 15.59 ± 1.67 cd | 19.76 ± 0.69 g–i | 4.60 ± 0.10 g–i | 52.78 ± 3.58 bc | 8.50 ± 0.76 ab | 2959.8 ± 346.8 o |
| EC6 | I | 7.79 ± 0.23 ij | 23.02 ± 0.57 de | 4.59 ± 0.11 g–i | 25.87 ± 0.91 f | 8.08 ± 0.22 bc | 4773.5 ± 489.4 cd |
| | II | 14.02 ± 0.64 de | 24.94 ± 0.68 bc | 5.00 ± 0.12 be | 47.26 ± 2.26 d | 6.47 ± 0.24 e–h | 2431.0 ± 103.4 p |
| | III | 17.53 ± 1.02 ab | 18.10 ± 0.51 ij | 4.74 ± 0.11 c–h | 54.99 ± 1.45 a–c | 5.47 ± 0.39 hi | 3098.8 ± 290.7 m–o |
| | IV | 18.28 ± 0.97 a | 18.37 ± 0.33 ij | 4.75 ± 0.07 c–h | 56.14 ± 1.17 ab | 4.64 ± 0.51 ij | 3204.1 ± 355.5 m–o |
| | V | 18.21 ± 1.33 ab | 18.07 ± 0.58 ij | 4.64 ± 0.14 f–h | 57.21 ± 2.65 a | 6.77 ± 0.71 d–g | 3364.1 ± 318.0 mn |
| | VI | 17.90 ± 2.15 ab | 17.61 ± 0.76 j | 4.26 ± 0.21 j | 57.30 ± 2.32 a | 7.75 ± 0.82 b–d | 2396.3 ± 229.9 p |

Different letters within each column indicate significant differences according to Tukey's honestly significant difference (HSD) test at $p = 0.05$. Roman numbers (I–VI) indicate the successive harvests. EC0: tap water (0 ds m$^{-1}$); EC2: 2.0 dS m$^{-1}$; EC4: 4.0 dS m$^{-1}$; and EC6: 6.0 dS m$^{-1}$. All data are expressed as mean ± SE (standard error), n = 3.

Similar results to our study have been reported by Malécange et al. [66], who studied the effect of a biostimulant product rich in free amino acids on lettuce crop performance and suggested an increase in nitrogen content in treated plants, regardless of the irrigation regime. On the other hand, considering that salinity stress is associated with the decreased water availability and increased osmotic potential of the nutrient solution, nitrate accumulation in plants subjected to salinity stress indicates its osmoregulatory activity [64]. This finding should be associated with protective mechanisms similar to those of halophytes, which tend to accumulate minerals under saline conditions as part of their defense against abiotic stressors [67].

On the other hand, K concentration significantly increased from harvest III to V in all of the biostimulant treatments, while a decrease was recorded for the last harvest (Figure 3A). Similar trends were suggested for Cl content, which gradually increased with harvesting time after harvest II (Figure 3B). Moreover, Cl content was higher in the leaves of plants untreated with biostimulants (41.01 vs. 34.89 g kg$^{-1}$, mean value of BA and BT). Finally, contrasting trends were recorded in terms of nitrate content (Supplementary Table S4). In particular, the progress in harvesting time resulted in a decrease in nitrate content in plants that were not sprayed with biostimulants. In contrast, plants treated with either Auxym®or Trainer®recorded a steep decrease at the second harvest, followed by a significant increase at harvest III and fluctuating trends thereafter until harvest VI.

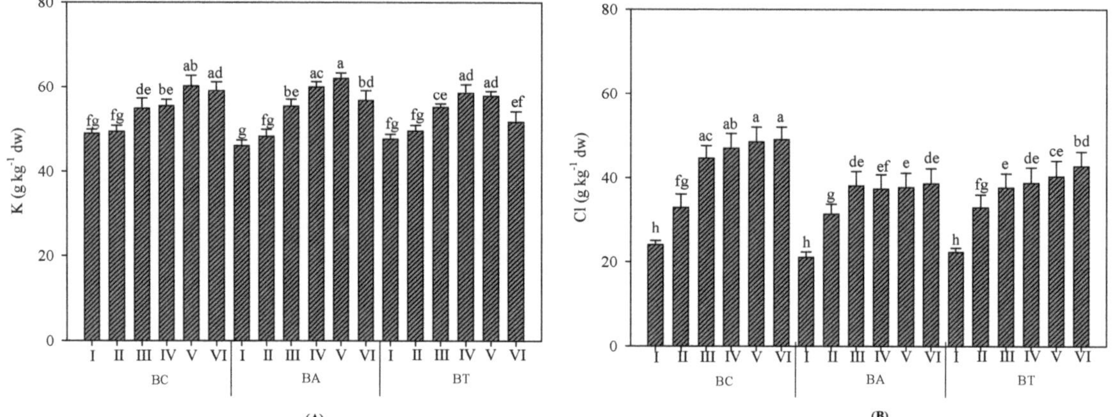

(A)  (B)

**Figure 3.** Effect of biostimulants and harvesting time on minerals K (**A**) and Cl (**B**) in *Diplotaxis tenuifolia* leaves, regardless of salinity level. Different letters indicate significant differences according to Tukey's honestly significant difference (HSD) test at *p* = 0.05. Roman numbers (I–VI) indicate the successive harvests. BC: no biostimulants added; BA: plants treated with Auxym®; and BT: plants treated with Trainer®. All data are expressed as mean ± SE (standard error), n = 3.

The increase in nitrate content with biostimulant application is already reported in the literature [16,30]; however, our findings indicate the importance of growing conditions in plant responses to nitrate accumulation, including soil properties, light intensity, and nitrogen form, among others [29,68]. Moreover, Bantis et al. [69] also suggested a higher nitrate content in the first harvest of rocket plants in an experiment where two successive harvests were implemented, a finding which is in agreement with our results for the plants that received no biostimulants. Regarding the nutrients profile, Caruso et al. [27] suggested a variable effect of growing conditions (winter and winter–spring cropping seasons) on wall rocket plants, while biostimulant application resulted in increased content for most of the nutrients (except for S, where no effects were recorded). According to the same authors, biostimulant application is associated with changes in root architecture that facilitate nutrient uptake, translocation, and assimilation, with the activity of signaling molecules or the expression of genes involved in macronutrient transportation through cell membranes [35]. Moreover, Giordano et al. [52], who tested the same biostimulant products in perennial wall rocket plants, reported that Trainer®and Auxym®contain bioactive compounds and peptides that may promote root growth as well as nutrient uptake, and therefore affect the mineral profile of leaves.

## 4. Conclusions

The results of the present work highlight the importance of biostimulant application in alleviating the negative effects of salinity stress on the chemical composition and mineral profile of *Diplotaxis tenuifolia* plants; however, a varied response in relation to harvesting time was recorded for most of the studied parameters, which indicates the pivotal effect of growing conditions in addition to the complexity of plants' responses to biostimulants and salinity stress. In conclusion, further research is needed in order to suggest those conditions that allow the alleviation of the negative effects of salinity stress through biostimulant application and prescheduled harvesting time. Moreover, special consideration is needed regarding the nitrate content of leaves, which tend to increase with biostimulant application, while moderate salinity and proper harvesting time seem to reduce the health risks associated with nitrate accumulation in *D. tenuifolia* leaves.

**Supplementary Materials:** The following supporting information can be downloaded at: https://www.mdpi.com/article/10.3390/app13031569/s1, Table S1: Analysis of variance of hydrophylic antioxidant activity (HAA), ABTS antioxidant activity, total phenols, total ascorbic acid (TAA), chlorophylls (a, b, and total chlorophyll), and total carotenoids; Table S2: Analysis of variance of minerals content; Table S3: Effect of biostimulant and harvesting time on hydrophylic antioxidant activity (HAA), ABTS antioxidant activity, total phenols, and total ascorbic acid (TAA) content in *Diplotaxis tenuifolia* leaves, regardless of salinity level; Table S4: Effect of biostimulants and harvest on minerals (Na, K, Ca, Mg, Cl, S, and P) and nitrate content in *Diplotaxis tenuifolia* leaves, regardless of salinity level.

**Author Contributions:** Conceptualization, I.D.M., E.C. and M.M.; methodology, S.A.P. and Y.R.; software, L.O. and C.E.-N.; validation, I.D.M., L.O. and C.E.-N.; formal analysis, S.A.P.; investigation, Y.R. and M.M.; resources, E.C.; data curation, L.O.; writing—original draft preparation, I.D.M. and S.A.P.; writing—review and editing, C.E.-N. and Y.R.; visualization, I.D.M. and S.A.P.; supervision, E.C. and M.M.; project administration, E.C.; funding acquisition, Y.R. and M.M. All authors have read and agreed to the published version of the manuscript.

**Funding:** This research received no external funding.

**Institutional Review Board Statement:** Not applicable.

**Informed Consent Statement:** Not applicable.

**Data Availability Statement:** The datasets generated for this study are available on request to the corresponding author.

**Acknowledgments:** The authors would like to thank Sabrina Nocerino, Maria Eleonora Pelosi, and Luca Scognamiglio for their support in the laboratory work.

**Conflicts of Interest:** The authors declare no conflict of interest.

# References

1. Jardim, S.; Sequeira, M.; Capelo, J.; Aguiar, C.; Costa, J.C.; Espírito-Santo, D.; Lousã, M. XXXVI: The vegetation of Madeira: IV—Coastal Vegetation of Porto Santo Island (Archipelag of Madeira). *Notas Herbário Estação Florest. Nac.* **1994**, *XVII*, 1–5.
2. Ashraf, M.; Foolad, M.R. Roles of glycine betaine and proline in improving plant abiotic stress resistance. *Environ. Exp. Bot.* **2007**, *59*, 206–216. [CrossRef]
3. Machado, R.M.A.; Serralheiro, R.P. Soil salinity: Effect on vegetable crop growth. Management practices to prevent and mitigate soil salinization. *Horticulturae* **2017**, *3*, 30. [CrossRef]
4. Shannon, M.C.; Grieve, C.M. Tolerance of vegetable crops to salinity. *HortScience* **1999**, *78*, 5–38. [CrossRef]
5. Rao, N.S.; Shivashankara, K.S.; Laxman, R.H. *Abiotic Stress Physiology of Horticultural Crops*; Rao, N.S., Shivashankara, K.S., Laxman, R.H., Eds.; Springer: New Delhi, India, 2016; ISBN 9788132227236.
6. Correa, R.C.G.; Di Gioia, F.; Ferreira, I.; SA, P. Halophytes for future horticulture: The case of small-scale farming in the Mediterranean basin. In *Halophytes for Future Horticulture: From Molecules to Ecosystems towards Biosaline Agriculture*; Grigore, M.-N., Ed.; Springer Nature: Cham, Switzerland, 2020; pp. 1–28; ISBN 9783030178543.
7. Hazell, P.; Poulton, C.; Wiggins, S.; Dorward, A. *The Future of Small Farms: Synthesis Paper*; World Bank: Washington, DC, USA, 2006; pp. 1–51.
8. Ventura, Y.; Sagi, M. Halophyte crop cultivation: The case for *Salicornia* and *Sarcocornia*. *Environ. Exp. Bot.* **2013**, *92*, 144–153. [CrossRef]
9. Mohammed, A.H. The Valuable Impacts of Halophytic Genus Suaeda; Nutritional, Chemical, and Biological Values. *Med. Chem.* **2020**, *16*, 1044–1057. [CrossRef]
10. Mohammed, H.A.; Ali, H.M.; Qureshi, K.A.; Alsharidah, M.; Kandil, Y.I.; Said, R.; Mohammed, S.A.A.; Al-omar, M.S.; Al Rugaie, O.; Abdellatif, A.A.H.; et al. Four Major Medicinal Halophytes from Qassim Flora. *Plants* **2021**, *10*, 2208. [CrossRef]
11. Amin, E.; Abdel-Bakky, M.S.; Mohammed, H.A.; Chigrupati, S.; Qureshi, K.A.; Hassan, M.H.A. Phytochemical Analysis and Evaluation of the Antioxidant and Antimicrobial Activities of Five Halophytes from Qassim Flora. *Polish J. Environ. Stud.* **2022**, *31*, 3005–3012. [CrossRef]
12. Battacharyya, D.; Babgohari, M.Z.; Rathor, P.; Prithiviraj, B. Seaweed extracts as biostimulants biostimulants in horticulture. *Sci. Hortic.* **2015**, *196*, 39–48. [CrossRef]
13. Shahrajabian, M.H.; Chaski, C.; Polyzos, N.; Petropoulos, S.A. Biostimulants Application: A Low Input Cropping Management Tool for Sustainable Farming of Vegetables. *Biomolecules* **2021**, *11*, 698. [CrossRef] [PubMed]
14. Andreotti, C. Management of abiotic stress in horticultural crops: Spotlight on biostimulants. *Agronomy* **2020**, *10*, 1514. [CrossRef]
15. Del Buono, D. Can biostimulants be used to mitigate the effect of anthropogenic climate change on agriculture? It is time to respond. *Sci. Total Environ.* **2021**, *751*, 141763. [CrossRef]

16. El-Nakhel, C.; Cozzolino, E.; Ottaiano, L.; Petropoulos, S.A.; Nocerino, S.; Pelosi, M.E.; Rouphael, Y.; Mori, M.; Di Mola, I. Effect of Biostimulant Application on Plant Growth, Chlorophylls and Hydrophilic Antioxidant Activity of Spinach (*Spinacia oleracea* L.) Grown under Saline Stress. *Horticulturae* **2022**, *8*, 971. [CrossRef]
17. Lucini, L.; Rouphael, Y.; Cardarelli, M.; Canaguier, R.; Kumar, P.; Colla, G. The effect of a plant-derived biostimulant on metabolic profiling and crop performance of lettuce grown under saline conditions. *Sci. Hortic.* **2015**, *182*, 124–133. [CrossRef]
18. Rouphael, Y.; Carillo, P.; Garcia-Perez, P.; Cardarelli, M.; Senizza, B.; Miras-Moreno, B.; Colla, G.; Lucini, L. Plant biostimulants from seaweeds or vegetal proteins enhance the salinity tolerance in greenhouse lettuce by modulating plant metabolism in a distinctive manner. *Sci. Hortic.* **2022**, *305*, 111368. [CrossRef]
19. Rouphael, Y.; Cardarelli, M.; Bonini, P.; Colla, G. Synergistic action of a microbial-based biostimulant and a plant derived-protein hydrolysate enhances lettuce tolerance to alkalinity and salinity. *Front. Plant Sci.* **2017**, *8*, 131. [CrossRef]
20. Van Oosten, M.J.; Pepe, O.; De Pascale, S.; Silletti, S.; Maggio, A. The role of biostimulants and bioeffectors as alleviators of abiotic stress in crop plants. *Chem. Biol. Technol. Agric.* **2017**, *4*, 5. [CrossRef]
21. Kim, H.J.; Fonseca, J.M.; Choi, J.H.; Kubota, C.; Dae, Y.K. Salt in irrigation water affects the nutritional and visual properties of romaine lettuce (*Lactuca sativa* L.). *J. Agric. Food Chem.* **2008**, *56*, 3772–3776. [CrossRef]
22. Rouphael, Y.; Petropoulos, S.A.; Cardarelli, M.; Colla, G. Salinity as eustressor for enhancing quality of vegetables. *Sci. Hortic.* **2018**, *234*, 361–369. [CrossRef]
23. Adhikari, B.; Olorunwa, O.J.; Wilson, J.C.; Barickman, T.C. Morphological and Physiological Response of Different Lettuce Genotypes to Salt Stress. *Stresses* **2021**, *1*, 285–304. [CrossRef]
24. Petropoulos, S.; Fernandes, Â.; Karkanis, A.; Ntatsi, G.; Barros, L.; Ferreira, I.C.F.R. Successive harvesting affects yield, chemical composition and antioxidant activity of *Cichorium spinosum* L. *Food Chem.* **2017**, *237*, 83–90. [CrossRef]
25. Abdalla, M.M. Boosting the growth of rocket plants in response to the application of *Moreinga oleifera* extracts as a biostimulant. *Life Sci. J.* **2014**, *11*, 1113–1121.
26. Ventura, Y.; Wuddineh, W.A.; Shpigel, M.; Samocha, T.M.; Klim, B.C.; Cohen, S.; Shemer, Z.; Santos, R.; Sagi, M. Effects of day length on flowering and yield production of *Salicornia* and *Sarcocornia* species. *Sci. Hortic.* **2011**, *130*, 510–516. [CrossRef]
27. Caruso, G.; El-Nakhel, C.; Rouphael, Y.; Comite, E.; Lombardi, N.; Cuciniello, A.; Woo, S.L. *Diplotaxis tenuifolia* (L.) DC. yield and quality as influenced by cropping season, protein hydrolysates, and *Trichoderma* applications. *Plants* **2020**, *9*, 697. [CrossRef] [PubMed]
28. Caruso, G.; Parrella, G.; Giorgini, M.; Nicoletti, R. Crop systems, quality and protection of Diplotaxis tenuifolia. *Agriculture* **2018**, *8*, 55. [CrossRef]
29. Weightman, R.M.; Huckle, A.J.; Roques, S.E.; Ginsburg, D.; Dyer, C.J. Factors influencing tissue nitrate concentration in field-grown wild rocket (*Diplotaxis tenuifolia*) in southern England. *Food Addit. Contam. Part A Chem. Anal. Control. Expo. Risk Assess.* **2012**, *29*, 1425–1435. [CrossRef] [PubMed]
30. Di Mola, I.; Ottaiano, L.; Cozzolino, E.; El-Nakhel, C.; Rippa, M.; Mormile, P.; Corrado, G.; Rouphael, Y.; Mori, M. Assessment of Yield and Nitrate Content of Wall Rocket Grown under Diffuse-Light-or Clear-Plastic Films and Subjected to Different Nitrogen Fertilization Levels and Biostimulant Application. *Horticulturae* **2022**, *8*, 138. [CrossRef]
31. Petropoulos, S.A.; Chatzieustratiou, E.; Constantopoulou, E.; Kapotis, G. Yield and quality of lettuce and rocket grown in floating culture system. *Not. Bot. Horti Agrobot. Cluj-Napoca* **2016**, *44*, 10611. [CrossRef]
32. Alexopoulos, A.A.; Marandos, E.; Assimakopoulou, A.; Vidalis, N.; Petropoulos, S.A.; Karapanos, I.C. Effect of Nutrient Solution pH on the Growth, Yield and Quality of *Taraxacum officinale* and *Reichardia picroides* in a Floating Hydroponic System. *Agronomy* **2021**, *11*, 1118. [CrossRef]
33. Kristensen, H.L.; Stavridou, E. Deep root growth and nitrogen uptake by rocket (*Diplotaxis tenuifolia* L.) as affected by nitrogen fertilizer, plant density and leaf harvesting on a coarse sandy soil. *Soil Use Manag.* **2017**, *33*, 62–71. [CrossRef]
34. Petropoulos, S.A.; Constantopoulou, E.; Karapanos, I.; Akoumianakis, C.A.; Passam, H.C. Diurnal variation in the nitrate content of parsley foliage. *Int. J. Plant Prod.* **2011**, *5*, 431–438.
35. Caruso, G.; De Pascale, S.; Cozzolino, E.; Giordano, M.; El-Nakhel, C.; Cuciniello, A.; Cenvinzo, V.; Colla, G.; Rouphael, Y. Protein Hydrolysate or Plant Extract-based Biostimulants Enhanced Yield and Quality Performances of Greenhouse Perennial Wall Rocket Grown in Different Seasons. *Plants* **2019**, *8*, 208. [CrossRef]
36. Rouphael, Y.; Giordano, M.; Cardarelli, M.; Cozzolino, E.; Mori, M.; Kyriacou, M.C.; Bonini, P.; Colla, G. Plant-and seaweed-based extracts increase yield but differentially modulate nutritional quality of greenhouse spinach through biostimulant action. *Agronomy* **2018**, *8*, 126. [CrossRef]
37. Vernieri, P.; Borghesi, E. Application of biostimulants in floating system for improving rocket quality. *J. Food* **2005**, *3*, 86–88.
38. Hargreaves, G.; Samani, Z. Reference crop evapotranspiration from temperature. *Appl. Eng. Agric.* **1985**, *1*, 96–99. [CrossRef]
39. Di Mola, I.; Ottaiano, L.; Cozzolino, E.; Senatore, M.; Giordano, M.; El-Nakhel, C.; Sacco, A.; Rouphael, Y.; Colla, G. Plant-Based Biostimulants Influence the Agronomical, Physiological, and Qualitative Responses of Baby Rocket Leaves under Diverse Nitrogen Conditions. *Plants* **2019**, *8*, 522. [CrossRef] [PubMed]
40. Re, R.; Pellegrini, N.; Proteggente, A.; Pannala, A.; Yang, M.; Rice-Evans, C. Antioxidant activity applying an improved ABTS radical cation decolorization assay. *Free Radic. Biol. Med.* **1999**, *26*, 1231–1237. [CrossRef] [PubMed]
41. Fogliano, V.; Verde, V.; Randazzo, G.; Ritieni, A. Method for measuring antioxidant activity and its application to monitoring the antioxidant capacity of wines. *J. Agric. Food Chem.* **1999**, *47*, 1035–1040. [CrossRef]

42. Kampfenkel, K.; Montagu, M.; Inzé, D. Extraction and determination of ascorbate and dehydroascorbate from plant tissue. *Anal. Biochem.* **1995**, *225*, 165–167. [CrossRef]
43. Singleton, V.; Orthofer, R.; Lamuela-Raventós, R.M. Analysis of Total Phenols and Other Oxidation Substrates and Antioxidants by Means of Folin-Ciocalteu Reagent. *Methods Enzymol.* **1999**, *299*, 152–178. [CrossRef]
44. Wellburn, A.R. The spectral determination of chlorophyls a and b, as well as tottal carotenoids, using various solvents with spectrophotometers of different resolution. *J. Plant Physiol.* **1994**, *144*, 307–313. [CrossRef]
45. Rouphael, Y.; Colla, G.; Graziani, G.; Ritieni, A.; Cardarelli, M.; De Pascale, S. Phenolic composition, antioxidant activity and mineral profile in two seed-propagated artichoke cultivars as affected by microbial inoculants and planting time. *Food Chem.* **2017**, *234*, 10–19. [CrossRef] [PubMed]
46. Mori, M.; Di Mola, I.; Chiarandà, F.Q. Salt stress and transplant time in snap bean: Growth and productive behaviour. *Int. J. Plant Prod.* **2011**, *5*, 49–64.
47. Schiattone, M.I.; Candido, V.; Cantore, V.; Montesano, F.F.; Boari, F. Water use and crop performance of two wild rocket genotypes under salinity conditions. *Agric. Water Manag.* **2017**, *194*, 214–221. [CrossRef]
48. Feng, G.; Zhang, Z.; Wan, C.; Lu, P.; Bakour, A. Effects of saline water irrigation on soil salinity and yield of summer maize (*Zea mays* L.) in subsurface drainage system. *Agric. Water Manag.* **2017**, *193*, 205–213. [CrossRef]
49. Chen, W.; Jin, M.; Ferré, T.P.A.; Liu, Y.; Xian, Y.; Shan, T.; Ping, X. Spatial distribution of soil moisture, soil salinity, and root density beneath a cotton field under mulched drip irrigation with brackish and fresh water. *Field Crops Res.* **2018**, *215*, 207–221. [CrossRef]
50. Schiattone, M.I.; Boari, F.; Cantore, V.; Castronuovo, D.; Denora, M.; Di Venere, D.; Perniola, M.; Renna, M.; Sergio, L.; Candido, V. Effects of nitrogen, azoxystrobin and a biostimulant based on brown algae and yeast on wild rocket features at harvest and during storage. *Agronomy* **2021**, *11*, 2326. [CrossRef]
51. Candido, V.; Boari, F.; Cantore, V.; Castronuovo, D.; Di Venere, D.; Perniola, M.; Sergio, L.; Viggiani, R.; Schiattone, M.I. Interactive Effect of Nitrogen and Azoxystrobin on Yield, Quality, Nitrogen and Water Use Efficiency of Wild Rocket in Southern Italy. *Agronomy* **2020**, *10*, 849. [CrossRef]
52. Giordano, M.; El-Nakhel, C.; Caruso, G.; Cozzolino, E.; De Pascale, S.; Kyriacou, M.C.; Colla, G.; Rouphael, Y. Stand-alone and combinatorial effects of plant-based biostimulants on the production and leaf quality of perennial wall rocket. *Plants* **2020**, *9*, 922. [CrossRef]
53. Carillo, P.; Giordano, M.; Raimondi, G.; Napolitano, F.; Di Stasio, E.; Kyriacou, M.C.; Sifola, M.I.; Rouphael, Y. Physiological and nutraceutical quality of green and red pigmented lettuce in response to NaCl concentration in two successive harvests. *Agronomy* **2020**, *10*, 1358. [CrossRef]
54. Corrado, G.; Chiaiese, P.; Lucini, L.; Miras-Moreno, B.; Colla, G.; Rouphael, Y. Successive harvests affect yield, quality and metabolic profile of sweet basil (*Ocimum basilicum* L.). *Agronomy* **2020**, *10*, 830. [CrossRef]
55. Ciriello, M.; Formisano, L.; El-Nakhel, C.; Kyriacou, M.C.; Soteriou, G.A.; Pizzolongo, F.; Romano, R.; De Pascale, S.; Rouphael, Y. Genotype and successive harvests interaction affects phenolic acids and aroma profile of Genovese basil for pesto sauce production. *Foods* **2021**, *10*, 278. [CrossRef]
56. Di Mola, I.; Rouphael, Y.; Colla, G.; Fagnano, M.; Paradiso, R.; Mori, M. Morphophysiological traits and nitrate content of greenhouse lettuce as affected by irrigation with saline water. *HortScience* **2017**, *52*, 1716–1721. [CrossRef]
57. Bonasia, A.; Lazzizera, C.; Elia, A.; Conversa, G. Nutritional, biophysical and physiological characteristics of wild rocket genotypes as affected by soilless cultivation system, salinity level of nutrient solution and growing period. *Front. Plant Sci.* **2017**, *8*, 300. [CrossRef]
58. Hamilton, J.M.; Fonseca, J.M. Effect of saline irrigation water on antioxidants in three hydroponically grown leafy vegetables: Diplotaxis tenuifolia, eruca sativa, and lepidium sativum. *HortScience* **2010**, *45*, 546–552. [CrossRef]
59. Corrado, G.; Vitaglione, P.; Soteriou, G.A.; Kyriacou, M.C.; Rouphael, Y. Configuration by osmotic eustress agents of the morphometric characteristics and the polyphenolic content of differently pigmented baby lettuce varieties in two successive harvests. *Horticulturae* **2021**, *7*, 264. [CrossRef]
60. Franzoni, G.; Cocetta, G.; Trivellini, A.; Ferrante, A. Transcriptional regulation in rocket leaves as affected by salinity. *Plants* **2020**, *9*, 20. [CrossRef]
61. Chaski, C.; Petropoulos, S.A. The Effects of Biostimulant Application on Growth Parameters of Lettuce Plants Grown under Deficit Irrigation Conditions. *Horticulturae* **2022**, *8*, 4. [CrossRef]
62. Bulgari, R.; Cocetta, G.; Trivellini, A.; Ferrante, A. Borage extracts affect wild rocket quality and influence nitrate and carbon metabolism. *Physiol. Mol. Biol. Plants* **2020**, *26*, 649–660. [CrossRef]
63. Grattan, S.R.; Grieve, C.M. Salinity-mineral nutrient relations in horticultural crops. *Sci. Hortic.* **1998**, *78*, 127–157. [CrossRef]
64. Chung, J.B.; Jin, S.J.; Cho, H.J. Low water potential in saline soils enhances nitrate accumulation of lettuce. *Commun. Soil Sci. Plant Anal.* **2005**, *36*, 1773–1785. [CrossRef]
65. Caparrotta, S.; Masi, E.; Atzori, G.; Diamanti, I.; Azzarello, E.; Mancuso, S.; Pandolfi, C. Growing spinach (*Spinacia oleracea*) with different seawater concentrations: Effects on fresh, boiled and steamed leaves. *Sci. Hortic.* **2019**, *256*, 108540. [CrossRef]
66. Malécange, M.; Pérez-Garcia, M.D.; Citerne, S.; Sergheraert, R.; Lalande, J.; Teulat, B.; Mounier, E.; Sakr, S.; Lothier, J. Leafamine®, a Free Amino Acid-Rich Biostimulant, Promotes Growth Performance of Deficit-Irrigated Lettuce. *Int. J. Mol. Sci.* **2022**, *23*, 7338. [CrossRef]

67. Mohammed, H.A.; Al-Omar, M.S.; Mohammed, S.A.A.; Alhowail, A.H.; Eldeeb, H.M.; Sajid, M.S.M.; Abd-Elmoniem, E.M.; Alghulayqeh, O.A.; Kandil, Y.I.; Khan, R.A. Phytochemical analysis, pharmacological and safety evaluations of halophytic plant, *Salsola cyclophylla*. *Molecules* **2021**, *26*, 2384. [CrossRef]
68. Chang, A.C.; Yang, T.Y.; Riskowski, G.L. Changes in nitrate and nitrite concentrations over 24 h for sweet basil and scallions. *Food Chem.* **2013**, *136*, 955–960. [CrossRef]
69. Bantis, F.; Kaponas, C.; Charalambous, C.; Koukounaras, A. Strategic successive harvesting of rocket and spinach baby leaves enhanced their quality and production efficiency. *Agriculture* **2021**, *11*, 465. [CrossRef]

**Disclaimer/Publisher's Note:** The statements, opinions and data contained in all publications are solely those of the individual author(s) and contributor(s) and not of MDPI and/or the editor(s). MDPI and/or the editor(s) disclaim responsibility for any injury to people or property resulting from any ideas, methods, instructions or products referred to in the content.

*applied*
*sciences*

*Article*

# Proximate Composition and Antioxidant Activity of Selected Morphological Parts of Herbs

Wioletta Biel [1], Urszula Pomietło [2], Robert Witkowicz [3], Ewa Piątkowska [2] and Aneta Kopeć [2,*]

[1]  Department of Monogastric Animal Sciences, Division of Animal Nutrition and Food, West Pomeranian University of Technology in Szczecin, Klemensa Janickiego 29, 71-270 Szczecin, Poland
[2]  Department of Human Nutrition and Dietetics, Faculty of Food Technology, University of Agriculture in Krakow, Balicka 122, 30-149 Krakow, Poland
[3]  Department of Agroecology and Crop Production, Faculty of Agriculture and Economics, University of Agriculture in Krakow, Mickiewicza 21, 31-120 Krakow, Poland
*  Correspondence: aneta.kopec@urk.edu.pl

**Abstract:** The aim of the study was to provide an analytical evaluation of the proximate composition, the total content of polyphenolic compounds and the antioxidant activity, of 27 selected plant materials collected in Poland (West Pomeranian). The basic chemical composition was determined in the ground samples according to the Association of Official Analytical Chemists methods. Antioxidant activity was tested using free radical methods $ABTS^{\bullet+}$, $DPPH^{\bullet+}$ and the FRAP method. The lowest concentration of dry matter (DM) was measured in black chokeberry (88.82 g/100 g) and the highest was found in milk thistle (94.65 g/100 g) as well as black cumin (95.09 g/100 g). The content of total polyphenols, assessed using the Folin–Ciocalteu method, ranged from 291.832–7565.426 mg of chlorogenic acid equivalent (CGA)/100 g of DM. Antioxidant activity measured sequentially against the radical $ABTS^{\bullet+}$, $DPPH^{\bullet+}$ and using the FRAP method was 26.334–1912.016 μM Trolox/g DM, 9.475–1061.068 μM Trolox/g DM and 26.252–1769.766 μM Trolox/g DM, respectively. The methanolic extract from milk thistle fruit in most assays was characterized by the lowest antioxidant activity and the lowest total content of polyphenolic compounds. Methanol extracts prepared from garlic, stinging nettle and cleavers showed the highest content of total polyphenols and antioxidant activity among the tested plant materials. The parts of plants with the highest antioxidant potential can be a source of new bioactive compounds, but further research is required to describe the profile of compounds harmful to human health.

**Keywords:** antioxidant activity; botanicals; herbs; polyphenols; principal component analysis (PCA); proximate composition

**Citation:** Biel, W.; Pomietło, U.; Witkowicz, R.; Piątkowska, E.; Kopeć, A. Proximate Composition and Antioxidant Activity of Selected Morphological Parts of Herbs. *Appl. Sci.* **2023**, *13*, 1413. https://doi.org/10.3390/app13031413

Academic Editors: Luca Mazzoni, Maria Teresa Ariza Fernández and Franco Capocasa

Received: 18 December 2022
Revised: 16 January 2023
Accepted: 18 January 2023
Published: 20 January 2023

**Copyright:** © 2023 by the authors. Licensee MDPI, Basel, Switzerland. This article is an open access article distributed under the terms and conditions of the Creative Commons Attribution (CC BY) license (https://creativecommons.org/licenses/by/4.0/).

## 1. Introduction

In recent years, the interest in antioxidants derived from herbal raw materials has increased significantly due to their health-promoting properties. Their preventive effect is appreciated in the context of chronic non-communicable diseases such as obesity, type II diabetes, atherosclerosis or neurodegenerative diseases, the common risk factor of which is oxidative stress [1]. In biological systems, the formation of free radicals is the result of many metabolic changes, including aerobic respiration and inflammatory reactions. Free radicals have proven useful in the fight against pathogens. This is due to their bacteriostatic and bactericidal effects, but they are also capable of removing cancer cells [2]. Free radicals are responsible for controlling blood flow through blood vessels, removal of xenobiotics from the body, and they are also responsible for transmitting signals within the cell [3–5]. Under normal conditions, there is a balance between the formation of free radicals and their removal. However, in conditions of disturbed homeostasis, the amount of free radicals increases, beyond the possibility of their systematic and efficient removal by enzymatic and non-enzymatic mechanisms [5]. During aerobic respiration, some of the electrons leave the complexes of the respiratory chain, reducing the oxygen molecule by a one-electron or two-electron redox reaction, which leads to the formation of reactive oxygen species (ROS) [6,7]. This causes disturbances at the metabolic level due to high reactivity, short lifetime and extraordinary ease of chemical reactions of free oxygen radicals with cell components [5,7]. Enzymatic systems located in the mitochondrial matrix and cytoplasm, i.e., superoxide dismutase (SOD), glutathione reductase (GR), glutathione peroxidase (GPx), catalase (CAT) are responsible for the removal of these compounds [5,8].

Apart from enzyme systems, there are also defense systems based on proteins located in the blood plasma (ceruloplasmin, ferritin, transferrin, or albumin) that remotely bind copper and iron ions, preventing free radical reactions. These proteins interact with uric acid, complementing each others antioxidant potential [8]. Non-enzymatic defense mechanisms against free radicals are some vitamins (C, E, A, including provitamin A–β-carotene), coenzyme Q10 or polyphenolic compounds found in plants [9]. The excess of free radicals, including reactive oxygen species, together with the inefficiency in their removal, leads to lipid peroxidation, damage to cell organelles, proteins, and DNA degradation and mutation, which results in chronic non-communicable diseases [10].

The main antioxidant compounds present in plant products are polyphenolic compounds, including flavonoids [11]. Antioxidant compounds contained in herbs have following effects: anti-inflammation, antibacterial, antifungal, antiviral and immunostimulation effects [12]. Their effect cannot be reproduced on the basis of individually isolated antioxidant compounds, because the health benefiting effect includes many compounds simultaneously appearing in the plant [13].

Herbal raw materials, long known in traditional folk medicine, are mostly well-tested in terms of antioxidant properties, but there are still raw materials that have not been widely distributed in the phytotherapeutic and pharmaceutical industries [14]. Such raw materials may turn out to be new nutraceuticals, helping pharmacological substances (medicines) in the fight against common diseases. Adequate dietary intake of a variety of antioxidants, such as polyphenols, vitamin C, vitamin E, carotenoids and selenium, is associated with a lower risk of developing chronic non-communicable diseases [14,15].

Increasing amount of research results suggest that antioxidants affecting cells may also trigger interactions with specific proteins crucial for intracellular signaling cascades, modulating their expression and activity [15]. These compounds also affect epigenetic mechanisms and modulate the intestinal microbiota [16].

In the available literature, many studies are devoted to the antioxidant properties of various plants and their morphological parts (e.g., fruits, seeds, leaves, roots) [17–20]. Usually, in this type of article, the assessment of antioxidant properties is most often performed using the methods of determining total polyphenols, ABTS, DPPH and FRAP [12,18,21,22].

It is very important that in the published literature there are studies on the antioxidant properties of herbs, but most often they concern popular herbs, e.g., sage, oregano and basil. In our publication, we decided to research less popular plants, which, however, are often used for their culinary and medicinal properties in many countries around the world.

The aim of the study was the analytical evaluation of the basic chemical composition, the total content of polyphenolic compounds (TPC) the antioxidant activity, of 27 selected morphological parts of plant materials collected in Poland.

## 2. Materials and Methods

### 2.1. Materials

Plant material was collected in 2019 from the collection of medicinal and useful plants of the Experimental Station in Lipnik, Poland (53°20′35″ N, 14°58′10″ E). The collection was conducted by a team of botanists and agrotechnicians from the West Pomeranian University of Technology in Szczecin (Poland).

Samples of tested plants weighed and dried at room temperature (18–22 °C) for 3–4 days were ground to 0.1 mm by use of a laboratory mill type KNIFETEC 1095 (Foss

Tecator, Höganäs, Sweden) and placed in sterile containers, according to the list presented in Table 1.

**Table 1.** Plant material.

| Plant | Raw Material |
| --- | --- |
| Black chokeberry (*Aronia melanocarpa* L.) | fruit |
| Plantain (*Plantago lanceolata* L.) | fruit |
| Common cumin (*Carum carvi* L.) | fruit |
| Fenugreek (*Trigonella foenum-graecum* L.) | fruit |
| Wild rose (*Rosa canina* L.) | fruit |
| Marigold (*Calendula officinalis* L.) | flowerheads |
| Common chamomile (*Matricaria chamomilla* L.) | flowerheads |
| Birch (*Betula* L.) | leaves |
| Raspberry (*Rubus idaeus* L.) | leaves |
| Marsh mallow (*Althaea officinalis* L.) | leaves |
| Psyllium (*Plantago afra* L.) | seedhusks |
| Purple coneflower (*Echinacea purpurea* Moench L.) | herb |
| Yarrow (*Achillea millefolium* L.) | herb |
| Marjoram (*Origanum majorana* L.) | herb |
| Lemon balm (*Melissa officinalis* L.) | herb |
| Mint (*Mentha* L.) | herb |
| Common dandelion (*Taraxacum officinale* F.H. Wigg) | herb |
| Knotgrass (*Polygonum aviculare* L.) | herb |
| Stinging nettle (*Urtica dioica* L.) | herb |
| Cleavers (*Galium aparine* L.) | herb |
| Field horsetail (*Equisetum arvense* L.) | herb |
| Thyme (*Thymus vulgaris* L.) | herb |
| Mezzanine (*Filipendulae ulmariae herba* L.) | herb |
| Willow (*Salix alba* L.) | bark |
| Black cumin (*Nigella sativa* L.) | seeds |
| Garlic (*Allium sativum* L.) | bulbs |

*2.2. Methods*

2.2.1. Proximate Composition

Before conducting analyses by the weight-dryer method, the dry matter content was determined and afterwards components in the air-dry mass were analysed. The proximate composition of the samples was determined according to the Association of Official Analytical Chemists (AOAC) methods [23]. To determine dry matter, samples were dried at 105 °C to constant weight (method 945.15). Crude fat (as ether extract EE); method 2003.06) was determined using the Soxhlet extraction method with diethyl ether as solvent; crude ash (CA; method 920.153 by incineration in a muffle furnace at 580 °C for 8 h; crude protein (CP; method 945.18) (N × 6.25) by Kjeldahl method using a Büchi B-324 distillation unit (Büchi Labortechnik AG, Switzerland). Crude fiber (CF) was determined as the residue after sequential treatment with 1.25% $H_2SO_4$ and with 1.25% NaOH using an ANKOM$_{220}$ Fibre Analyser (ANKOM Technology, New York, NY, USA). Total carbohydrates were calculated as: nitrogen free extract (NFE) (%) = 100 − % (moisture + crude protein + crude fat + crude ash + crude fiber).

2.2.2. Extraction

Methanolic extracts were prepared by weighing a 1.0 g sample of the material and extracting it for 1.5 h with 40 mL of 70% methanol (analytical grade) in a water bath with a shaker at 31 °C ± 1 °C. Then, the cooled extract was filtered using Munktell filter paper (84 g/m$^2$) and funnels into 50 mL containers. These extracts were used for subsequent polyphenols content determination and antioxidant analysis. The extracts were stored in a freezer at −18 °C ± 1 °C.

### 2.2.3. Total Polyphenols Content

The content of polyphenols was determined using the method described by Swain and Hillis using the Folin–Ciocalteu reagent [24]. The absorbance of the obtained colored solutions was then measured using a spectrophotometer (Specto 2000 RD, LaboMed, We Los Angeles, CA, USA) at $\lambda$ = 760 nm against 70% methanol. Total polyphenol content is expressed as chlorogenic acid equivalents (mg CGA/100 g DM).

### 2.2.4. Antioxidant Activity

Antioxidant activity was measured using the method of Re et al. [25] with the ABTS$^{\bullet+}$ radical (2,2′-azino-bis-(3-ethylbenzothiazoline-6-sulfonic acid)). The absorbance of the colored solution was measured using a spectrophotometer (Specto 2000 RD, LaboMed, Inc., Los Angeles, CA, USA) at a wavelength of 734 nm in the presence of 70% methanol used to prepare the extracts. The result was expressed in $\mu$M Trolox/g DM.

Antioxidant activity was determined by the method of Brand-Williams et al. [26] using the free radical DPPH$^{\bullet+}$. To a 1.5 mL sample suitably diluted with methanol, 3 mL of the prepared DPPH$^{\bullet+}$ solution (diluted to an absorbance between 0.900–1.000) was added and the contents of the tube were mixed. The samples were left for 10 min of incubation protected from light and at room temperature. After this time, the absorbance was measured with a spectrophotometer (Specto 2000 RD, LaboMed, Inc. Los Angeles, CA, USA) at 515 nm against 99% pure undiluted methanol. The result was expressed in $\mu$M Trolox/g DM.

Antioxidant activity was determined by the FRAP method according to Benzie and Strain [27]. The absorbance at 593 nm against 70% methanol was measured using a spectrophotometer (Specto 2000 RD, LaboMed, Inc., Los Angeles, CA, USA). The results obtained are expressed in $\mu$M Trolox/g DM.

### 2.3. Statistical Analysis

All analyses were carried out in duplicate (proximate composition) and triplicate (total polyphenols and antioxidant activity). One factorial analysis of variance (ANOVA) and principal component analysis (PCA) were carried out using the STATISTICA v13.3 software. The significance of differences between the means was assessed using the Tukey test at $p$ = 0.05.

## 3. Results

### 3.1. Proximate Composition

Seeds of black cumin and milk thistle were characterized by the highest level of dry matter (respectively, 95.09 g/100 g, 94.65 g/100 g) while the lowest level was found in black chokeberry fruit (88.82 g/100 g). The highest content of protein was found in fruit of fenugreek (26.164 g/100 g DM). Herb of stinging nettle was the best source of fiber (42.661 g/100 g DM), and herb of common dandelion (24.506 g/100 g DM) was the best source of crude ash (Table 2).

### 3.2. Total Polyphenols Content

The tested herbs were characterized by a varied content of TPC (Table 3). The three herbs with the highest total polyphenolic content are marjoram (6956.584 mg CGA/100 g DM), chamomile (7240.002 mg CGA/100 g DM) and cleavers (7565.426 mg CGA/100 g DM). It is worth mentioning that birch (6585.825 mg CGA/100 g DM), field horsetail (6545.378 mg CGA/100 g DM) and garlic (6512.095 mg CGA/100 g DM), contained more than 6000 mg CGA/100 g DM phenolic compounds. The lowest content of total polyphenols, among the tested plants, was shown in the milk thistle (291.832 mg CGA/100 g DM) test, and the difference between the other tested samples was statistically significant at $p$ = 0.05.

**Table 2.** Proximate composition of analysed plant materials.

| Plant | DM *<br>(After Drying, g/100 g) | CP * | CF * | EE *<br>(g/100 g DM) | CA * | NFE * |
|---|---|---|---|---|---|---|
| Black chokeberry (*Aronia melanocarpa* L.) | 88.82 [a] ± 0.19 | 3.901 [a] ± 0.06 | 6.60 [c] ± 0.09 | 2.08 [bc] ± 0.03 | 2.48 [a] ± 0.06 | 73.76 [r] ± 0.07 |
| Plantain (*Plantago lanceolata* L.) | 91.66 [fg] ± 0.01 | 16.28 [i] ± 0.14 | 15.11 [g] ± 0.07 | 2.69 [cde] ± 0.10 | 12.59 [l] ± 0.12 | 44.99 [lm] ± 0.04 |
| Common cumin (*Carum carvi* L.) | 91.04 [cd] ± 0.03 | 20.57 [m] ± 0.40 | 28.00 [o] ± 0.11 | 14.91 [n] ± 0.06 | 5.68 [e] ± 0.05 | 21.87 [c] ± 0.49 |
| Fenugreek (*Trigonella foenum-graecum* L.) | 90.79 [bc] ± 0.01 | 26.16 [n] ± 0.26 | 10.25 [e] ± 0.13 | 4.47 [h] ± 0.04 | 3.05 [a] ± 0.01 | 46.85 [no] ± 0.35 |
| Milk thistle (*Silybum marianum* (L.) | 94.65 [n] ± 0.03 | 17.17 [j] ± 0.37 | 32.20 [q] ± 0.34 | 20.63 [o] ± 0.610 | 4.73 [d] ± 0.02 | 19.93 [b] ± 0.62 |
| Wild rose (*Rosa canina* L.) | 92.62 [ij] ± 0.03 | 5.76 [c] ± 0.05 | 37.05 [t] ± 0.14 | 5.40 [i] ± 0.01 | 4.18 [bcd] ± 0.04 | 40.22 [gh] ± 0.16 |
| Marigold (*Calendula officinalis* L.) | 90.37 [b] ± 0.13 | 19.53 [l] ± 0.09 | 17.25 [j] ± 0.02 | 7.60 [l] ± 0.07 | 9.77 [j] ± 0.08 | 36.24 [f] ± 0.35 |
| Common chamomile (*Matricaria chamomilla* L.) | 91.73 [fg] ± 0.16 | 14.21 [h] ± 0.09 | 23.10 [m] ± 0.07 | 3.40 [fg] ± 0.02 | 8.80 [i] ± 0.19 | 42.22 [ijk] ± 0.09 |
| Birch (*Betula* L.) | 93.13 [kl] ± 0.11 | 15.86 [i] ± 0.11 | 15.25 [g] ± 0.14 | 11.56 [m] ± 0.13 | 4.50 [cd] ± 0.01 | 45.96 [mno] ± 0.01 |
| Raspberry (*Rubus idaeus* L.) | 92.69 [ijk] ± 0.14 | 12.32 [g] ± 0.06 | 15.22 [g] ± 0.14 | 6.39 [jk] ± 0.09 | 6.53 [f] ± 0.04 | 52.22 [p] ± 0.21 |
| Marsh mallow (*Althaea officinalis* L.) | 93.29 [l] ± 0.04 | 18.86 [kl] ± 0.07 | 22.62 [n] ± 0.09 | 3.60 [g] ± 0.07 | 16.41 [n] ± 0.05 | 31.79 [e] ± 0.08 |
| Psyllium (*Plantago afra* L.) | 91.11 [cde] ± 0.01 | 5.52 [bc] ± 0.06 | 4.94 [b] ± 0.06 | 2.96 [efg] ± 0.06 | 2.59 [a] ± 0.05 | 75.11 [r] ± 0.09 |
| Purple coneflower (*Echinacea purpurea Moench* L.) | 91.64 [fg] ± 0.05 | 10.58 [f] ± 0.19 | 16.73 [ij] ± 0.09 | 2.79 [def] ± 0.02 | 14.18 [m] ± 0.15 | 47.37 [o] ± 0.09 |
| Yarrow (*Achillea millefolium* L.) | 92.11 [gh] ± 0.16 | 9.57 [e] ± 0.03 | 31.98 [q] ± 0.03 | 3.37 [fg] ± 0.16 | 8.06 [h] ± 0.12 | 39.14 [g] ± 0.38 |
| Marjoram (*Origanum majorana* L.) | 93.81 [m] ± 0.17 | 14.20 [h] ± 0.01 | 18.29 [j] ± 0.09 | 5.83 [ij] ± 0.06 | 12.39 [l] ± 0.04 | 43.11 [k] ± 0.02 |
| Lemon balm (*Melissa officinalis* L.) | 92.32 [hij] ± 0.03 | 17.25 [j] ± 0.59 | 13.62 [f] ± 0.11 | 3.51 [g] ± 0.06 | 11.67 [k] ± 0.21 | 46.27 [mno] ± 0.20 |
| Mint (*Mentha* L.) | 91.51 [def] ± 0.09 | 14.00 [h] ± 0.53 | 19.23 [l] ± 0.12 | 3.25 [efg] ± 0.18 | 11.41 [k] ± 0.05 | 43.62 [kl] ± 0.64 |
| Common dandelion (*Taraxacum officinale* F.H Wigg) | 92.75 [jk] ± 0.09 | 19.24 [l] ± 0.01 | 15.68 [gh] ± 0.04 | 4.32 [h] ± 0.05 | 24.51 [o] ± 0.36 | 29.01 [d] ± 0.46 |
| Knotgrass (*Polygonum aviculare* L.) | 91.45 [def] ± 0.41 | 12.44 [g] ± 0.06 | 27.89 [o] ± 0.38 | 1.93 [b] ± 0.08 | 6.60 [f] ± 0.09 | 42.58 [jk] ± 0.72 |
| Stinging nettle (*Urtica dioica* L.) | 90.90 [c] ± 0.05 | 17.36 [j] ± 0.27 | 42.66 [u] ± 0.17 | 2.10 [bc] ± 0.01 | 12.81 [l] ± 0.55 | 15.97 [a] ± 0.38 |
| Cleavers (*Galium aparine* L.) | 92.30 [hij] ± 0.06 | 9.85 [ef] ± 0.19 | 30.14 [p] ± 0.22 | 2.17 [bcd] ± 0.06 | 9.58 [j] ± 0.04 | 40.56 [gh] ± 0.37 |
| Field horsetail (*Equisetum arvense* L.) | 91.47 [def] ± 0.09 | 15.91 [i] ± 0.28 | 19.98 [m] ± 0.04 | 2.96 [efg] ± 0.07 | 11.49 [k] ± 0.05 | 41.14 [hij] ± 0.38 |
| Thyme (*Thymus vulgaris* L.) | 93.14 [kl] ± 0.03 | 14.15 [h] ± 0.13 | 16.23 [hi] ± 0.15 | 6.86 [k] ± 0.05 | 10.03 [j] ± 0.06 | 45.86 [mn] ± 0.32 |
| Mezzanine (*Filipendulae ulmariae herba* L.) | 91.59 [ef] ± 0.01 | 8.36 [d] ± 0.07 | 33.10 [r] ± 0.07 | 3.07 [efg] ± 0.02 | 4.52 [cd] ± 0.26 | 42.54 [jk] ± 0.24 |
| Willow (*Salix alba* L.) | 92.24 [hi] ± 0.19 | 4.92 [b] ± 0.03 | 35.82 [s] ± 0.09 | 3.41 [fg] ± 0.07 | 7.33 [g] ± 0.01 | 40.78 [hi] ± 0.24 |
| Black cumin (*Nigella sativa* L.) | 95.09 [n] ± 0.07 | 21.27 [m] ± 0.22 | 8.22 [d] ± 0.07 | 42.28 [p] ± 0.44 | 3.97 [bc] ± 0.01 | 19.35 [b] ± 0.65 |
| Garlic (*Allium sativum* L.) | 91.47 [def] ± 0.08 | 18.19 [k] ± 0.01 | 2.55 [a] ± 0.16 | 0.74 [a] ± 0.01 | 3.74 [b] ± 0.01 | 66.25 [q] ± 0.07 |

Means with at least same letter not differ statistically at $p = 0.05$. * dry matter (DM), crude protein (CP), crude fiber (CF), ether extract (EE), crude ash (CA), nitrogen free extract (NFE).

**Table 3.** Total polyphenol content and antioxidant activity of selected raw materials.

| Plants | Polyphenols * | ABTS ** | DPPH ** | FRAP ** |
|---|---|---|---|---|
| Black chokeberry (*Aronia melanocarpa* L.) | 595.738 [ab] ± 50.91 | 52.327 [ab] ± 0.52 | 13.364 [a] ± 0.00 | 41.047 [ab] ± 3.14 |
| Plantain (*Plantago lanceolata* L.) | 4578.407 [ij] ± 147.95 | 891.833 [k] ± 65.56 | 394.914 [e] ± 5.28 | 697.550 [k] ± 6.82 |
| Common cumin (*Carum carvi* L.) | 4144.998 [hi] ± 130.86 | 489.577 [hi] ± 32.01 | 245.704 [cde] ± 28.75 | 449.973 [hi] ± 32.12 |
| Fenugreek (*Trigonella foenum-graecum* L.) | 736.449 [ab] ± 20.06 | 81.057 [ab] ± 1.96 | 20.000 [a] ± 0.73 | 48.932 [ab] ± 3.16 |
| Milk thistle (*Silybum marianum* (L.) | 291.832 [a] ± 28.41 | 26.334 [a] ± 0.96 | 14.271 [a] ± 0.11 | 26.252 [a] ± 1.14 |
| Wild rose (*Rosa canina* L.) | 5072.939 [jk] ± 242.88 | 512.909 [i] ± 44.75 | 385.150 [de] ± 1.06 | 622.873 [jk] ± 43.06 |
| Marigold (*Calendula officinalis* L.) | 656.669 [ab] ± 5.00 | 48.923 [ab] ± 0.95 | 40.634 [ab] ± 13.76 | 114.776 [abcd] ± 6.84 |
| Common chamomile (*Matricaria chamomilla* L.) | 7240.002 [op] ± 134.46 | 1188.944 [l] ± 32.07 | 666.845 [f] ± 61.34 | 1699.766 [o] ± 48.26 |
| Birch (*Betula* L.) | 6585.825 [no] ± 233.64 | 1201.158 [l] ± 27.80 | 608.443 [f] ± 48.01 | 1030.362 [l] ± 25.28 |
| Raspberry (*Rubus idaeus* L.) | 5653.171 [kl] ± 53.93 | 896.536 [k] ± 34.30 | 712.134 [f] ± 128.36 | 930.384 [l] ± 73.74 |
| Marsh mallow (*Althaea officinalis* L.) | 3181.631 [fg] ± 42.20 | 375.841 [fgh] ± 78.47 | 230.800 [cde] ± 15.97 | 371.949 [gh] ± 17.76 |
| Psyllium (*Plantago afra* L.) | 1159.538 [bc] ± 74.92 | 111.030 [abc] ± 1.69 | 9.475 [a] ± 1.00 | 147.121 [abcde] ± 8.19 |
| Purple coneflower (*Echinacea purpurea* Moench L.) | 5778.254 [klm] ± 21.84 | 881.856 [k] ± 25.65 | 610.147 [f] ± 56.00 | 950.445 [l] ± 27.57 |
| Yarrow (*Achillea millefolium* L.) | 3322.056 [fg] ± 28.35 | 332.152 [efg] ± 12.73 | 124.108 [abc] ± 14.29 | 267.674 [efg] ± 18.70 |
| Marjoram (*Origanum majorana* L.) | 6956.584 [nop] ± 98.87 | 1396.437 [m] ± 6.40 | 664.860 [f] ± 98.39 | 1357.709 [n] ± 48.12 |
| Lemon balm (*Melissa officinalis* L.) | 2577.805 [ef] ± 203.09 | 303.079 [def] ± 14.82 | 61.489 [abc] ± 1.06 | 230.477 [def] ± 4.10 |
| Mint (*Mentha* L.) | 3466.203 [gh] ± 152.11 | 449.340 [ghi] ± 20.81 | 206.872 [bcd] ± 9.86 | 518.388 [ij] ± 0.00 |
| Common dandelion (*Taraxacum officinale* F.H Wigg) | 3132.016 [fg] ± 107.63 | 367.012 [fgh] ± 31.60 | 235.068 [cde] ± 7.36 | 354.451 [fgh] ± 0.00 |
| Knotgrass (*Polygonum aviculare* L.) | 2651.438 [ef] ± 148.12 | 319.487 [efg] ± 8.47 | 164.562 [abc] ± 20.07 | 285.465 [efg] ± 21.85 |
| Stinging nettle (*Urtica dioica* L.) | 6205.064 [lmn] ± 365.78 | 1912.016 [o] ± 17.08 | 1034.725 [g] ± 87.86 | 1698.517 [o] ± 18.35 |
| Cleavers (*Galium aparine* L.) | 7565.426 [p] ± 639.80 | 1890.523 [o] ± 76.71 | 1061.068 [g] ± 74.42 | 1769.766 [o] ± 119.09 |
| Field horsetail (*Equisetum arvense* L.) | 6545.378 [mno] ± 146.08 | 736.158 [j] ± 29.90 | 613.430 [f] ± 23.98 | 1202.939 [m] ± 27.55 |
| Thyme (*Thymus vulgaris* L.) | 1710.696 [cd] ± 92.73 | 176.528 [bcd] ± 6.43 | 112.595 [abc] ± 1.61 | 169.851 [bcde] ± 5.53 |
| Mezzanine (*Filipendulae ulmariae herba* L.) | 2200.496 [de] ± 124.08 | 272.093 [def] ± 32.00 | 102.573 [abc] ± 5.32 | 198.428 [cde] ± 16.05 |
| Willow (*Salix alba* L.) | 778.397 [ab] ± 19.09 | 82.138 [ab] ± 12.14 | 23.797 [ab] ± 0.81 | 59.523 [abc] ± 0.98 |
| Black cumin (*Nigella sativa* L.) | 2306.676 [de] ± 160.32 | 216.174 [cde] ± 15.03 | 161.562 [abc] ± 17.14 | 233.329 [defg] ± 7.38 |
| Garlic (*Allium sativum* L.) | 6512.095 [mno] ± 324.82 | 1751.490 [n] ± 0.00 | 933.050 [g] ± 66.27 | 1471.586 [n] ± 23.26 |

Means with the same letter do not differ statistically at $p = 0.05$; * mg CGA/100 g DM; ** µM Trolox/g DM; ABTS•+—2,2′-azino-bis-(3-ethylbenzothiazoline-6-sulfonic acid), DPPH•+—2,2′-Diphenyl-1-picrylhydrazyl, FRAP ferric ion reducing antioxidant parameter.

### 3.3. Antioxidant Activity Measured by the ABTS$^{+\bullet}$ Method

The values of antioxidant activity against the ABTS$^{\bullet+}$ radical ranged from 26.3–1912.0 μM Trolox/g DM, as shown in Table 3. The three highest results of antioxidant activity against the ABTS$^{\bullet+}$ radical were found in extracts from stinging nettle (1912.016 μM Trolox/g DM), cleavers (1890.523 μM Trolox/g DM), and garlic bulbs (1751.490 μM Trolox/g DM). In the tested plant material, the lowest result was obtained in milk thistle (26.334 μM Trolox/g DM; Table 3).

### 3.4. Antioxidant Activity Measured by the DPPH$^{+\bullet}$ Method

The values of antioxidant activity in the tested herbs against the DPPH$^{\bullet+}$ radical ranged from 9.5–1061.1 μM Trolox/g DM, which is shown in Table 3. The highest antioxidant activity against the DPPH$^{\bullet+}$ radical was shown by garlic bulbs (933.050 μM Trolox/g DM), stinging nettle herb (1034.725 μM Trolox/g DM) and cleavers herb (1061.068 μM Trolox/g DM). The lowest antioxidant activity against the DPPH$^{\bullet+}$ radical was found in the extract from psyllium seed husk, (9.475 μM Trolox/g DM); (Table 3).

### 3.5. Antioxidant Activity Measured by the FRAP Method

The three herbs with the highest antioxidant activity determined using the FRAP method were stinging nettle herb (1698.517 μM Trolox/g DM), chamomile flower heads (1699.766 μM Trolox/g DM) and cleavers herb (1769.766 μM Trolox/g DM), as shown in Table 3. The lowest antioxidant activity determined by the FRAP method was found in milk thistle extract (26.252 μM Trolox/g DM), which corresponded to the results obtained earlier in the study among others, total polyphenol content and antioxidant activity against the ABTS$^{\bullet+}$ radical (Table 3).

### 3.6. PCA Analysis

The PCA analysis showed that the first component is essentially related to the antioxidant activity, as the factor describing FRAP, DPPH$^{\bullet+}$, ABTS$^{\bullet+}$ and polyphenols. The second component, with slightly smaller factor loads, describes the variability of the basic composition of the tested raw materials (Figure 1A).

This analysis also confirmed a very high degree of correlation between the antioxidant activity (determined by various methods) and the content of polyphenols, and confirmed the correlation of the dry matter content with the protein content. In turn, the content of DM and CP was negatively correlated with NFE. A thesis can also be formulated about the lack of correlation between antioxidant activity, the content of polyphenols and the proximate composition of the tested raw materials.

The analysis of the factorial coordinates of the cases (Figure 1B) indicates the existence of four clusters of points (surrounded by ellipses). Raw materials in clusters in the second and third quadrants of the coordinate system are characterized by the greatest activity of the antioxidant activity.

Herbs that proved the highest antioxidant activity include: stinging nettle herb, marjoram herb, cleavers herb, common chamomile flower heads and garlic bulbs. Other raw materials are characterized by a smaller antioxidant activity and a significant differentiation of the basic composition. This is evidenced by the indicated extreme groups, i.e., two-element groups in the first and fourth quarters. The first one includes seeds of black cumin and fruits of milk thistle, whose raw materials were characterized by a low content of NFE, in contrast to the husks of psyllium seeds and black chokeberry fruits, of course, with significantly lower antioxidant activity.

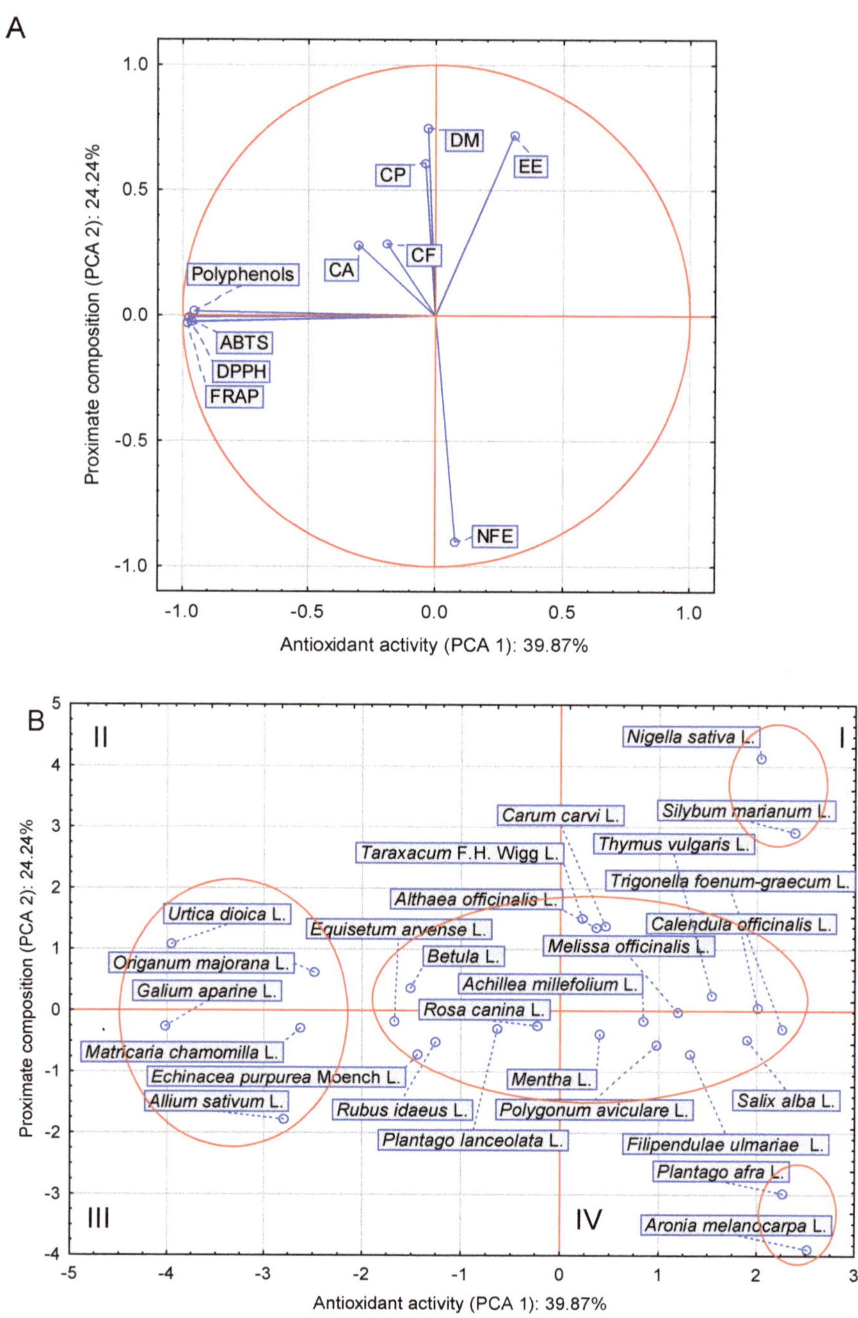

**Figure 1.** Biplot based on first two principal component axes for nutritional value and antioxidant activity (**A**) and distribution of 27 herbs on the first two components obtained from principal component analysis (**B**). Abbreviation crude protein (CP), crude fibre (CF), ether extract (EE), crude ash (CA), nitrogen free extract (NFE), 2,2′-azino-bis(3-ethylbenzothiazoline-6-sulfonic acid (ABTS$^{•+}$ radical), 2,2′-Diphenyl-1-picrylhydrazyl (DPPH$^{•+}$), ferric ion reducing antioxidant parameter (FRAP).

## 4. Discussion

### 4.1. Proximate Composition

The oldest method of herb preservation is drying, which is associated with the loss of water and inactivation of enzymes when begun immediately after harvest. Properly dried herbs do not ferment and do not turn moldy. Moreover, the levels of their active substances do not change over a long time.

The available literature contains little information about the proximate composition of many of the tested plants presented in this study. What is more, there is little information concerning proximate composition of morphological parts of herbs and medical plants.

It should be emphasized that some of the plants we analyzed are a rich source of CP (fenugreek), CF (stinging nettle), EE (black cumin) and NFE (psyllium). However, taking into account their consumption in the average daily diet, which is small and results from the use of these plants as an addition to dishes, their nutritional significance is minimal. Albeit they can be considered for use as raw materials for the production of food for special nutritional purposes.

Concerning protein content, Sadowska et al. [18] reported that whole plant of dried thyme contained 6.48 g of protein per DM; these results are lower than ours.

Some authors reported in their researches that black cumin contains about 32–40% of oil, 16–19.9% of protein, 1.79–3.74% of minerals, 5.5% of fiber [28]. These results are lower than we presented in our study (about 42% of fat, 21% of protein, almost 4% of ash, and 8% of fiber). On the other hand, Boskabady and Shirmohammadi [29] showed that this plant contains 26.7% of protein, 28.5% of fat, 8.4% of crude fiber, and 4.8% of total ash.

Sójka et al. [30] examined basic composition of chokeberry pomace fractions achieved as a result of industrial-scale processing of fruit into juice. They showed that this valuable raw material is rich in fat (13.9%), and proteins (24%). Chokeberry fruit are good sources of dietary fiber, i.e., at the level of 5.6% of fresh mass [31]. The lower content was determined in the present study (6.6%).

Sulieman et al. [32] showed that the contents of fiber, ash, protein, fat in fenugreek seed were 6.50%, 3.20%, 28.55% and 4%, respectively. Other authors analyzed chemical composition of different varieties of fenugreek and the ash content was in the range 3–7%, protein content 23.1–26.8%, fat content 8.8–21.0% and fiber content 5.1–7.1% [33]. These results are similar to our results, except fat content. In our study, the content of fat is much lower (4.5%).

Piątkowska et al. [34] determined proximate composition of common dandelion. The amount of protein was 15.25%, the crude fiber 13.09%, fat 6.81% and ash 9.11%. These results are similar to ours, except ash content. The content of ash was established as 24.5%.

### 4.2. Total Polyphenols Content

The tested herbs were characterized by a varied content of total polyphenols. Antioxidants, especially polyphenolic compounds (secondary metabolites), are produced by the protective systems of various plants in response to the destructive effects of free radicals. Higher plants produce secondary metabolites that protect them against environmental stress, pathogens or herbivores [35]. Depending on the number of aromatic rings and the way they are bound, they are divided into the following classes: flavonoids, phenolic acids, stilbenes and lignans [35–41]. In this study, selected dried herbs and plant materials were compared, from which various morphological parts of the plant were collected. These were plants commonly found in Polish meadows and gardens, easily available and cheap to obtain. The total content of polyphenolic compounds is presented in Table 3. Herbs with the highest content of these components in the conducted research turned out to be marjoram herb, flower heads of chamomile and cleavers herb. It is important to note that in many studies, the total polyphenol content is expressed as gallic acid, and in our research we used chlorogenic acid as the standard. Certainly, this affects the differences in the level of total polyphenols in the discussed material and may affect the interpretation of the results.

Our results are different from data reported by Bieżanowska-Kopeć and Piątkowska [38]. These authors showed lower content of total polyphenols (mg CGA/100 g DM) in leaves of lemon balm, marjoram, and thyme compared to our study. Similarly, the results of Tsivelika et al. [39] in common chamomile are lower than ours, but these results are calculated per mg of gallic acid/g DM). Common cleavers in the study by Milić et al. [40] was characterized by a higher content of total polyphenols. This plant is unknown, and any research on the content of polyphenolic compounds may turn out to be important in the context of further study of the common cleavers profile.

### 4.3. Antioxidant Activity Measured by the $ABTS^{+\bullet}$ Method

The highest antioxidant activity measured using the method with the $ABTS^{+\bullet}$ radical was found in extracts from marjoram herb, garlic bulbs, stinging nettle herb and cleavers herb (Table 3). Marjoram is a commonly used spice herb, which, in addition to a high content of polyphenolic compounds, is also characterized by high antioxidant activity. Gramza-Michałowska et al. [41] determined the activity against $ABTS^{\bullet+}$ radical in ethanolic solutions of marjoram herb at the level of 14.09 mg Trolox/g DM. In literature sources, the $ABTS^{\bullet+}$ values in garlic bulbs (47.7 µM Trolox/g DM) are lower than those observed in this study [36]. The leaves of the cleavers in the study by Csepregi et al. [42] showed activity against the $ABTS^{\bullet+}$ radical at the level of about 1.0 µM Trolox/µg of FM. Stinging nettle in the assessment of Rasa et al. [43] was characterized by activity against the $ABTS^{\bullet+}$ radical at the level of 18 mM Trolox/100 g DM. The $ABTS^{\bullet+}$ values obtained in the study (in the case of garlic, cleavers and stinging nettle) are, therefore, much higher than those observed in literature sources [42–44]. As in the case of the total polyphenol content analysis, the lowest antioxidant activity against the $ABTS^{\bullet+}$ radical was found in the milk thistle sample.

### 4.4. Antioxidant Activity Measured by the $DPPH^{+\bullet}$ Method

The highest antioxidant activity against the $DPPH^{\bullet+}$ radical was shown by garlic bulbs, stinging nettle herb and cleavers herb, and these values were not statistically different (Table 3). In the methanol extracts studied by Wojdyło et al. [12], there was reported lower antioxidant activity in fenugreek, (3.64 µmol Trolox/g DM; coneflower leaves (0.75 µmol Trolox/g DM, and in knotgrass 1.41 µmol Trolox/g DM. In the stinging nettle herb, Belmaghraoui et al. [45] determined the activity against the $DPPH^{\bullet+}$ radical at the level of 483.98 IC50 µg/mL. Vlase et al. [46], in studies on different varieties of cleavers, determined the activity against the $DPPH^{\bullet+}$ radical in this herb at the level of 107.45 IC50 µg/mL. There are no sources in the literature describing the antioxidant activity of cleavers in µM Trolox/g DM. Gorinstein et al. [47] determined the lower antioxidant activity with the DPPH method in cloves of Polish garlic varieties, amounting to 14.81–34.86 µmol Trolox/g DM, as compared to our study.

### 4.5. Antioxidant Activity Measured by the FRAP Method

The highest results of antioxidant activity assessed by the FRAP method were recorded in extracts from flower heads of chamomile, stinging nettle and cleavers, and they did not differ significantly regarding statistics with respect to $p = 0.05$. Kukric et al. [48] determined the antioxidant activity in stinging nettle leaves using the FRAP method at the level of 7.5 mM $Fe^{2+}$/g DM. Mărghitaş et al. [49] examined pollen collected from common chamomile, in which the $Fe^{3+}$ reducing capacity was close to 5.35 mM $Fe^{2+}$/g DM. In the available literature, there is no information on the antioxidant activity in the cleavers determined using the FRAP method. As in the previous determinations, samples containing marjoram and garlic bulbs were characterized by high antioxidant activity. Hossain et al. [50], in his study on the extraction of polyphenolic compounds from the marjoram herb, assessed the antioxidant activity using the FRAP method at 18.96 g Trolox/100 g DM. In methanolic extracts of garlic cloves tested by Gorinstein et al. [47], the antioxidant activity measured with FRAP method was lower 6.63–11.95 µmol Trolox/g DM. Again, the lowest antioxidant activity was determined in the milk thistle sample.

Determination of antioxidant activity by several methods is crucial in determining the profile of antioxidant compounds. $DPPH^{\bullet+}$ dissolves only in organic solvents and does not allow the determination of hydrophilic antioxidants [51]. The $ABTS^{\bullet+}$ radical method and the FRAP method are used to determine the activity of both hydrophobic and hydrophilic antioxidant samples [52]. A difference can be seen in the antioxidant activity of chamomile flower heads, where this activity assessed using the FRAP method is much higher than that assessed using the $DPPH^{\bullet+}$ radical.

## 5. Conclusions

The conducted research turns out to be innovative due to the discovery of potential sources of new antioxidant compounds in selected herbal materials. Cleavers can be used as a source of new antioxidant compounds. However, further research is required in the context of its profile of anti-nutritional and potentially harmful compounds, due to the very limited literature sources analyzing this plant. Some analyzed plants are also a rich source of nutrients. They can be used as ingredients in functional food products. Selected herbs show high antioxidant activity, but only their systematic use combined with a dosage appropriate for the individual can show a beneficial effect in maintaining the health of the body. In addition, it seems beneficial to use herbal mixtures with high antioxidant potential; however, possible adverse interactions should be taken into further consideration.

**Author Contributions:** Conceptualization, W.B. and E.P.; methodology, W.B. and U.P.; software, R.W.; investigation, W.B., E.P. and U.P.; data curation, R.W.; writing—original draft preparation, W.B. and U.P., writing—review and editing, W.B., E.P., R.W. and A.K. All authors have read and agreed to the published version of the manuscript.

**Funding:** The study was financed by the Ministry of Science and Higher Education of the Republic of Poland for University of Agriculture in Kraków, Poland.

**Institutional Review Board Statement:** Not applicable.

**Informed Consent Statement:** Not applicable.

**Data Availability Statement:** Not applicable.

**Conflicts of Interest:** The authors declare no conflict of interest.

## References

1. Ayoka, T.O.; Ezema, B.O.; Eze, C.N.; Nnadi, C.O. Antioxidants for the Prevention and Treatment of Non-communicable Diseases. *J. Explor. Res. Pharmacol.* **2022**, *7*, 178–188. [CrossRef]
2. Pisoschi, A.M.; Pop, A.; Iordache, F.; Stanca, L.; Predoi, G.; Serban, A.I. Oxidative stress mitigation by antioxidants-an overview on their chemistry and influences on health status. *Eur. J. Med. Chem.* **2021**, *209*, 112891. [CrossRef] [PubMed]
3. Heunks, L.M.A.; Dekhuijzen, P.N.R. Respiratory muscle function and free radicals: From cell to COPD. *Thorax* **2000**, *55*, 704–716. [CrossRef]
4. Trinity, J.D.; Broxterman, R.M.; Richardson, R.S. Regulation of exercise blood flow: Role of free radicals. *Free Radic. Biol. Med.* **2016**, *98*, 90–102. [CrossRef]
5. Patlevič, P.; Vašková, J.; Švorc, P., Jr.; Vaško, L.; Švorc, P. Reactive oxygen species and antioxidant defense in human gastrointestinal diseases. *Integr. Med. Res.* **2016**, *5*, 250–258. [CrossRef] [PubMed]
6. Geto, Z.; Molla, M.D.; Challa, F.; Belay, Y.; Getahun, T. Mitochondrial dynamic dysfunction as a main triggering factor for inflammation associated chronic non-communicable diseases. *J. Inflamm. Res.* **2020**, *13*, 97. [CrossRef]
7. Vakifahmetoglu-Norberg, H.; Ouchida, A.T.; Norberg, E. The role of mitochondria in metabolism and cell death. *Biochem. Biophys. Res. Commun.* **2017**, *482*, 426–431. [CrossRef]
8. Mirończuk-Chodakowska, I.; Witkowska, A.M.; Zujko, M.E. Endogenous non-enzymatic antioxidants in the human body. *Adv. Med. Sci.* **2018**, *63*, 68–78. [CrossRef]
9. Kurutas, E.B. The importance of antioxidants which play the role in cellular response against oxidative/nitrosative stress: Current state. *Nutr. J.* **2016**, *15*, 71. [CrossRef]
10. Seyedsadjadi, N.; Grant, R. The potential benefit of monitoring oxidative stress and inflammation in the prevention of non-communicable diseases (NCDs). *Antioxidants* **2020**, *10*, 15. [CrossRef]
11. Embuscado, M.E. Spices and herbs: Natural sources of antioxidants—A mini review. *J. Funct. Foods* **2015**, *18*, 811–819. [CrossRef]
12. Wojdyło, A.; Oszmiański, J.; Czemerys, R. Antioxidant activity and phenolic compounds in 32 selected herbs. *Food Chem.* **2007**, *105*, 940–949. [CrossRef]

13. Jacobs, D.R.; Tapsell, L.C.; Temple, N.J. Food synergy: The key to balancing the nutrition research effort. *Public Health Rev.* **2012**, *33*, 507–529. [CrossRef]
14. Jayedi, A.; Rashidy-Pour, A.; Parohan, M.; Zargar, M.S.; Shab-Bidar, S. Dietary antioxidants, circulating antioxidant concentrations, total antioxidant capacity, and risk of all-cause mortality: A systematic review and dose-response meta-analysis of prospective observational studies. *Adv. Nutr.* **2018**, *9*, 701–716. [CrossRef]
15. Banjarnahor, S.D.; Artanti, N. Antioxidant properties of flavonoids. *Med. J. Indones.* **2015**, *23*, 239–244. [CrossRef]
16. Okoh, M.P. Antioxidants an Epigenetics Regulator for the Prevention of Diseases and Aging Process. *J. Genet. Eng. Biotechnol. Res.* **2019**, *1*, 1–5.
17. Chandra, S.; Khan, S.; Avula, B.; Lata, H.; Yang, M.H.; Elsohly, M.A.; Khan, I.A. Assessment of total phenolic and flavonoid content, antioxidant properties, and yield of aeroponically and conventionally grown leafy vegetables and fruit crops: A comparative study. *Evid. Based Complement. Altern. Med.* **2014**, *2014*, 253875. [CrossRef]
18. Sadowska, U.; Kopeć, A.; Kourimska, L.; Zarubova, L.; Koloucek, P. The effect of drying methods on the concentration of compounds in sage and thyme. *J. Food Process. Preserv.* **2017**, *41*, e13286. [CrossRef]
19. Paunović, S.M.; Mašković, P.; Nikolić, M.; Miletić, R. Bioactive compounds and antimicrobial activity of black currant (Ribes nigrum L.) berries and leaves extract obtained by different soil management system. *Sci. Hortic.* **2017**, *222*, 69–75. [CrossRef]
20. Dziadek, K.; Kopeć, A.; Tabaszewska, M. Potential of sweet cherry (Prunus avium L.) by-products: Bioactive compounds and antioxidant activity of leaves and petioles. *Eur. Food Res. Technol.* **2019**, *245*, 763–772. [CrossRef]
21. Roby, M.H.H.; Sarhan, M.A.; Selim, K.A.H.; Khalel, K.I. Evaluation of antioxidant activity, total phenols and phenolic compounds in thyme (*Thymus vulgaris* L.), sage (*Salvia officinalis* L.), and marjoram (*Origanum majorana* L.) extracts. *Ind. Crops Prod.* **2013**, *43*, 827–831. [CrossRef]
22. Viuda-Martos, M.; Ruiz Navajas, Y.; Sánchez Zapata, E.; Fernández-López, J.; Pérez-Álvarez, J.A. Antioxidant activity of essential oils of five spice plants widely used in a Mediterranean diet. *Flavour Fragr. J.* **2010**, *25*, 13–19. [CrossRef]
23. AOAC. *Official Methods of Analysis of the AOAC*, 21st ed.; AOAC: Gaithersburg, MD, USA, 2019.
24. Swain, T.; Hillis, W.E. The phenolic constituents of *Prunus Domesticus* (L.). The quantity of analisys of phenolic constituents. *J. Food Sci. Agric.* **1959**, *10*, 63–68. [CrossRef]
25. Re, R.; Pellegrini, N.; Proteggente, A.; Pannala, A.; Yang, M.; Rice-Evans, C. Antioxidant activity applying an improved ABTS radical cation decolorization assay. *Free. Radic. Biol. Med.* **1999**, *26*, 1231–1237. [CrossRef] [PubMed]
26. Brand-Williams, W.; Cuvelier, M.E.; Berset, C.L.W.T. Use of a free radical method to evaluate antioxidant activity. *LWT-Food Sci. Technol.* **1995**, *28*, 25–30. [CrossRef]
27. Benzie, I.F.; Strain, J.J. The ferric reducing ability of plasma (FRAP) as a measure of "antioxidant power": The FRAP assay. *Anal. Biochem.* **1996**, *239*, 70–76. [CrossRef]
28. Datta, A.K.; Saha, A.; Bhattacharya, A.; Mandal, A.; Paul, R.; Sengupta, S. Black cumin (*Nigella sativa* L.)—A review. *J. Plant Dev. Sci.* **2012**, *4*, 1–43.
29. Boskabady, M.H.; Shirmohammadi, B. Effect of Nigella sativa on isolated guinea pig trachea. *Arch. Iran. Med.* **2002**, *5*, 103–107.
30. Sójka, M.; Kołodziejczyk, K.; Milala, J. Polyphenolic and basic chemical composition of black chokeberry industrial by-products. *Ind. Crops Prod.* **2013**, *51*, 77–86. [CrossRef]
31. Tanaka, T.; Tanaka, A. Chemical components and characteristics of black chokeberry. *J. Jpn. Soc. Food Sci. Technol.* **2001**, *48*, 606–610. [CrossRef]
32. Sulieman, A.M.E.; Ahmed, H.E.; Abdelrahim, A.M. The chemical composition of fenugreek (*Trigonella foenum graceum* L.) and the antimicrobial properties of its seed oil. *Gezira J. Eng. Appl. Sci.* **2008**, *3*, 52–71.
33. Bouhenni, H.; Doukani, K.; Şekeroğlu, N.; Gezici, S.; Tabak, S. Comparative study on chemical composition and antibacterial activity of fenugreek (*Trigonella foenum graecum* L.) and cumin (*Cuminum cyminum* L.) seeds. *Ukr. Food J.* **2019**, *8*, 755–767. [CrossRef]
34. Piątkowska, E.; Biel, W.; Witkowicz, R.; Kępińska-Pacelik, J. Chemical Composition and Antioxidant Activity of Asteraceae Family Plants. *Appl. Sci.* **2022**, *12*, 12293. [CrossRef]
35. Quan, T.H.; Benjakul, S.; Sae-leaw, T.; Balange, A.K.; Maqsood, S. Protein–polyphenol conjugates: Antioxidant property, functionalities and their applications. *Trends Food Sci. Technol.* **2019**, *91*, 507–517. [CrossRef]
36. Rudrapal, M.; Khairnar, S.J.; Khan, J.; Dukhyil, A.B.; Ansari, M.A.; Alomary, M.N.; Alshabrmi, F.M.; Palai, S.; Deb, P.K.; Devi, R. Dietary polyphenols and their role in oxidative stress-induced human diseases: Insights into protective effects, antioxidant potentials and mechanism(s) of action. *Front. Pharmacol.* **2022**, *14*, 806470. [CrossRef]
37. Cory, H.; Passarelli, S.; Szeto, J.; Tamez, M.; Mattei, J. The role of polyphenols in human health and food systems: A mini-review. *Front. Nutr.* **2018**, *5*, 87. [CrossRef]
38. Bieżanowska-Kopeć, R.; Piątkowska, E. Total Polyphenols and Antioxidant Properties of Selected Fresh and Dried Herbs and Spices. *Appl. Sci.* **2022**, *12*, 4876. [CrossRef]
39. Tsivelika, N.; Irakli, M.; Mavromatis, A.; Chatzopoulou, P.; Karioti, A. Phenolic Profile by HPLC-PDA-MS of Greek Chamomile Populations and Commercial Varieties and Their Antioxidant Activity. *Foods* **2021**, *10*, 2345. [CrossRef]
40. Milić, P.S.; Stanojević, L.P.; Rajković, K.M.; Milić, S.M.; Nikolić, V.D.; Nikolić, L.B.; Veljković, V.B. Antioxidant activity of Galium mollugo L. extracts obtained by different recovery techniques. *Hem. Ind.* **2013**, *67*, 89–94. [CrossRef]

41. Gramza-Michałowska, A.; Abramowski, Z.; Jovel, E.; Hes, M. Antioxidant potential of herbs extracts and impact on HEPG2 cells viability. *Acta Sci. Pol. Technol. Aliment.* **2008**, *7*, 61–72.
42. Csepregi, R.; Bencsik, T.; Papp, N. Examination of secondary metabolites and antioxidant capacity of Anthyllis vulneraria, Fuchsia sp., Galium mollugo and Veronica beccabunga. *Acta Biol. Hung.* **2016**, *67*, 442–446. [CrossRef] [PubMed]
43. Rasa, K.; Jurga, B.; Daiva, M.; Katerina, D.; Ruta, M.; Jan, M.; Arunas, S. Investigation of antiradical activity of Salvia officinalis L., Urtica dioica L., and Thymus vulgaris L. extracts as potential candidates for a complex therapeutic preparation. *J. Med. Plants Res.* **2011**, *5*, 6090–6096.
44. Najman, K.; Sadowska, A.; Buczak, K.; Leontowicz, H.; Leontowicz, M. Effect of heat-treated garlic (*Allium sativum* L.) on growth parameters, plasma lipid profile and histological changes in the ileum of atherogenic rats. *Nutrients* **2022**, *14*, 336. [CrossRef] [PubMed]
45. Belmaghraoui, W.; Manni, A.; Harir, M.; Filali-Maltouf, A.; Fatni, O.K.E.; Hajjaji, S.E. Phenolic compounds quantification, antioxidant and antibacterial activities of different parts of Urtica dioica and Chenopodium murale. *Res. J. Pharm. Technol.* **2018**, *11*, 5490–5496. [CrossRef]
46. Vlase, L.; Mocan, A.; Hanganu, D.; Benedec, D.; Gheldiu, A.; Crisan, G. Comparative study of polyphenolic content, antioxidant and antimicrobial activity of four Galium species (*Rubiaceae*). *Dig. J. Nanomater. Biostruct.* **2014**, *9*, 1085–1094.
47. Gorinstein, S.; Leontowicz, H.; Leontowicz, M.; Namiesnik, J.; Najman, K.; Drzewiecki, J.; Cvikrová, M.; Martincová, O.; Katrich, E.; Trakhtenberg, S. Comparison of the main bioactive compounds and antioxidant activities in garlic and white and red onions after treatment protocols. *J. Agric. Food Chem.* **2008**, *25*, 4418–4426. [CrossRef]
48. Kukrić, Z.Z.; Topalić-Trivunović, L.N.; Kukavica, B.M.; Matoš, S.B.; Pavičić, S.S.; Boroja, M.M.; Savić, A.V. Characterization of antioxidant and antimicrobial activities of nettle leaves (*Urtica dioica* L.). *Acta Period. Technol.* **2012**, *43*, 257–272. [CrossRef]
49. Mărghitaş, L.A.; Stanciu, O.G.; Dezmirean, D.S.; Bobiş, O.; Popescu, O.; Bogdanov, S.; Campos, M.G. In vitro antioxidant capacity of honeybee-collected pollen of selected floral origin harvested from Romania. *Food Chem.* **2009**, *115*, 878–883. [CrossRef]
50. Hossain, M.B.; Brunton, N.P.; Patras, A.; Tiwari, B.; O'donnell, C.P.; Martin-Diana, A.B.; Barry-Ryan, C. Optimization of ultrasound assisted extraction of antioxidant compounds from marjoram (*Origanum majorana* L.) using response surface methodology. *Ultrason. Sonochemistry* **2012**, *19*, 582–590. [CrossRef]
51. Sanchez-Moreno, C. Method used to evaluate the free radical scavenging activity in food and biological systems. *Food Sci. Technol.* **2000**, *8*, 419–421. [CrossRef]
52. Witkowicz, R.; Biel, W.; Skrzypek, E.; Chłopicka, J.; Gleń-Karolczyk, K.; Krupa, M.; Prochownik, E.; Galanty, A. Microorganisms and Biostimulants Impact on the Antioxidant Activity of Buckwheat (*Fagopyrum esculentum* Moench) Sprouts. *Antioxidants* **2020**, *9*, 584. [CrossRef] [PubMed]

**Disclaimer/Publisher's Note:** The statements, opinions and data contained in all publications are solely those of the individual author(s) and contributor(s) and not of MDPI and/or the editor(s). MDPI and/or the editor(s) disclaim responsibility for any injury to people or property resulting from any ideas, methods, instructions or products referred to in the content.

*Article*

# Stability of Strawberry Fruit (*Fragaria x ananassa* Duch.) Nutritional Quality at Different Storage Conditions

Rohullah Qaderi, Bruno Mezzetti, Franco Capocasa and Luca Mazzoni *

Department of Agriculture, Food and Environmental Sciences, Università Politecnica delle Marche, Via Brecce Bianche 10, 60131 Ancona, Italy
* Correspondence: l.mazzoni@staff.univpm.it; Tel.: +39-071-2204640

**Abstract:** Strawberry fruit is a very rich source of vitamins and phenolic compounds, which determine its nutritional properties. Strawberries are a highly perishable non-climacteric fruit, and their perishable nature can lead to physical and chemical damage during storage. Therefore, the large market of fresh fruit relies on the capacity of fast distribution and marketing under a continuous cold-storage chain. In this study, we applied different cold-storage temperatures (domestic $-20\ ^{\circ}$C and industrial $-80\ ^{\circ}$C) on different treatments (whole fruits and dried fruits) of three strawberry cultivars (Arianna, Francesca, and Silvia), for up to seven months, and evaluated the influence of different storage conditions and lengths on the stability of the fruits' nutritional compounds (vitamin C, phenolic acids, anthocyanins, and folate). The results show that the nutritional quality of the fruits was significantly affected by storage temperature (with $-80\ ^{\circ}$C storage preserving more nutritional compounds), while storage time did not greatly affect the composition of the nutritional compounds in the whole or dried fruits. Oven drying the fruits dramatically affected their vitamin C content, almost completely degrading this compound (from 731.8 to 23.2 mg/kg at time 0 for fresh Arianna fruit, the cultivar with the highest amount). The amount of folate was increased during storage (from 126.17 at time 0 to 190.61 µg/kg at time 7 for fresh whole Arianna fruit). The interesting results obtained in this study are worth considering in future studies, to better plan fruit-storage conditions and time, for maintaining better fruit nutritional quality.

**Keywords:** folate; polyphenols; storage; strawberry; vitamin C

**Citation:** Qaderi, R.; Mezzetti, B.; Capocasa, F.; Mazzoni, L. Stability of Strawberry Fruit (*Fragaria x ananassa* Duch.) Nutritional Quality at Different Storage Conditions. *Appl. Sci.* **2023**, *13*, 313. https://doi.org/10.3390/app13010313

Academic Editor: Maria Kanellaki

Received: 1 December 2022
Revised: 20 December 2022
Accepted: 25 December 2022
Published: 27 December 2022

**Copyright:** © 2022 by the authors. Licensee MDPI, Basel, Switzerland. This article is an open access article distributed under the terms and conditions of the Creative Commons Attribution (CC BY) license (https://creativecommons.org/licenses/by/4.0/).

## 1. Introduction

Strawberries are the most cultivated berry fruit worldwide, with an annual production of 13.3 million tons, on a surface area of 522,527 ha [1]. Consumer health expectations are linked to the intrinsic characteristics of the product, such as the presence of bioactive compounds with a nutraceutical effect. Many epidemiologic studies have shown that a diet rich in fruits and vegetables is often associated with a lower incidence of several chronic pathologies, including obesity, infections, cancer, and cardiovascular and neurologic diseases [2,3]. Berries, including strawberries, have an important role among fruits because of their high phytochemical content [4,5]. Numerous scientific studies confirm that the strawberry contains bioactive molecules with antioxidant power, such as ascorbic acid, polyphenolic compounds such as ellagic acid, ferulic acid, and some flavonoids (anthocyanins, catechins, phenolic acids, etc.). These compounds exhibit a nutraceutical effect, exerting beneficial and protective properties on the human body [6].

Strawberries are a highly perishable non-climacteric fruit, which can lead to physical damage during storage and transportation. Therefore, the large market of fresh fruit relies on the capacity of fast distribution and marketing under a continuous cold-storage chain. Fruit overproduction, especially during off-season periods, is addressed mostly toward the processing industry to produce bakery products such as jam, jellies, juice, puree, flavor additives, etc. [4,7], with a consequence being a price reduction for the product.

Strawberries' post-harvest decay can be due to physical, physiological, or pathological factors that may happen in the pre-harvest period (and then take place during storage) or directly in the post-harvest period, and their shelf life is diminished by firmness loss, fruit desiccation, and the growth of spoilage microorganisms [7].

Freezing is a simple but common and effective method for long-term preservation and storage of fruits and vegetables, while maintaining many of their fresh-like qualities [8,9]. In addition, freezing is less destructive than other available preserving methods [10]. Given the nutritional value of fruits and vegetables, it is essential that bioactive compounds and nutrients are considered healthy when frozen and stored. These methods support the food supply chain and producer income when fresh food is not available. Long-term storage is becoming a more common in-home consumer behavior, and labs freeze food prior to analysis, both for convenience. Analytical labs and home kitchens typically have $-20\,°C$ freezers. Although $-80\,°C$ can be used in a laboratory, this equipment is expensive and has high operating costs, mainly due to energy consumption [11]. The freezing rate is the most important factor in the freezing process to prevent fruit-tissue damage, loss of bioactive compounds, and drip loss in thawing. Faster freezing results in small ice crystals and better frozen-fruit quality [12]. The effect of freezing fruits on microbial and enzymatic activity can influence the chemical composition, but storage temperature and storage period could also have an effect on the chemical composition of the strawberry [13].

Dehydration is another method for long-term preserving and increases the shelf life of delicate fruits. Oven drying fruits has been the most common preserving method used for many years [14], and, in the past few decades, considerable efforts have been made to understand the chemical and biochemical changes that occur during dehydration and to develop methods for preventing undesirable quality losses. Drying reduces water activity, and avoids microbial growth and deteriorative chemical reactions. The effects of heat on microorganisms and on the activity of enzymes are also important when drying fruits [15].

In this study, we evaluate the effect of fruit treatment on whole fruits and dried fruits (WF and DF) at different storage times (from 0 to 7 months) and temperatures ($-20$ and $-80\,°C$) on the nutritional quality (vitamin C, anthocyanins, phenolic acids, and folate) of three strawberry cultivars.

## 2. Materials and Methods

### 2.1. Plant Material

Three commercial cultivars (Arianna, Francesca, and Silvia) were planted in July 2020 in non-fumigated soil in "P. Rosati" experimental farm of Università Politecnica delle Marche, sited in Agugliano (Ancona, Italy), with the following main characteristics: pH 7.9, active calcium 9%, and texture composed of 40% clay, 25% sand, and 35% silt, following the procedure described in Mezzetti et al. [16]. Plants were grown in open field conditions according to the plastic hill culture production system. Fruit samples were harvested at fully red stage, at the second, third, and fourth main seasonal pickings [7], and immediately treated according to the experimental storage design. The environmental data of the picking season (April–June 2021) were registered and are reported in Table 1.

**Table 1.** Monthly rainfall (sum) and average daily maximum, average, and minimum temperatures registered in April, May, and June 2021.

| Month | Rainfall (mm) | Maximum Temperature (°C) | Average Temperature (°C) | Minimum Temperature (°C) |
|---|---|---|---|---|
| April | 33.0 | 17.9 | 12.2 | 7.1 |
| May | 23.8 | 24.1 | 18.2 | 12.7 |
| June | 7.8 | 30.8 | 25.0 | 19.1 |

*2.2. Experimental Storage Design*

For all cultivars analysed, two fruit treatments were taken into consideration: whole fruits (WF) and oven-dried fruits (DF). Regarding the fruit treatments, for the WF, we collected fresh ripe fruit and directly put them into a normal $-20\,°C$ domestic refrigerator (Zoppas, Vittorio Veneto, Italy) and a laboratory $-80\,°C$ refrigerator (HFU B series, Thermo Fisher Scientific, Milan, Italy). For the DF, after the collection of fruit from the field, we put them in a laboratory oven (Pbi international, Milan, Italy) at $65\,°C$ for 1 week. The loss of water and the complete dryness of fruits was monitored by checking the fruit weight at intervals of 24 h, until no further weight reductions were detected. Then, dried fruits were placed at $-20\,°C$ and $-80\,°C$, as indicated before for WF.

Fruits were stored for five different storage times: 0, 1, 3, 5, and 7 months after the harvest day. During each storage time (except at 0, when the fruits were immediately analyzed), fruits of three cultivars were stored at $-20\,°C$ and $-80\,°C$ temperature conditions. For each combination of cultivar/storage time/treatment/storage temperature, three repetitions of 5 fruits were made. A summary of the experimental storage design is described in Table 2.

**Table 2.** Storage experimental design applied to each of the three studied cultivars. WF: whole fruit; DF: dried fruit.

| Storage Time | Fruit Treatment | Storage Temperature | Number of Fruits |
|---|---|---|---|
| 0 M | WF | - | 5 |
| | DF | | 5 |
| 1 M | WF | −20 | 5 |
| | | −80 | 5 |
| | DF | −20 | 5 |
| | | −80 | 5 |
| 3 M | WF | −20 | 5 |
| | | −80 | 5 |
| | DF | −20 | 5 |
| | | −80 | 5 |
| 5 M | WF | −20 | 5 |
| | | −80 | 5 |
| | DF | −20 | 5 |
| | | −80 | 5 |
| 7 M | WF | −20 | 5 |
| | | −80 | 5 |
| | DF | −20 | 5 |
| | | −80 | 5 |

For the analyses of fruit nutritional quality, the WF and DF were extracted and analyzed after 1, 3, 5, and 7 months of refrigerated storage; for the "time 0", WF were immediately extracted and analyzed after collection, while DF were immediately extracted and analyzed after oven-drying. The analyzed nutritional parameters were anthocyanins, phenolic acids, vitamin C, and folate. All the parameters were analyzed by HPLC-UV-FD.

*2.3. Methanolic Extraction*

Fruit extracts were prepared as described by Diamanti et al. [17]. First, 10 g of WF and 1 g of DF were homogenized with Ultraturrax T25 homogenizer (Janke and Kunkel, IKA Labortechnik, Staufen, Denmark) in 20 mL of methanol and agitated for 30 min in the dark. The suspension was centrifuged at 4500 rpm for 10 min at $4\,°C$, the supernatant was collected, and the pellet of the fruits was extracted for a second time by adding another 20 mL of methanol and repeating the procedure. The second supernatant was added to the first one and then immediately injected to HPLC.

### 2.4. Vitamin C Extraction

Vitamin C of fruits was extracted with an ultrasound-assisted extraction protocol, as described by Tulipani et al. [18]. The extraction was carried out with the use of an ultrasound bath (Bioblock/ELMA 88155, Stuttgart, Germany), an instrument that generates ultrasound waves inside a tank containing water, using high frequency electric current produced by a generator. The process is useful to speed up the dissolution of solutes in certain solvents.

The analysis requires homogenizing 1 g of frozen strawberry for WF fruit and 0.1 g for DF with an aliquot of 4 mL taken from the extraction buffer solution, containing 5% metaphosphoric acid and 1 mM DTPA, followed by 5 min of sonication and centrifugation at 4000 rpm for 10 min at 4 °C. The supernatants obtained from each sample were filtered (filter pore size 0.45 μm) and inserted into a vial to perform analysis on an HPLC system.

### 2.5. Extraction of Vitamin B9 (Folate)

Following the method described by Mezzetti et al. [19], with slight modifications, 8 mL of the extracting solution (0.1 mol/L sodium acetate containing 10% (*w/v*) sodium chloride, 1% (*w/v*) ascorbic acid, and 0.1% 2-mercaptoethanol) were pipetted in 2 g of the frozen strawberry for WF samples and 0.2 g for DF samples, then homogenized with Ultraturrax T25 homogenizer (Janke and Kunkel, IKA Labortechnik, Staufen, Denmark). The falcon tube was loosely capped, boiled in water for 12 min, and rapidly cooled in the freezer for 10 min. Hog kidney folate conjugate enzyme was prepared, and about 1.5 mL of the enzyme was added to the cooled solution and incubated in a shaking oven at 37 °C for 3 h. Afterwards, the enzyme was inactivated by boiling in water for 5 min followed by cooling for 10 min in the freezer. The samples were then centrifuged at 4500 rpm for 30 min at 4 °C, and the supernatant was transferred into a new labeled falcon tube. Moreover, another 8 mL of the extracting solution was added to the resulting pellet and centrifuged again for 30 min. The second supernatant was added to the first one, then extracting solution was added to top up the supernatants to 25 mL. The final supernatant of 25 mL was then filtered using 0.45 μm filter pore size, 25 mm inner diameter, nylon disposable syringe filters, and the filtrates were purified through solid-phase extraction on strong anion-exchange isolate cartridges, as described by Iniesta et al. [20].

### 2.6. HPLC Determination of Vitamin C Content

Vitamin C content was measured as described in Helsper et al. [21]. Extracts were subjected to HPLC analysis after the extraction procedure. The HPLC system comprised a Jasco PU-2089 plus controller (Jasco, Easton, MD, USA), a Jasco UV-2070 plus ultraviolet (UV) detector (Jasco Easton, MD, USA) set at an absorbance of 260 nm, and an autosampler Jasco AS-4050 (Jasco, Easton, MD, USA). The HPLC column used was Ascentis Express C18 150 × 4.6 mm (Supelco, Bellefonte, PA, USA), protected by a Phenomenex 4.0 × 3.0 mm C18 ODS guard column (Phenomenex, Torrance, CA, USA). The gradient program consisted of two mobile phases: A (50mM phosphate buffer with pH 3.2) and B (Acetonitrile), which started with 100% of A until 6 min, then decreased to 50% for 2 min, and again increased to 100% until the end.

The quantification of vitamin C content was carried out through calibration curve prepared by running standard concentration of vitamin C, and the results were expressed as mg vit-C per 1 kg fresh weight of strawberries (mg/kg FW).

### 2.7. HPLC Determination of Phenolic Acid Content

Phenolic acids were analyzed as previously described in Schieber et al. [22] and Fredericks et al. [23]. The HPLC system comprised a Jasco PU-2089 plus controller (Jasco, Easton, MD, USA), a Jasco UV-2070 plus ultraviolet (UV) detector (Jasco Easton, MD, USA), and an autosampler Jasco AS-4050 (Jasco, Easton, MD, USA). The HPLC UV detector was set at 320 nm and the column used was an Aqua Luna C18 250 × 4.6 mm (Phenomenex, Torrance, CA, USA) protected by a Phenomenex 4.0 × 3.0 mm C18 ODS guard column

(Phenomenex, Torrance, CA, USA). The gradient program consisted of two mobile phases: A (2% Acetic acid) and B (acetic acid, acetonitrile and $H_2O$ 1:50:49). It started with 55% A and 45% B for 50 min, followed by 10 min 100% of B, and then decreased to 10 % B until the end.

For the quantification of phenolic acid content, external chlorogenic acid (CHA), caffeic acid (CA), and ellagic acid (EA) calibration curves were used. Values were expressed as mg corresponding to phenolic acid per kilogram of fresh weight of strawberries (mg/kg FW).

### 2.8. HPLC Determination of Anthocyanin Content

Anthocyanin content was analyzed following the method of Fredericks et al. [23]. The HPLC system comprised a Jasco PU-2089 plus controller (Jasco, Easton, MD, USA), a Jasco UV-2070 plus ultraviolet (UV) detector (Jasco Easton, MD, USA), and an autosampler Jasco AS-4050 (Jasco, Easton, MD, USA). The compounds were separated on an Aqua Luna C18 (2) (250 × 4.6 mm) reverse-phase column with a particle size of 5 μm (Phenomenex, Torrance, CA, USA) protected by a Phenomenex 4.0 × 3.0 mm C18 ODS guard column (Phenomenex, Torrance, CA, USA), and monitoring was performed at 520 nm. The gradient program consisted of two mobile phases: A (formic acid, acetonitrile, and $H_2O$ 10:3:87) and B (formic acid, acetonitrile, and $H_2O$ 10:50:40). It started with 75% A for 10 min, decreased to 69% A for 5 min, decreased again to 60% A for 5 min, and later continued 50% A for 10 min, followed by 90% A for 16 min.

Anthocyanins were quantified using calibration curves made with external standards of cyanidin-3-glucoside, pelargonidin-3-glucoside, and pelargonidin-3-rutinoside, and were calculated as mg per 1 kg of fresh weight of strawberries (mg/kg FW).

### 2.9. Quantification of Folate Content

Folate was quantified using the HPLC as cited by Stralsjo et al. [24]. The HPLC system comprised a pump model Jasco PU- 2089 (Jasco, Easton, MD, USA), a fluorescence detector (FLD) Jasco FP-2020 Plus (Jasco, Easton, MD, USA) set at wavelengths of 290 nm excitation and 360 nm emission, and an autosampler Jasco AS-4050 (Jasco, Easton, MD, USA). The analytical column was a Luna C18, 250 × 4.6, 5 μm (Phenomenex, Torrance, CA, USA), protected by a Phenomenex 4.0 × 3.0 mm C18 ODS guard column (Phenomenex, Torrance, CA, USA). Quantification of folate content was determined through a calibration curve prepared by running standard concentrations of 5-methyl-tetrahydrofolic acid (5-$CH_3$-$H_4$ folate). The gradient program consisted of two mobile phases: A (30 mM phosphate buffer with 2.3 pH) and B (Acetonitrile). It started with 94%A for 8 min, then decreased to 75% A for 27 min, and increased again to 94% A for 15 min.

Results are expressed as μg 5-$CH_3$-$H_4$ folate per 1 kg of fresh weight of strawberries (μg 5-$CH_3$-$H_4$folate/kg FW).

### 2.10. Data Analyses

The results are presented as the values ± standard error and were subjected to one-way analysis of variance (ANOVA), at a confidence level of 95%. Significant differences were calculated according to Tukey's tests, and differences at $p < 0.05$ were considered to be significant. A correlation matrix has also been developed among the nutritional parameters to check their inter-relationship, with $p < 0.05$. Statistical analyses were performed by using Statistica 7 software (StatSoft, TIBCO Software, Palo Alto, CA, USA).

## 3. Results and Discussion

### 3.1. Vitamin C

The vitamin C amounts of the Arianna, Francesca, and Silvia fruits at different storage times with different treatments are shown in Table 3. The highest vitamin C amount was detected in Arianna WF (731.8 mg/1 kg FW) at time 0, and the oven-drying process caused a high loss of total vitamin C (almost 97% at time 0 for DF). The vitamin C amount in the Arianna and Silvia WF during storage at −20 °C decreased from 1 M until the end of storage, with the lowest amount detected at 5 M (436.1 mg/kg FW and 388.1 mg/kg FW, respectively). However, in Francesca WF, vitamin C slightly increased after 1 M (599.8 mg/kg FW) and was maintained until the end of storage. Regarding the DF in $-20\,^{\circ}$C storage, there was a slight increasing trend in the vitamin C amount for all cultivars until the end of storage, but the values remained dramatically lower than those of the WF.

**Table 3.** Average of vitamin C content (mg/kg FW) in fruits of Arianna, Francesca, and Silvia cultivars. The data are expressed as mean ± standard error. Different lowercase letters indicate significant differences for $p \leq 0.05$ in each cultivar during storage. Different uppercase letters indicate significant differences for $p \leq 0.05$ for all cultivars for a specific storage time. WF: whole fruit, DF: dried fruit. 0 M: 0 months of storage; 1 M: 1 month of storage; 3 M: 3 months of storage; 5 M: 5 months of storage; 7 M: 7 months of storage.

| Storage Time | Storage Temperature | Treatment | Arianna | Francesca | Silvia |
|---|---|---|---|---|---|
| 0 M | | WF | 731.8 ± 6.5aA | 580.4 ± 3.6fB | 519.3 ± 0.3cC |
| | | DF | 23.2 ± 0.3jD | 18.8 ± 0.3mD | 16.5 ± 0.3lD |
| 1 M | −20 | WF | 650.6 ± 3.9bB | 599.8 ± 2.8dC | 402.2 ± 1.8gE |
| | −80 | WF | 737.4 ± 4aA | 736 ± 2aA | 592.1 ± 0.9aD |
| | −20 | DF | 26.8 ± 0.7jF | 22.8 ± 0.4lmF | 16.4 ± 0.3lG |
| | −80 | DF | 27.9 ± 0.1jF | 25.1 ± 0.3ijklF | 14.6 ± 0.1lG |
| 3 M | −20 | WF | 554.4 ± 0.1dD | 594.3 ± 3.9eB | 423.4 ± 0.2fF |
| | −80 | WF | 571.7 ± 13.4cC | 609.5 ± 1.6cA | 523.7 ± 2.3cE |
| | −20 | DF | 24.2 ± 0.3jHI | 28 ± 0.1ijkH | 19.4 ± 0.2lHI |
| | −80 | DF | 93.5 ± 5.3hG | 24.3 ± 0.3jklHI | 14.1 ± 0.1lI |
| 5 M | −20 | WF | 436.1 ± 0.8gD | 517.2 ± 0.2gA | 388.1 ± 0.3hF |
| | −80 | WF | 507.1 ± 0.5eB | 500.3 ± 0hC | 430.1 ± 0.1eE |
| | −20 | DF | 25.4 ± 0.2jI | 23.1 ± 0.2klmI | 39.3 ± 0.8jG |
| | −80 | DF | 29.2 ± 0.7jH | 30 ± 0.7iH | 31.3 ± 2.3kH |
| 7 M | −20 | WF | 544.6 ± 1.2dC | 576.8 ± 1fB | 509.4 ± 0.3dD |
| | −80 | WF | 473 ± 0.5fE | 664.1 ± 1.5bA | 537.5 ± 8.2bC |
| | −20 | DF | 46.3 ± 7.6iF | 29.3 ± 0.5ijG | 26.7 ± 0.1kG |
| | −80 | DF | 27.3 ± 1jG | 26.1 ± 2ijklG | 49.6 ± 0.7iF |

At $-80\,^{\circ}$C, similar to $-20\,^{\circ}$C, the Arianna and Silvia WF had a decreasing trend for vitamin C during storage, but presented a higher value than at $-20\,^{\circ}$C; the Francesca WF showed an increasing trend for vitamin C until the end of storage, except at 5 M. The DF also presented higher vitamin C values (with some exceptions) at $-80\,^{\circ}$C until the end of storage for all cultivars compared to $-20\,^{\circ}$C, but with dramatically lower values than the WF. There was a significant difference in fruit vitamin C content between fruits stored at $-20\,^{\circ}$C and $-80\,^{\circ}$C during storage and between treatments.

Ancos et al. [25] reported that vitamin C was highly preserved after freezing fruits, but the degradation increased during storage, confirming some of our results. The chemical changes that may occur after the freezing of berries, as a result of oxidation and enzymatic activity, might influence the degradation and loss of vitamin C content [26]. The concentration of reagents in the non-frozen phase of frozen fruits and crystallization can favor chemical and enzymatic oxidation reactions [27]. Regarding the drying process, vitamin C is highly sensitive to and unstable in heat, so it was degraded during the sample preparation and oven-drying process. Vitamin C is also hydrosoluble and, as 90% of strawberries are water, the loss of vitamin C occurs during the drying process [28].

*3.2. Phenolic Acids*

Phenolic acids are a part of the large group of phenolic compounds, widely distributed in strawberry fruits. They are considered important ingredient of strawberries, contributing to taste, color, and nutritional properties [29].

The phenolic acids' content in the WF of all three cultivars during storage is shown in Table 4. The Arianna, Francesca, and Silvia fruits had a content of 396.4 mg/kg FW, 569 mg/kg FW, and 170 mg/kg FW phenolic acids at time 0, respectively. The oven-drying process reduced the amounts of phenolic acids in the fruits at time 0 to 192.6 mg/kg FW, 288.1 mg/kg FW, and 138.2 mg/kg FW, respectively.

**Table 4.** Average of phenolic acids' content (mg/kg FW) in cultivars of Arianna, Francesca, and Silvia fruits. The data are expressed as mean ± standard error. Different lowercase letters indicate significant differences for $p \leq 0.05$ in each cultivar during storage. Different uppercase letters indicate significant differences for $p \leq 0.05$ for all cultivars for a specific storage time. WF: whole fruit, DF: dried fruit. 0 M: 0 months of storage; 1 M: 1 month of storage; 3 M: 3 months of storage; 5 M: 5 months of storage; 7 M: 7 months of storage.

| Storage Time | Storage Temperature | Treatment | Arianna | Francesca | Silvia |
|---|---|---|---|---|---|
| 0 M | | WF | 396.4 ± 15.3abB | 569 ± 1.9cA | 170.7 ± 9.7dE |
| | | DF | 192.6 ± 1efD | 288.1 ± 4.3gC | 138.2 ± 1.3fghF |
| 1 M | −20 | WF | 305.2 ± 7.5cB | 643.2 ± 14.6aA | 206.9 ± 3.3bEF |
| | −80 | WF | 273.4 ± 5.4cC | 631.7 ± 11.5aA | 222.5 ± 1.9aDE |
| | −20 | DF | 188.2 ± 2.2efF | 313.2 ± 9.5efgB | 129.2 ± 0.1ghijkG |
| | −80 | DF | 201.2 ± 2.3deEF | 237 ± 9.5hD | 131.2 ± 6.3ghijG |
| 3 M | −20 | WF | 427.2 ± 6.6aC | 601.2 ± 25.6bB | 191.5 ± 2.5cEF |
| | −80 | WF | 400.2 ± 6.6abC | 646.8 ± 14aA | 191.6 ± 5.5cEF |
| | −20 | DF | 230.6 ± 23.7dE | 386.2 ± 5.3dC | 164 ± 12.8deFG |
| | −80 | DF | 387 ± 11.7bC | 292.2 ± 8.9fgD | 140.9 ± 2.2fgG |
| 5 M | −20 | WF | 162 ± 3fgE | 321.4 ± 4.8efA | 149.8 ± 3.3efE |
| | −80 | WF | 191.5 ± 4.4efC | 325.5 ± 2.8eA | 129.7 ± 2.8ghijkF |
| | −20 | DF | 91.4 ± 0hG | 177.1 ± 2.9jD | 123.1 ± 4.7hijkF |
| | −80 | DF | 151.8 ± 10.4gE | 226.5 ± 2hiB | 117.5 ± 2.3jkF |
| 7 M | −20 | WF | 190.9 ± 6.1efB | 294.3 ± 5fgA | 134.5 ± 2.4ghiCDE |
| | −80 | WF | 186.8 ± 5.7efB | 291.6 ± 7gA | 126.1 ± 1.6ghijkDEF |
| | −20 | DF | 138.9 ± 4gCD | 198.7 ± 2.4ijB | 122.3 ± 4.9ijkEF |
| | −80 | DF | 144 ± 3.3gC | 191.6 ± 5.2jB | 114.6 ± 2.3kF |

The phenolic acids' content in Francesca and Silvia WF increased at 1M during storage at −20 °C (643 mg/kg FW and 206.9 mg/kg FW, respectively) and had a decreasing trend until the end of storage. Differently, in Arianna, the highest amount was detected after 3 M of storage at −20 °C, with the value of 427.2 mg/kg FW. Regarding the DF, there was a decreasing trend until the end of storage, with a peak of phenolic acid content in fruits after 3 M of storage in all the cultivars.

At −80 °C storage temperature, the WF showed a similar reaction to −20 °C, with increased phenolic acids' content at the early stage of storage, then a decreasing trend until the end of storage. The highest amounts of phenolic acids were detected in Francesca fruit at 3M of storage, with the value of 646 mg/kg FW. Regarding the DF, the phenolic content of Francesca and Silvia fruits increased after 1M of storage, decreasing until the end of storage; Arianna showed a slightly different behavior, with increased phenolic acids' content from 1 M to 3 M, then decreasing until the end of storage. There was not a significant difference between the different temperatures during storage, but there was a significant difference between the treatments in all the cultivars, with the WF phenolic acids' content always being higher than those of the DF. The same effect was observed in a storage study of frozen strawberries by Oszmainski et al. [30]. They reported that the degradation of phenolic acids may be due to enzymatic oxidation. The degradation of phenolic acids during the oven-drying process were also confirmed in studies by Coklar et al. [31] and Semenov et al. [32]. The oxidative and thermal degradation of phenolic compounds due to the increasing heat treatment led to decreases in the phenolic compounds in oven-dried fruits [33].

### 3.3. Anthocyanins

Anthocyanins are the most important phenolic compounds of the strawberry [34], and their concentration increases during the ripening progress [35]. The Arianna, Francesca, and Silvia WF had a content of 202.2 mg/kg FW, 108.8 mg/kg FW, and 300.7 mg/kg FW of total anthocyanins at time 0, respectively (Table 5). The oven-drying process greatly reduced these amounts to 98 mg/kg FW, 77.4 mg/kg FW, and 108.6 mg/kg FW, respectively.

**Table 5.** Average of total anthocyanins' content (mg/kg FW) in cultivars of Arianna, Francesca, and Silvia fruits. The data are expressed as mean ± standard error. Different lowercase letters indicate significant differences for $p \leq 0.05$ in each cultivar during storage. Different uppercase letters indicate significant differences for $p \leq 0.05$ for all cultivars for a specific storage time. WF: whole fruits, DF: dried fruits. 0 M: 0 months of storage; 1 M: 1 month of storage; 3 M: 3 months of storage; 5 M: 5 months of storage; 7 M: 7 months of storage.

| Storage Time | Storage Temperature | Treatment | Arianna | Francesca | Silvia |
|---|---|---|---|---|---|
| 0 M |  | WF | 202.4 ± 2.7Eb | 108.8 ± 5.5Cc | 300.7 ± 3.3Ba |
|  |  | DF | 98 ± 1.8Fd | 77.4 ± 1.2Ee | 108.6 ± 0.5Hc |
| 1 M | −20 | WF | 236.9 ± 1.1abcC | 127.3 ± 1.8abD | 266.6 ± 9.1Cb |
|  | −80 | WF | 224.7 ± 14.4Dc | 123.3 ± 1.1Bde | 300.7 ± 7.5Ba |
|  | −20 | DF | 61.1 ± 0.8Gfg | 52.9 ± 1.5Jg | 122.1 ± 0Gde |
|  | −80 | DF | 50.8 ± 0.7ghG | 74.9 ± 1.5efF | 109.4 ± 3.8He |
| 3 M | −20 | WF | 249 ± 0.1Aa | 101.4 ± 0.6Cf | 187.6 ± 0.4Fd |
|  | −80 | WF | 244 ± 1.2abB | 89.4 ± 2.6Dg | 207.4 ± 1Ec |
|  | −20 | DF | 45.8 ± 0.8Hj | 61.2 ± 0.6hiH | 111.9 ± 1.1ghE |
|  | −80 | DF | 56.4 ± 1.5ghI | 64.8 ± 0.4ghH | 92.1 ± 1.9Ig |
| 5 M | −20 | WF | 203.6 ± 2.9Ec | 132.7 ± 0.3Ad | 259.4 ± 4.8Cb |
|  | −80 | WF | 200.1 ± 5.7Ec | 108.9 ± 6.3Ce | 335 ± 2.6Aa |
|  | −20 | DF | 19.5 ± 0Ig | 69.2 ± 1.4fgI | 72.1 ± 0.8klG |
|  | −80 | DF | 56.1 ± 1.2ghH | 57.1 ± 0.5ijH | 86.6 ± 1ijF |
| 7 M | −20 | WF | 232.1 ± 2bcdB | 124 ± 1.5Bd | 298.7 ± 2.6Ba |
|  | −80 | WF | 224.9 ± 3cdC | 126.3 ± 4.1abD | 220.2 ± 1.4Dc |
|  | −20 | DF | 48.1 ± 1.1Hgh | 49.6 ± 0.2Jgh | 77.8 ± 2.8jkE |
|  | −80 | DF | 44.7 ± 0.8Hh | 51.1 ± 0.6Jg | 62 ± 1Lf |

Storage duration and temperature do not significantly affect the anthocyanins amount in the WF. There was a slight increase in the amounts of the anthocyanins in the Arianna and Francesca WF from 1 month to 7 months, while in Silvia fruit the anthocyanins amount was maintained during storage. Regarding the oven-dried fruits, the anthocyanin content of Arianna and Silvia fruits dramatically decreased after 1 month, and the values remained very low until 7 months of storage; however, in Silvia fruit, the anthocyanin content increased after the first month of storage, and, by increasing storage time, the anthocyanin content decreased in parallel. The lowest amount of anthocyanin was found in the Arianna DF after 5 months of storage at −20 °C. There was also no significant effect of storage temperature on the anthocyanins' content of the dried fruit of all cultivars.

Freezing is commonly regarded as a technique that has a less deleterious effect on the anthocyanins of strawberries for long-term storage [36]. Kamiloglu [9] studied the effect of different freezing methods on the bioaccessibility of strawberry polyphenols and concluded that freezing retains the bioactive compounds of strawberries and might enhance the total amounts of bioaccessible anthocyanins. Freezing rate and storage time are the most important parameters in the losses of anthocyanin in strawberries during the freezing process. It has been determined that a higher freezing rate is essential for the better preservation of the bioactive compounds in strawberries. Freezing should be done at an appropriate freezing rate to preserve the cell structure and the nutritional content of strawberries [37]. The effects of heat on microorganisms and the activity of enzymes are important when drying fruits [15]. Some important properties of fruits change during dehydration such as texture, chemical changes affecting flavor and nutrients, and color, as the evolution of the color is strongly correlated with the anthocyanin concentration across various fruits and vegetables [38–40]. A high temperature during the drying process is an important factor for the loss of quality. Lowering the process temperature has great potential to improve the quality of dried products [41].

### 3.4. Folate

The folate content of all three cultivars during storage is shown in Table 6. The folate content of the Arianna, Francesca, and Silvia WF at time 0 was 126 µg/kg FW, 171.55 µg/kg FW, and 113.72 µg/kg FW, respectively. The oven-drying process slightly decreased the folate content of the fruits of all cultivars, with the folate content being 120.96 µg/kg FW, 115.19 µg/kg FW, and 103.09 µg/kg FW in the DF at time 0, respectively. The folate content of the WF during storage at −20 °C had an increasing trend from 1M until the end of storage for all cultivars, and, at the end of storage, the folate contents of the Arianna, Francesca, and Silvia WF were detected to be 190.61 µg/kg FW, 208.4 µg/kg FW, and 132.06 µg/kg FW, respectively. In the DF, similarly to the WF, the folate content increased with the extension of storage for all cultivars.

**Table 6.** Average of folate content (µg/kg FW) in cultivars of Arianna, Francesca, and Silvia fruits. The data are expressed as mean ± standard error. Different lowercase letters indicate significant differences for $p \leq 0.05$ in each cultivar during storage. Different uppercase letters indicate significant differences for $p \leq 0.05$ for all cultivars for a specific storage time. WF: whole fruits, DF: dried fruits. 0 M: 0 months of storage; 1 M: 1 month of storage; 3 M: 3 months of storage; 5 M: 5 months of storage; 7 M: 7 months of storage.

| Storage Time | Storage Temperature | Treatment | Arianna | Francesca | Silvia |
|---|---|---|---|---|---|
| 0 M | | WF | 126.17 ± 4.8cdB | 171.55 ± 1.39abcdA | 113.72 ± 0.79cdD |
| | | DF | 120.96 ± 2.73dC | 115.19 ± 1.27dD | 103.09 ± 1.78dE |
| 1 M | −20 | WF | 133.58 ± 1.61cdB | 156.39 ± 1.82bcdA | 114.88 ± 1.44cdCD |
| | −80 | WF | 132.27 ± 2.92cdB | 116.97 ± 0.54dCD | 117.92 ± 1.45cdCD |
| | −20 | DF | 112.37 ± 0.88dDE | 131.47 ± 3.77dB | 107.35 ± 0.91cdEF |
| | −80 | DF | 119.88 ± 3.17dC | 134.83 ± 0.62dB | 104.04 ± 0.3dF |
| 3 M | −20 | WF | 155.39 ± 2.85bcdC | 141.68 ± 3.53cdD | 126.53 ± 1.7bcdEF |
| | −80 | WF | 229.67 ± 0.9aA | 223.59 ± 5.4aA | 123.39 ± 1.08cdEF |
| | −20 | DF | 127.6 ± 2.99cdE | 164.8 ± 2.94abcdB | 106.5 ± 0.13cdG |
| | −80 | DF | 139.74 ± 2.94bcdD | 119.27 ± 1.52dF | 110.13 ± 0.54cdG |
| 5 M | −20 | WF | 147.71 ± 7.25bcdCDE | 220.68 ± 3.37aA | 121.9 ± 24.31cdEF |
| | −80 | WF | 131.23 ± 2.11cdDEF | 132.08 ± 2.18dDEF | 177.65 ± 5.68aBC |
| | −20 | DF | 157.83 ± 12.88bcdCD | 150.22 ± 2.41bcdCDE | 121.43 ± 15.66cdEF |
| | −80 | DF | 188.49 ± 7.18abB | 109.92 ± 5.94dF | 138.88 ± 8.78bcDEF |
| 7 M | −20 | WF | 190.61 ± 3.66abAB | 208.4 ± 7.24abA | 132.06 ± 1.2bcdAB |
| | −80 | WF | 175.58 ± 7.32bcAB | 199.44 ± 51.7abcAB | 157.55 ± 1.75abAB |
| | −20 | DF | 174.64 ± 40.41bcAB | 156.41 ± 54.41bcdAB | 124.8 ± 32.44cdAB |
| | −80 | DF | 188.04 ± 58.92abAB | 166.46 ± 38.51abcdAB | 102.66 ± 1.63dB |

At $-80\ °C$ storage, the folate content of the WF for all cultivars increased until the end of storage, as registered for the $-20\ °C$ storage. The Arianna and Francesca fruits had a peak of folate content after 3M of storage, with a content of 229.67 µg/kg FW and 223.59 µg/kg FW respectively, while the Silvia fruit had a peak after 5M of storage, with a content of 177.65 µg/kg FW. In the DF, there was also an increasing trend for folate content until the end of storage, with the highest content detected at the end of storage in the Arianna and Francesca fruits (188.04 µg/kg FW and 166.46 µg/kg FW, respectively); in the Silvia fruit, the highest folate content was detected after 5M of storage, with a value of 138.88 µg/kg FW. There was not a significant difference in folate content between temperature and treatments, but the WF presented more folate content than the DF during storage.

The strawberry is one of the most important sources of antioxidants and is considered a functional fruit due to the presence of a diverse range of bioactive components and high levels of vitamin C, vitamin E, folate, phenolic compounds, and fiber [42]. In this respect, variations of the amount of vitamin C in strawberries can affect the folate content, as a higher vitamin C content can lead to increased stability of folate [43]. The retention of folate and ascorbic acid was affected by the same factors, and a high content of ascorbic acid could provide possible protection against folate degradation [44]. This assumption was also confirmed by Ringling and Rychlik [45], who performed in vivo studies to simulate food folate digestion and found out that ascorbic acid stabilizes folate, particularly 5-CH3-H4 folate during digestion. The addition of ascorbic acid in physiological amounts improved the stability of some types of folate, depending on the food matrix [45].

### 3.5. Correlation Matrix

The correlation matrix among the analyzed nutritional parameters indicated that there was a good positive correlation between anthocyanins and vitamin C content (0.70, $p < 0.05$). This means that these two classes of compounds presented a similar behavior in the tested conditions, decreasing dramatically with the oven-drying treatment, while they did not change greatly regarding storage temperature or length. Vitamin C also presented a medium correlation with phenolic acids (0.52) and a lower but still significant correlation with folate content (0.27) (Table 7).

**Table 7.** Correlation matrix among the nutritional parameters analyzed. * indicates a significant correlation for $p \leq 0.05$.

|  | Anthocyanins | Vitamin C | Folate | Phenolic Acids |
|---|---|---|---|---|
| Anthocyanins | 1.00 | 0.70 * | 0.04 | −0.03 |
| Vitamin C | 0.70 * | 1.00 | 0.31 * | 0.52 * |
| Folate | 0.04 | 0.31 * | 1.00 | 0.27 * |
| Phenolic acids | −0.03 | 0.52 * | 0.27 * | 1.00 |

## 4. Conclusions

In this study, some interesting results were obtained, describing how the nutritional compounds of strawberries react during storage at different temperatures with different storage times and treatments.

- Higher amounts of nutritional compounds were detected in the WF compared to the DF.
- For preserving nutritional quality, the WF treatment was optimum, and it showed good results in $-80\ °C$ storage compared to $-20\ °C$. This is an important indication for application in the laboratory analysis and processing industry.
- The anthocyanins' content in the WF seemed to not decrease during 7 months of $-80\ °C$ storage, with some exceptions.
- Oven drying was not an ideal treatment for preserving vitamin C, almost completely degrading this compound in strawberry fruits.

- In the WT and DF, the amount of folate increased during storage. More folate was detected in the WF.
- The different strawberry cultivars presented different amounts of nutritional compounds: the Arianna and Francesca fruits had more vitamin C, phenolic acids, and folates, but the Silvia fruit had more anthocyanins.
- At the end of storage (7M), there was more loss of vitamin C in the Arianna WF compared to the other two cultivars' WF.
- There was a slight increase in the amount of anthocyanin in the Arianna and Francesca WF after 7M of storage, but in the Silvia WF the amount of anthocyanin was retained.

Generally, it can be concluded that oven drying is not a recommended technique to treat fruits for preserving nutritional quality during storage. Storage time did not greatly affect the nutritional quality in whole fruits, though they presented a higher amount when stored at −80 °C.

**Author Contributions:** Conceptualization, L.M. and B.M.; methodology, R.Q.; software, R.Q.; validation, L.M., F.C. and B.M.; formal analysis, R.Q.; investigation, L.M. and B.M.; resources, B.M. and F.C.; data curation, L.M. and F.C.; writing—original draft preparation, R.Q.; writing—review and editing, L.M. and B.M.; visualization, R.Q.; supervision, B.M.; project administration, B.M.; funding acquisition, B.M. All authors have read and agreed to the published version of the manuscript.

**Funding:** This research was funded by the Partnership for Research and Innovation in the Mediterranean Area (PRIMA)'s 2019–2022 MEDBERRY project.

**Data Availability Statement:** Not applicable.

**Conflicts of Interest:** The authors declare no conflict of interest.

# References


1. FAO. 2019. Available online: http://www.fao.org/faostat/en/#data/QC (accessed on 15 December 2022).
2. Giampieri, F.; Islam, M.S.; Greco, S.; Gasparrini, M.; Forbes Hernandez, T.Y.; Delli Carpini, G.; Giannubilo, S.R.; Ciavattini, A.; Mezzetti, B.; Mazzoni, L.; et al. Romina: A powerful strawberry with in vitro efficacy against uterine leiomyoma cells. *J. Cell. Physiol.* **2019**, *234*, 7622–7633. [CrossRef] [PubMed]
3. Gasparrini, M.; Forbes-Hernandez, T.Y.; Cianciosi, D.; Quiles, J.L.; Mezzetti, B.; Xiao, J.; Giampieri, F.; Battino, M. The efficacy of berries against lipopolysaccharide-induced inflammation: A review. *Trends Food Sci. Technol.* **2021**, *117*, 74–91. [CrossRef]
4. Giampieri, F.; Tulipani, S.; Alvarez-Suarez, J.M.; Quiles, J.L.; Mezzetti, B.; Battino, M. The strawberry: Composition, nutritional quality, and impact on human health. *Nutrition* **2012**, *28*, 9–19. [CrossRef] [PubMed]
5. Sabbadini, S.; Capocasa, F.; Battino, M.; Mazzoni, L.; Mezzetti, B. Improved nutritional quality in fruit tree species through traditional and biotechnological approaches. *Trends Food Sci. Technol.* **2021**, *117*, 125–138. [CrossRef]
6. Capocasa, F.; Balducci, F.; Di Vittori, L.; Mazzoni, L.; Stewart, D.; Williams, S.; Hargreaves, R.; Bernardini, D.; Danesi, L.; Zhong, C.-F.; et al. Romina and Cristina: Two new strawberry cultivars with high sensorial and nutritional values. *Int. J. Fruit Sci.* **2016**, *16* (Suppl. 1), 207–219. [CrossRef]
7. Bhat, R.; Stamminger, R. Preserving strawberry quality by employing novel food preservation and processing techniques–Recent updates and future scope—An overview. *J. Food Process Eng.* **2015**, *38*, 536–554. [CrossRef]
8. Prochaska, L.J.; Nyugen, X.T.; Donat, N.; Piekutowski, W.V. Effects of food processing on the thermodynamic and nutritive value of foods: Literature and databese survey. *Med. Hypotheses* **2000**, *54*, 254–262. [CrossRef]
9. Kamiloglu, S. Effect of different freezing methods on the bioaccessibility of strawberry polyphenols. *Int. J. Food Sci. Technol.* **2019**, *54*, 2652–2660. [CrossRef]
10. Chaves, A.; Zaritzky, N. Cooling and Freezing of Fruits and Fruit Products. In *Fruit Preservation*; Rosenthal, A., Deliza, R., Welti-Chanes, J., Barbosa-Cánovas, G., Eds.; Food Engineering Series; Springer: New York, NY, USA, 2018.
11. Dias, M.G.; Camões, M.F.G.; Oliveira, L. Carotenoid stability in fruits, vegetables and working standards—Effect of storage temperature and time. *Food Chem.* **2014**, *156*, 37–41. [CrossRef]
12. Alexandre, E.M.C.; Brandao, T.R.S.; Silvia, C.L.M. Impact of non-thermal technologies and sanitizer solutions on microbial load reduction and quality factor retention of frozen red bell peppers. *Innov. Food Emerg. Technol.* **2013**, *17*, 99–105. [CrossRef]
13. Sahari, M.A.; Boostani, F.M.; Hamidi, E.Z. Effect of low temperature on the ascorbic acid content and quality characteristics of frozen strawberry. *Food Chem.* **2004**, *86*, 357–363. [CrossRef]
14. Erbersdobler, H.F. *Concentration and Drying of Foods*; MacCarthy, D., Ed.; Elsevier: London, UK, 1986; pp. 69–87.
15. Rahman, M.S. Drying and food preservation. In *Handbook of Food Preservation*, 2nd ed.; CRC Press: Boca Raton, FL, USA, 2007.

16. Mezzetti, B.; Mazzoni, L.; Qaderi, R.; Balducci, F.; Marcellini, M.; Capocasa, F. Generating novel strawberry pre-breeding material from a Fragaria × ananassa backcrossing program with *F. virginiana* subsp. *glauca* inter-specific hybrids. *Acta Hortic.* **2021**, *1309*, 197–204.

17. Diamanti, J.; Capocasa, F.; Balducci, F.; Battino, M.; Hancock, J.; Mezzetti, B. Increasing strawberry fruit sensorial and nutritional quality using wild and cultivated germplasm. *PLoS ONE* **2012**, *7*, e46470. [CrossRef] [PubMed]

18. Tulipani, S.; Mezzetti, B.; Capocasa, F.; Bompadre, S.; Beekwilder, J.; De Vos, C.R.; Capanoglu, E.; Bovy, A.; Battino, M. Antioxidants, phenolic compounds, and nutritional quality of different strawberry genotypes. *J. Agric. Food Chem.* **2008**, *56*, 696–704. [CrossRef]

19. Mezzetti, B.; Balducci, F.; Capocasa, F.; Zhong, C.F.; Cappelletti, R.; Di Vittori, L.; Mazzoni, L.; Giampieri, F.; Battino, M. Breeding strawberry for higher phytochemicals content and claim it: Is it possible? *Int. J. Fruit Sci.* **2016**, *16*, 194–206. [CrossRef]

20. Iniesta, M.D.; Perez-Conesa, J.D.; Garcia-Alonso, G.R.; Periago, M.J. Folate content in tomato (*Lycopersicon esculentum*): Influence of cultivar, ripeness, year of harvest, and pasteurization and storage temperatures. *J. Agric. Food Chem.* **2009**, *57*, 4739–4745. [CrossRef]

21. Helsper, J.P.F.G.; de Vos, C.H.R.; Maas, F.M.; Jonker, H.H.; van den Broeck, H.C.; Jordi, W.; Pot, C.S.; Keizer, L.C.P.; Schapendonk, A.H.C.M. Response of selected antioxidants and pigments in tissues of Rosa hybrida and Fuchsia hybrida to supplemental UV-A exposure. *Physiol. Plantlantarum* **2003**, *117*, 171–187. [CrossRef]

22. Schieber, A.; Keller, P.; Carle, R. Determination of phenolic acids and flavonoids of apple and pear by high-performance liquid chromatography. *J. Chromatogr. A* **2001**, *910*, 265–273. [CrossRef]

23. Fredericks, C.H.; Fanning, K.J.; Gidley, M.J.; Netzel, G.; Zabaras, D.; Herrington, M.; Netzel, M. High-anthocyanin strawberries through cultivar selection. *J. Sci. Food Agric.* **2013**, *93*, 846–852. [CrossRef]

24. Strålsjö, L.; Åhlin, H.; Witthöft, C.M.; Jastrebova, J. Folate determination in Swedish berries by radioprotein-binding assay (RPBA) and high-performance liquid chromatography (HPLC). *Eur. Food Res. Technol.* **2003**, *216*, 264–269. [CrossRef]

25. Ancos, B.; Gonzalez, E.M.; Pilar Cano, M. Ellagic acid, vitamin C, and total phenolic contents and radical scavenging capacity affected by freezing and frozen storage in raspberry fruit. *J. Agric. Food Chem.* **2000**, *48*, 565–4570. [CrossRef] [PubMed]

26. Zhao, Y. Freezing process of berries. In *Berry Fruit, Value-Added Products for Health Promotion*; Zhao, Y., Ed.; CRC, Taylor and Francis Group: Abingdon, UK, 2007; pp. 291–312.

27. Stevanovic, S.M.; Petrovic, T.S.; Markovic, D.D.; Milovancevic, U.M.; Stevanovic, S.V.; Urosevic, T.M.; Kozarski, M.S. Changes of quality and free radical scavenging activity of strawberry and raspberry frozen under different conditions. *J. Food Process. Preserv.* **2021**, *46*, e15981. [CrossRef]

28. Phan, A.D.T.; Adoamo, O.; Akter, S.; Netzel, M.E.; Cozzolino, D.; Sultanbawa, Y. Effects of drying methods and maltodextrin on vitamin C and quality of Terminalia ferdinandiana fruit powder, an emerging Australian functional food ingredient. *J. Sci. Food Agric.* **2021**, *101*, 5132–5141. [CrossRef] [PubMed]

29. Tomás-Barberán, F.A.; Espín, J.C. Phenolic compounds and related enzymes as determinants of quality in fruits and vegetables. *J. Sci. Food Agric.* **2001**, *81*, 853–876. [CrossRef]

30. Oszmainski, J.; Wohdylo, A.; Kolnaik, J. Effect of L-ascorbic acid, sugar, pectin and freeze-thaw treatment on polyphenol content of frozen strawberries. *Food Sci. Technol.* **2009**, *42*, 581–586.

31. Coklar, H.; Akbulut, M.; Kilinc, S.; Yildirim, A.; Alhassan, I. Effect of Freeze, Oven and Microwave Pretreated Oven Drying on Color, Browning Index, Phenolic Compounds and Antioxidant Activity of Hawthorn (*Crataegus orientalis*) Fruit. *Not. Bot. Horti Agrobot.* **2018**, *46*, 449–456. [CrossRef]

32. Semenov, G.V.; Krasnova, I.S.; Suvorov, O.A.; Shuvalova, I.D.; Posokhov, N.D. Influence of freezing and drying on phytochemical properties of various fruit. *Biosci. Biotechnol. Res. Asia* **2015**, *12*, 1311–1320. [CrossRef]

33. Wojdylo, A.; Figiel, A.; Oszmianski, J. Effect of drying methods with application of vacuum microwaves on the bioactive compounds, color and antioxidant activity of strawberry fruits. *J. Agric. Food Chem.* **2009**, *57*, 1337–1343. [CrossRef]

34. Lopes-da-Silva, F.; de Pascual-Teresa, S.; Rivas-Gonzalo, J.; Santos-Buelga, C. Identification of anthocyanin pigments in strawberry (cv. Camarosa) by LC using DAD and ESI-MS detection. *Eur. Food Res. Technol.* **2002**, *214*, 248–253. [CrossRef]

35. Tosun, I.; Ustun, N.S.; Tekguler, B. Physical and chemical changes during ripening of blackberry fruits. *Sci. Agric* **2008**, *65*, 87–90. [CrossRef]

36. Celli, G.B.; Ghanem, A.; Brooks, M.S.L. Influence of Freezing Process and Frozen Storage on Fruits and Fruit Products Quality. *Food Rev. Int.* **2015**, *32*, 280–304. [CrossRef]

37. Yanat, M.; Baysal, T. Effect of freezing rate and storage time on quality parameters of strawberry frozen in modified and home type freezer. *Hrvat. Časopis Za Prehrambenu Tehnol. Biotehnol. I Nutr.* **2018**, *13*, 154–158. [CrossRef]

38. Goncalves, B.; Silva, A.P.; Moutinho-Pereira, J.; Bacelar, E.; Rosa, E.; Meyer, A.S. Effect of ripeness and postharvest storage on the evolution of colour and anthocyanins in cherries (*Prunus avium* L.). *J. Food Chem.* **2007**, *103*, 976–984. [CrossRef]

39. Borochov-Neori, H.; Judeinstein, S.; Harati, M.; Bar-Yaakov, I.; Patil, B.S.; Lurie, S.; Holland, D. Climate Effects on Anthocyanin Accumulation and Composition in the Pomegranate (*Punica granatum* L.) Fruit Arils. *J. Agric. Food Chem.* **2011**, *59*, 5325–5334. [CrossRef]

40. Shete, Y.V.; Chavan, S.M.; Champawat, P.S.; Jain, S.K. Reviews on osmotic dehydration of fruits and vegetables. *J. Pharmacogn. Phytochem.* **2018**, *7*, 1964–1969.

41. Kumar, S.; Sagar, V.R. Influence of packaging materials and storage temperature on quality of osmo-vac dehydrated aonla segments. *J. Food Sci. Technol.* **2009**, *46*, 259–262.

42. Giampieri, F.; Forbes-Hernandez, T.Y.; Gasparrini, M.; Alvarez-Suarez, J.M.; Afrin, S.; Bompadre, S.; Quiles, J.L.; Mezzetti, B.; Battino, M. Strawberry as a health promoter: An evidence based review. *Food Funct.* **2015**, *6*, 1386–1398. [CrossRef]
43. Wilson, S.D.; Horne, D.W. Evaluation of ascorbic-acid in protecting labile folic-acid derivatives. *Proc. Natl. Acad. Sci. USA* **1983**, *80*, 6500–6504. [CrossRef]
44. Strålsjö, L.M.; Witthöft, C.M.; Sjöholm, I.M.; Jägerstad, M.I. Folate content in strawberries (Fragaria x ananassa): Effects of cultivar, ripeness, year of harvest, storage, and commercial processing. *J. Agric. Food Chem.* **2003**, *51*, 128–133. [CrossRef]
45. Ringling, C.; Rychlik, M. Simulation of food folate digestion and bioavailability of an oxidation product of 5-methyltetrahydrofolate. *Nutrients* **2017**, *9*, 969. [CrossRef]

**Disclaimer/Publisher's Note:** The statements, opinions and data contained in all publications are solely those of the individual author(s) and contributor(s) and not of MDPI and/or the editor(s). MDPI and/or the editor(s) disclaim responsibility for any injury to people or property resulting from any ideas, methods, instructions or products referred to in the content.

*applied*
*sciences*

*Article*

# Supplementation of an Anthocyanin-Rich Elderberry (*Sambucus nigra* L.) Extract in FVB/n Mice: A Healthier Alternative to Synthetic Colorants

Tiago Azevedo [1,2], Tiago Ferreira [1,2], João Ferreira [3,4], Filipa Teixeira [5], Diana Ferreira [5], Rita Silva-Reis [1,2,6], Maria João Neuparth [7,8,9], Maria João Pires [1,2], Maria de Lurdes Pinto [4], Rui M. Gil da Costa [1,2,10,11,12,13], Margarida M. S. M. Bastos [11,12], Rui Medeiros [10,14,15,16,17], Luís Félix [1,2,*], Carlos Venâncio [1,2,4], Maria Inês Dias [18,19], Isabel Gaivão [3,4], Lillian Barros [18,19] and Paula A. Oliveira [1,2,*]

1. Centre for the Research and Technology of Agro-Environmental and Biological Sciences (CITAB), University of Trás-os-Montes and Alto Douro, 5000-801 Vila Real, Portugal
2. Inov4Agro—Institute for Innovation, Capacity Building and Sustainability of Agri-Food Production, University of Trás-os-Montes and Alto Douro, 5000-801 Vila Real, Portugal
3. Department of Genetics and Biotechnology, School of Life and Environmental Sciences, University of Trás-os-Montes and Alto Douro, 5000-801 Vila Real, Portugal
4. Animal and Veterinary Research Centre (CECAV), University of Trás-os-Montes and Alto Douro, Quinta de Prados, 5000-801 Vila Real, Portugal
5. School of Life and Environmental Sciences, University of Trás-os-Montes and Alto Douro, 5000-801 Vila Real, Portugal
6. LAQV-REQUIMTE, Department of Chemistry, University of Aveiro, 3810-193 Aveiro, Portugal
7. Research Center in Physical Activity, Health and Leisure (CIAFEL), Faculty of Sports, University of Porto, 4200-319 Porto, Portugal
8. TOXRUN—Toxicology Research Unit, University Institute of Health Sciences, CESPU, 4585-116 Gandra, Portugal
9. Laboratory for Integrative and Translational Research in Population Health (ITR), 4050-600 Porto, Portugal
10. Molecular Oncology and Viral Pathology Group, Research Center of IPO Porto (CI-IPOP)/RISE@CI-IPOP (Health Research Network), Portuguese Oncology Institute of Porto (IPO Porto), Porto Comprehensive Cancer Center (Porto.CCC), 4200-072 Porto, Portugal
11. Laboratory for Process Engineering, Environment, Biotechnology and Energy (LEPABE), Faculty of Engineering, University of Porto, 4200-465 Porto, Portugal
12. Associate Laboratory in Chemical Engineering (ALiCE), Faculty of Engineering, University of Porto, 4200-465 Porto, Portugal
13. Postgraduate Programme in Adult Health (PPGSAD), Department of Morphology, Federal University of Maranhão (UFMA), UFMA University Hospital (HUUFMA), São Luís 65020-070, Brazil
14. Research Department of the Portuguese League against Cancer Regional Nucleus of the North (LPCC-NRN), 4200-177 Porto, Portugal
15. Faculty of Medicine, University of Porto, 4200-319 Porto, Portugal
16. Institute of Biomedical Sciences Abel Salazar (ICBAS), 4050-313 Porto, Portugal
17. Biomedical Research Center (CEBIMED), Faculty of Health Sciences of Fernando Pessoa University (UFP), 4249-004 Porto, Portugal
18. Centro de Investigação de Montanha (CIMO), Instituto Politécnico de Bragança, Campus de Santa Apolónia, 5300-253 Bragança, Portugal
19. Laboratório Associado para a Sustentabilidade e Tecnologia em Regiões de Montanha (SusTEC), Instituto Politécnico de Bragança, Campus de Santa Apolónia, 5300-253 Bragança, Portugal
* Correspondence: lfelix@utad.pt (L.F.); pamo@utad.pt (P.A.O.)

**Citation:** Azevedo, T.; Ferreira, T.; Ferreira, J.; Teixeira, F.; Ferreira, D.; Silva-Reis, R.; Neuparth, M.J.; Pires, M.J.; Pinto, M.d.L.; Gil da Costa, R.M.; et al. Supplementation of an Anthocyanin-Rich Elderberry (*Sambucus nigra* L.) Extract in FVB/n Mice: A Healthier Alternative to Synthetic Colorants. *Appl. Sci.* **2022**, *12*, 11928. https://doi.org/10.3390/app122311928

Academic Editor: Monica Gallo

Received: 31 October 2022
Accepted: 19 November 2022
Published: 23 November 2022

**Publisher's Note:** MDPI stays neutral with regard to jurisdictional claims in published maps and institutional affiliations.

**Copyright:** © 2022 by the authors. Licensee MDPI, Basel, Switzerland. This article is an open access article distributed under the terms and conditions of the Creative Commons Attribution (CC BY) license (https://creativecommons.org/licenses/by/4.0/).

**Featured Application: Due to anthocyanins' potential as colorants, this anthocyanin-rich elderberry extract presents a healthier alternative to commercially available colorants and has the potential to be used at industry level.**

**Abstract:** *Sambucus nigra* L., popularly known as elderberry, is renowned for its amazing therapeutic properties, as well as its uses as a food source, in nutraceuticals, and in traditional medicine. This study's aim was to investigate the effects of an elderberry extract (EE) on mice for 29 days, as well as the safety of the extract when used as a natural colorant. Twenty-four FVB/n female mice (n = 6) were randomly assigned to one of four groups: control, 12 mg/mL EE (EE12), 24 mg/mL EE (EE24), or 48 mg/mL EE (EE48). The predominant anthocyanins detected were cyanidin-3-*O*-sambubioside and cyanidin-3-*O*-glucoside. Food and drink intake were similar between groups, with the exception of EE48, who drank significantly less compared with the Control. Biochemical analysis of the liver showed that the changes observed in histological analysis had no pathological significance. The EE, at doses of 24 and 48 mg/mL, significantly reduced the oxidative DNA damage compared with the non-supplemented group. The *S. nigra* extract showed a favorable toxicological profile, affording it potential to be used in the food industry.

**Keywords:** *Sambucus nigra*; in vivo; oral administration; colorant

## 1. Introduction

The genus *Sambucus* belong to the Adoxaceae family and consists of 5 to 30 species, with *Sambucus nigra* L. being the most commonly occurring species [1–3]. This species, which is native to the northern hemisphere and may be found on practically every continent, is sometimes referred to as "black elder, European elder, or elderberry" [4]. It is a 6 m tall deciduous shrub that thrives in locations with direct sunlight. Between the spring and summer, the elderberry produces white hermaphrodite flowers, and in the late summer, the fruits mature [5]. The fruits of *S. nigra* L. are dark purple berries in a cluster with a diameter of up to 6 mm [5], and are a good source of vitamins, sugars, organic acids, fatty acids, protein, and essential oils [3,6–8]. Elderberry is also rich in phenolic compounds, especially anthocyanins. In folk medicine, elderberry has been widely consumed for many years due to its therapeutic effects; the berries have been used for the preparation of juice and tea to treat several illnesses, such as constipation, common cold, and diarrhea [8–10].

In the food industry, elderberry fruits and flowers are used to produce several products, such as liqueurs, jams, and juices [11]. Moreover, *S. nigra* can be used as a functional food for the prevention and the treatment of numerous diseases, as it has demonstrated antioxidant, anti-inflammatory, immune-stimulating, anti-cancer, and atheroprotective properties [12,13].

The largest class of water-soluble pigments, anthocyanins, afford many fruits and vegetables their red, purple, and blue hues [14]. These natural pigments can be found in flowers, roots, and vegetables, and are mostly associated with red fruits [14,15]. Furthermore, the food industry's interest in natural dyes has been increasing in order to replace synthetic dyes, as the latter have demonstrated disadvantages, such as allergenic, toxic, and even carcinogenic effects [16]. Cyanidin-3-glucoside and cyanidin-3-sambubioside are the most abundant anthocyanins found in elderberry juice [11,17]. Furthermore, anthocyanin cyanidin-3-glucoside has demonstrated anti-cancer [18,19], anti-angiogenic [20] and anti-obesity properties [21]. As such, anthocyanins have the added benefit of being used to prevent a variety of diseases, in addition to being of particular interest due to their great colorant properties, and could be used as a potential natural colorant [14,15]. The aim of this research was to evaluate the effect of an anthocyanin-rich elderberry extract (EE) supplementation on mice's physiological parameters.

## 2. Materials and Methods

### 2.1. Sample Preparation

*Sambucus nigra* L. fruits were collected in mid-September 2019 in Braganza, Portugal, when the maturation process was complete (Figure 1). The harvested fruits were immediately separated from the stems, washed, and frozen at −20 °C. The frozen fruits were crushed using a knife mill (model A327R1, Moulinex, Madrid, Spain) with a small amount of water to obtain juice; this process was facilitated by the previous freezing of the fruits, and later centrifuging them (K24OR refrigerated centrifuge, Centurion, West Sussex, UK) to separate the husks and small seeds. The supernatant obtained was frozen, lyophilized, reduced to a fine, dried powder (35 mesh), and stored for further analysis.

**Figure 1.** Elderberry tree with the ripe fruit used to obtain the extract.

Stability of the Aqueous Extract

The stability of the aqueous extract was assessed taking into consideration the anthocyanin profile of these fruits and the respective percentage of degradation over four days. The anthocyanin phenolic profile was previously described by other authors [17]; nonetheless, a new identification was performed for this sample following the procedure previously described by Bastos et al. (2015) [22]. Two anthocyanins were identified, cyanidin-3-*O*-sambubioside and cyanidin-3-*O*-glucoside, and, as such, the stability of the aqueous extract was assessed by measuring the percentage of anthocyanin loss by high-performance liquid chromatography coupled with a diode array detector (HPLC-DAD) using 520 nm as preferred wavelength.

The dried lyophilized extract of *S. nigra* was re-hydrated with water (100%) at a final concentration of 5 mg/mL and stored for 96 h at room temperature (~25 °C, mimicking the conditions under which the animals would be supplemented with these extracts, described below) and protected from light (to avoid the maximum degradation of the extracts). An aliquot of the samples was collected at 24, 72, and 96 h, injected into the chromatography system, and quantified using a 7-level calibration curve of the most similar standard compound available in the laboratory, cyanidin-3-*O*-glucoside (y = 105,078x − 12,437; R2: 0.9993; LOD = 0.28 µg/mL; LOQ = 0.84 µg/mL). The results were expressed as the percentage of anthocyanin loss (%).

### 2.2. Experimental Design

The University of Trás-os-Montes and Alto Douro Ethics Committee (approval no. 10/2013) and the Portuguese Veterinary Authorities (approval no. 0421/000/000/2014) approved this study. The national law (Decree-Law 113/2013) and European Directive 2010/63/EU on the protection of animals used in scientific research were both followed in all animal procedures.

Animals

As this is a preliminary study, to best apply the 3Rs for animal experiments (replacement, reduction, and refinement) and obtain enough data, we estimated the minimum number of animals required using a power analysis. Twenty-four FVB/n 8-week-old female mice (*Mus musculus*), obtained from a colony at University of Trás-os-Montes and Alto

Douro's animal facility, were randomly divided into four different groups (n = 6/group) using a computer-generated randomization sequence.

The control group drank tap water, while EE12, EE24, and EE48 were supplemented with increasing concentrations of a *S. nigra* extract, 12 mg/mL, 24 mg/mL, and 48 mg/mL, respectively, dissolved in normal tap water and changed every 2–3 days due to the compounds' stabilities. Drink and food (Diet Standard 4RF21 Certificate, 4RF21, Mucedola, Milan, Italy) were available *ad libitum* for every group. Under controlled conditions of temperature (19 ± 2 °C), 12 h:12 h light–dark cycle, and relative humidity (50 ± 10%), the animals were housed in separate open polycarbonate cages with bedding made of corn cob and provided with environmental enrichment.

Weekly, several murinometric parameters were evaluated, such as the body weight and temperature of each animal, as well as drink and food consumption by each group. The same researcher examined humane endpoints on a weekly basis using a previously published grading sheet [23]. Animals with a cumulative score of four or higher at any point in time were designated for euthanasia.

Twenty-nine days after the beginning of the experiment, the animals were sacrificed by intraperitoneal administration of ketamine (Imalgene 1000, Vetoquinol, Barcarena, Portugal) and xylazine (Rompun® 2% Bayer, Healthcare S.A., Kiel, Germany), followed by cardiac puncture and exsanguination according to the Federation of European Laboratory Animal Science Associations guidelines [24].

Complete necropsies were performed, and organs were collected and weighed on a precision balance (KERN® PLT 6200-2A, Dias de Sousa S.A., Alcochete, Portugal). These organs were fixed by immersion in 10% neutral-buffered formalin. Liver and kidney samples were stored at −80 °C for further analysis of oxidative stress.

*2.3. Hematological Analysis*

2.3.1. Microhematocrit

Microhematocrit values were obtained after the blood samples were centrifuged at $4500\times g$ for 5 min (PrO-Vet, Centurion Scientific Limited, Chichester, UK) in capillary tubes, and the column of red blood cells was measured with a ruler.

2.3.2. Serum Biochemistry

Blood collected into lithium–heparin tubes (FL MEDICAL, Torreglia, Italy) was centrifuged at $1400\times g$ (Heraeus Labofuge™ 400R, Thermo Fischer Scientific, Waltham, MA, USA) for 15 min, at 4 °C, and plasma was stored at −80 °C. Using an autoanalyzer (Prestige 24i, Cormay PZ, Warsaw, Poland), spectrophotometric methods were used to measure the concentrations of creatinine, urea, aspartate aminotransferase (ASAT), and alanine aminotransferase (ALAT).

*2.4. Comet Assay*

Mononuclear blood cells were used to perform the alkaline (pH > 13) comet assay, following the methods of Collins (2004) [25], and a system of twelve gels per slide, as adopted by Marques et al. (2021) [26], was used to increase the yield. Six slides precoated with normal-melting-point agarose were used for each treatment, with three replicates per animal. One set was used to run the assay with the repair enzyme, formamidopyrimidine DNA glycosylase (Fpg), and the other was used to conduct the assay without it. In order to precisely identify oxidative damage to DNA, particularly 8-oxoguanines and other changed purines, Fpg converted oxidized purines into DNA single-strand breaks. Peripheral blood was diluted in ice-cold phosphate-buffered saline (PBS), and this cell suspension was combined with 1% low-melting-point agarose. Twelve drops were placed on each of the twelve precoated slides, and the slides were refrigerated for solidification. The samples were then incubated in a lysis solution (2.5 M NaCl, 0.1 M EDTA, 10 mM Tris, 1% Triton X-100, pH 10) and rinsed (40 mM HEPES, 0.1 M KCl, 0.5 mM EDTA, 0.2 mg/mL bovine serum albumin, pH 8.0).

The slides with and without Fpg treatment were then incubated for 30 min at 4 °C in an alkaline electrophoresis solution (0.3 M NaOH and 1 mM EDTA, pH > 13) and electrophoresed for 30 min at 25 V and 300 mA. The cells were then neutralized with PBS, followed by distilled water, and then dehydrated in 70% and absolute ethanol. A fluorescent microscope at 400× magnification (Olympus BX41, Olympus America Inc., Hauppauge, NY, USA) was used to view the DNA stained with 4,6-diamidino-2-phenylindole (DAPI).

The comets were then classified visually based on a five-class scale of severity (class 0, no damage, and class 4, highest damage) [25]. Three hundred nucleoids were classified per animal (100 per mini gel) and, using the following formula, the total score was expressed as a genetic damage index (*GDI*) with a range of 0 to 400 arbitrary units:

$$GDI = \sum \% \text{ nucleoids class } i \times i$$

where *I* represents the number of each class.

Untreated *GDI* values were subtracted from the Fpg incubation scores ($GDI_{Fpg}$) to calculate the net enzyme-sensitive sites (NSSFpg).

### 2.5. Hepatic and Renal Histology

The organs underwent routine processing for paraffin embedding, including sectioning. Hematoxylin and eosin (H&E)-stained tissue sections that were 3 μm thick were examined under a light microscope for histological analysis.

Liver histological analysis included the recording of the presence or absence of hydropic changes, their distribution, as well as the presence of inflammatory cells. Kidneys were examined to assess the presence of inflammatory lesions.

### 2.6. Hepatic and Renal Oxidative Stress

Samples of the liver and kidney were homogenized in a cold buffer solution (0.32 mM sucrose, 20 mM HEPES, 1 mM $MgCl_2$, and 0.5 mM phenylmethylsulfonylfluoride (PMSF), pH 7.4) and the supernatants were collected following centrifugation and analyzed as previously described [27]. Briefly, excitation at 485 nm and emission at 530 nm were employed to assess the production of reactive oxygen species (ROS) using the probe $2',7'$-dichlorofluorescein diacetate (DCFH-DA). The inhibition of nitroblue tetazolium (NBT) reduction was used to assess the superoxide dismutase (SOD) activity at 560 nm. The $H_2O_2$ decay at 240 nm was used to evaluate the catalase (CAT) activity. The oxidation of NADPH to $NADP^+$ at 340 nm was employed to assess the glutathione peroxidase (GPx) activity. The reaction of glutathione's thiol group with 1-chloro-2,4-dinitrobenzene (CDNB) was used to determine the activity of glutathione S-transferase (GST) at 340 nm. Derivatization with ortho-phthalaldehyde at 320 nm and 420 nm (excitation and emission, respectively) was used to assess the quantities of reducing glutathione (GSH) and oxidized glutathione (GSSG). The ratio of GSH to GSSG was used to calculate the oxidative stress index (OSI). Malondialdehyde (MDA), a lipid peroxidation (LPO) biomarker, was measured at 530 nm using a thiobarbituric acid (TBA)-based technique.

### 2.7. Statistical Analysis

Statistical analysis was performed using IBM SPSS version 20 (Statistical Package for the Social Sciences, Chicago, IL, USA). The Shapiro–Wilk test was used to ensure that the data followed a normal distribution. A statistical one-way ANOVA was conducted, followed by the Bonferroni's multiple-comparison test. For histological analysis, the Chi-square test was performed. For oxidative stress analysis, statistical analyses were performed using one-way ANOVA followed by Tukey's multiple-comparison test or the Kruskal–Wallis test followed by Dunn's test. At $p \leq 0.05$ (*), $p \leq 0.01$ (**), $p \leq 0.001$ (***), or $p < 0.0001$ (****), data were deemed statistically significant.

## 3. Results

### 3.1. General Findings

The anthocyanin extract stability was assessed by HPLC-DAD (*data not shown*), revealing a maximum percentage of anthocyanin loss of 41.7% after 96 h of storage. In the first 48 and 72 h, the anthocyanin losses were 4.2 and 23.1%, respectively. Given these results, replacement of the supplementation water enriched with the anthocyanin-enriched extract occurred every two to three days during the experimental trial.

The animals in this study were monitored daily, and there were no phenotypic and/or behavioral changes in the mice in this study, as well as no deaths during the experiment. The aforementioned humane endpoint table was used to assess animal well-being, with no animal achieving the required score for euthanasia or any other notable behavioral change.

This study lasted for 29 days, during which almost no alteration was observed regarding the murinometric parameters evaluated. However, animals from EE12 ($p = 0.012$) lost a significant amount of body weight when compared with the control group (in weeks 1–4) and EE48 (in weeks 1 and 3), whereas animals from EE24 and EE48 did not show any significant differences when compared with the control group (Figure 2).

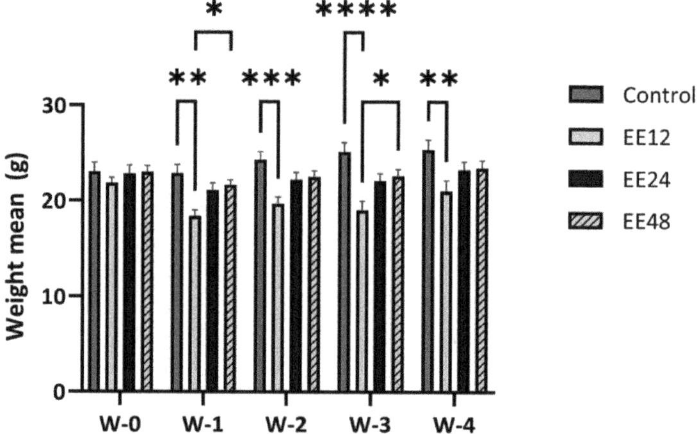


**Figure 2.** Body weight (g) mean [±Standard error (SE)] throughout the experiment. Statistically significant differences are denoted by asterisks: * $p < 0.05$; ** $p < 0.01$; *** $p < 0.001$; **** $p < 0.0001$.

Regarding food and drink intake, all of the groups had relatively similar average daily food and drink intake values (Table 1).

**Table 1.** Mean daily values of food and drink intake (g) per experimental group during the four weeks.

| Groups | Food (g) | | Drink (mL) | |
|---|---|---|---|---|
| | Initial | Final | Initial | Final |
| Control | 22.14 | 23.02 | 27.90 | 29.33 |
| EE12 | 19.99 | 25.27 | 26.12 | 26.60 |
| EE24 | 22.51 | 21.57 | 27.09 | 28.12 |
| EE48 | 21.47 | 21.60 | 22.08 | 27.12 |

The relative organ weight of the heart (Table 2) was significantly higher in EE12 compared with the control ($p = 0.011$), EE24 ($p = 0.02$), and EE48 ($p = 0.01$). Additionally, the left kidney's relative organ weight was significantly higher in the control than EE12 ($p = 0.013$) and EE24 ($p = 0.013$).

**Table 2.** Mean (±SE) relative organ weights (organ weight (mg)/body weight of the animal (g)) from the experimental groups.

| Group/Organ | Control | EE12 | EE24 | EE48 |
|---|---|---|---|---|
| Thymus | 1.92 ± 0.129 | 1.71 ± 0.259 | 2.02 ± 0.194 | 2.29 ± 0.255 |
| Heart | 4.69 ± 0.148 [a] | 5.50 ± 0.186 [b] | 4.50 ± 0.147 [a] | 4.49 ± 0.156 [a] |
| Lungs | 6.92 ± 0.349 | 7.37 ± 0.314 | 7.13 ± 0.257 | 7.78 ± 0.494 |
| Spleen | 3.89 ± 0.279 | 4.15 ± 0.239 | 4.08 ± 0.245 | 3.94 ± 0.344 |
| Liver | 54.56 ± 1.293 | 49.61 ± 0.904 | 50.50 ± 2.287 | 51.64 ± 2.171 |
| Left Kidney | 6.00 ± 0.118 [a] | 6.99 ± 0.114 [b] | 6.95 ± 0.196 [b] | 6.51 ± 0.271 [a,b] |
| Right Kidney | 6.05 ± 0.308 | 7.36 ± 0.134 | 6.33 ± 0.456 | 6.77 ± 0.330 |
| Left Adrenal | 0.42 ± 0.102 | 0.46 ± 0.126 | 0.47 ± 0.076 | 0.48 ± 0.101 |
| Right Adrenal | 0.32 ± 0.075 | 0.45 ± 0.064 | 0.29 ± 0.069 | 0.34 ± 0.060 |

Significant statistical differences between groups are indicated by different letters ([a] and [b]) ($p < 0.05$).

### 3.2. Haematological Analysis

The microhematocrit values per animal group are shown in Figure 3a (Supplementary Table S1), which were not significantly different between the experimental groups. Regarding the serum biochemical parameters (Figure 3c–e), there were no significant differences between the groups, except for the urea values (Figure 3e) in the control group and EE12 ($p = 0.011$).

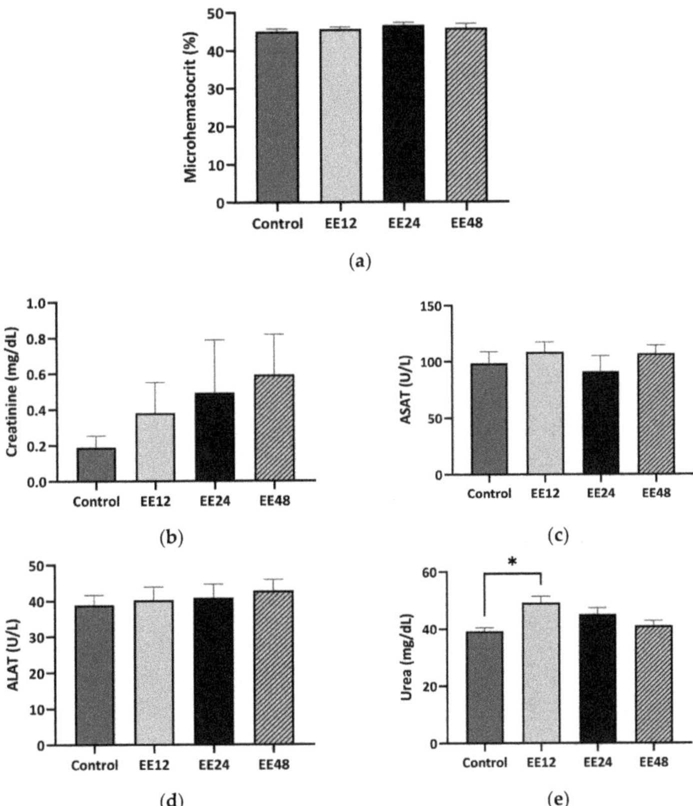

**Figure 3.** Hematological parameters evaluated (mean ± SE): (**a**) microhematocrit, (**b**) creatinine, (**c**) ASAT (aspartate aminotransferase), (**d**) ALAT (alanine aminotransferase), and (**e**) urea. * Statistically significant differences ($p < 0.05$).

### 3.3. Hepatic and Renal Histology

A significant difference between the control group and EE12 was seen in the liver's histology ($p = 0.036$), with the control group's histology displaying overall hydropic changes (Table 3). When compared with the other groups, the control mice showed a generalized distribution of these changes ($p = 0.006$), while the other groups showed hydropic changes at the centrilobular zone. With at least one animal in each group displaying chronic focal hepatitis, liver inflammation was also evaluated; the findings did not statistically differ between groups. Except for one instance of chronic interstitial nephritis in EE24, no histological alterations were seen in the experimental animals (Figure 4).

**Table 3.** Numbers of animals with hepatic and renal histological lesions per experimental group (%).

| | Control | EE12 | EE24 | EE48 |
|---|---|---|---|---|
| **Liver** | | | | |
| Normal | 1/5 (20.0%) [a] | 5/6 (83.3%) [b] | 4/6 (66.7%) [a,b] | 4/6 (66.7%) [a,b] |
| Hydropic changes (HC) | 4/5 (80.0%) [a] | 1/6 (16.7%) [b] | 2/6 (33.3%) [a,b] | 2/6 (33.3%) [a,b] |
| HC general | 4/5 (80.0%) [a] | 0/6 (0.0%) [b] | 0/6 (0.0%) [b] | 0/6 (0.0%) [b] |
| HC centrilobular | 0/5 (0.0%) | 1/6 (16.7%) | 2/6 (33.3%) | 2/6 (33.3%) |
| Inflammation | | | | |
| Absent | 3/5 (60.0%) | 5/6 (83.3%) | 3/6 (50.0%) | 3/6 (50.0%) |
| Chronic focal hepatitis | 2/5 (40.0%) | 1/6 (16.7%) | 3/6 (50.0%) | 3/6 (50.0%) |
| **Kidney** | | | | |
| Normal | 5/5 (100.0%) | 6/6 (100.0%) | 5/6 (83.3%) | 6/6 (100.0%) |
| Chronic interstitial nephritis | 0/0 (0.0%) | 0/0 (0.0%) | 1/6 (16.7%) | 0/0 (0.0%) |

Significant statistical differences between groups are indicated by different letters ([a] and [b]) ($p < 0.05$). HC: hydropic changes.

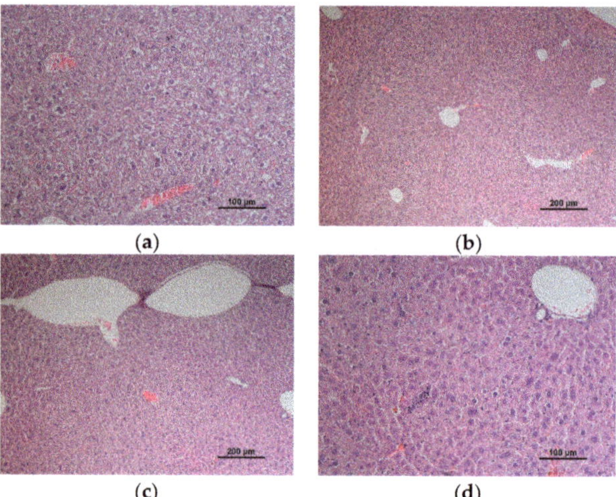

**Figure 4.** Microscopic images (staining with H&E) of liver sections from the different groups under study. (**a**) Cellular tumefaction and hydropic changes in a mouse's liver from the control group, 200×. (**b**) Hydropic changes observable only in the centrilobular region of an animal's liver from EE12, 100×. (**c**) Cellular tumefaction and centrilobular hydropic changes in a mouse's liver from EE24, 100×. (**d**) Cellular tumefaction, centrilobular hydropic changes, and multifocal inflammatory infiltration in an animal from EE48, 200×.

### 3.4. Comet Assay

The parameters evaluated with the comet assay (GDI, $GDI_{Fpg}$, and $NSS_{Fpg}$) were measured in peripheral blood mononuclear cells (Figure 5). Regarding the GDI values, there were no statistically significant differences between the experimental groups. However, the $GDI_{Fpg}$ values were significantly greater in the control group than EE24 ($p = 0.037$) and EE48 ($p = 0.047$). Regarding $NSS_{Fpg}$, the value in the control group was also significantly higher compared with those in EE24 ($p = 0.043$) and EE48 ($p = 0.033$).

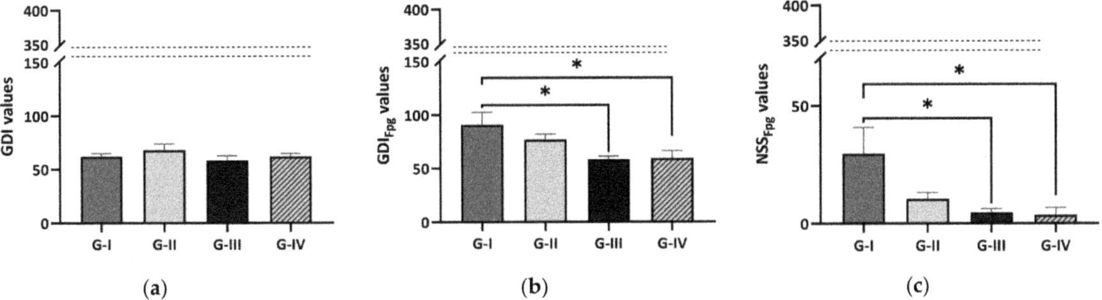

Figure 5. Genetic damage index values (mean ± SE and expressed as arbitrary units) determined by the comet assay, corresponding to untreated slides (**a**) and slides treated (**b**) with Fpg and the resulting $NSS_{Fpg}$ (**c**). * Statistically significant differences ($p < 0.05$).

### 3.5. Liver and Kidney Oxidative Stress

The liver and kidney oxidative stress analysis (Table 4) showed significant differences in the antioxidant enzymes SOD and CAT. The liver SOD activity increased in the EE48 group in comparison with the control and EE12 groups ($p < 0.01$). In the opposite direction, liver CAT activity decreased in the EE48 group in comparison with the control group ($p < 0.01$). Concerning kidney oxidative parameters, the only significant differences observed between groups were the increased activity of SOD in the EE48 group in comparison with the control ($p < 0.01$).

**Table 4.** Values of liver and kidney oxidative parameters. Data are expressed as mean ± standard deviation for parametric data distribution or median (25th–75th quartile) for non-parametric data.

| Biochemical Parameters | Control | EE12 | EE24 | EE48 |
|---|---|---|---|---|
| Liver | | | | |
| ROS | 1212 (1115–1623) | 1649 (1330–1729) | 2051 (1678–3046) | 1896 (1408–2308) |
| SOD | 231 (169–271) [a] | 250 (239–255) [a] | 274 (233–322) [a,b] | 333 (290–382) [b] |
| CAT | 5956 (5047–6608) [a] | 4001 (3197–4482) [a,b] | 3906 (2960–5081) [a,b] | 2925 (2241–3612) [b] |
| GPx | 0.41 ± 0.2 | 0.46 ± 0.13 | 0.61 ± 0.19 | 0.3 ± 0.27 |
| GST | 0.17 ± 0.01 | 0.18 ± 0.02 | 0.2 ± 0.03 | 0.19 ± 0.02 |
| GSH | 14.58 ± 3.4 | 14.71 ± 1.86 | 13.33 ± 5.55 | 12.95 ± 2.68 |
| GSSG | 270 ± 63.6 | 272.9 ± 138.5 | 339.7 ± 109 | 318.8 ± 141.6 |
| OSI | 0.05 ± 0.01 | 0.03 ± 0.02 | 0.04 ± 0.004 | 0.04 ± 0.01 |
| MDA | 205.1 ± 45.1 | 189.8 ± 23.8 | 207.0 ± 60.3 | 176.0 ± 28.6 |
| Kidney | | | | |
| ROS | 1460 ± 323 | 1545 ± 338 | 1429 ± 443 | 1337 ± 322 |
| SOD | 161 (139–188) [a] | 203 (178–271) [a,b] | 333 (267–432) [a,b] | 341 (305–374) [b] |
| CAT | 78.55 ± 7.19 | 76.05 ± 10.12 | 77.37 ± 12.21 | 77.89 ± 13.25 |
| GPx | 0.15 (0.13–0.22) | 0.08 (0.04–0.13) | 0.07 (0.05–0.1) | 0.08 (0.04–0.11) |
| GST | 0.058 ± 0.01 | 0.045 ± 0.002 | 0.058 ± 0.01 | 0.045 ± 0.002 |
| GSH | 148.1 ± 31.60 | 131.4 ± 16.69 | 146.2 ± 35.49 | 132.8 ± 21.63 |
| GSSG | 18.18 ± 13.02 | 23.07 ± 14.43 | 18.54 ± 8.78 | 14.93 ± 6.07 |
| OSI | 5.83 ± 1.92 | 8.01 ± 2.75 | 9.57 ± 5.12 | 10.6 ± 5.21 |
| MDA | 5600 ± 1189 | 4960 ± 728 | 5501 ± 1378 | 5040 ± 864 |

Significant statistical differences between groups are indicated by different letters ([a] and [b]) ($p < 0.05$). ROS: reactive oxygen species; SOD: superoxide dismutase; CAT: catalase; GPx: glutathione peroxidase; GST: glutathione S-transferase; GSH: reducing glutathione; GSSG: oxidized glutathione (GSSG); OSI: oxidative stress index; MDA: malondialdehyde.

## 4. Discussion

A previous study evaluated the extract's cytotoxicity in five different cell lines: human breast adenocarcinoma (MCF-7), human non-small lung carcinoma (NCI-H460), human cervical carcinoma (HeLa), human hepatocellular carcinoma (HepG2), and porcine liver cells (PLP2) [17]. These authors reported that the extract had significant capacity to inhibit the proliferation of these tumor cell lines, while low concentrations of EE had no effect on the normal cell line. After cell studies, a correct approach is to switch to an experimental animal protocol. To the best of our knowledge, this is the first study to assess the safety of different concentrations of an elderberry colorant extract in mice. Although the FVB/n strain is normally used for genetic and transgenic experiments, wild-type FVB/n mice have also been employed in metabolic and toxicology studies [28–31].

Animals supplemented with EE had considerably lower mean body weight values, particularly EE12, which was significantly inferior to the control group. *In vitro* assays have revealed that the primary components of this extract have an anti-obesity activity [32]. In a before-and-after clinical trial, Chrubasik et al. (2008) reported that a combination of *S. nigra* and *Asparagus officinalis* was significantly effective in lowering the mean body weight and blood pressure while improving the participants' physical and emotional well-being and quality of life [33]. The same occurrence appeared to happen to the EE-supplemented animals, especially at the lowest dose. Given that these animals consumed identical amounts of food and that the humane endpoint analysis revealed no changes in behavior associated with toxicity, the inclusion of anthocyanins in their diet appeared to influence their body weight gain. Another study discovered that dietary supplementation with purified mulberry (*Morus australis* Poir) anthocyanins suppressed body weight gain in high-fat-diet-fed C57BL/6 mice by reducing the epididymal fat weight and adipocyte size [34]. A future study should explore this extract's effect on adipose tissue and how it can affect body weight gain. Food consumption was standard compared with other studies developed by our group with the same diet [27,30]. The mice's initial reactions to the extract were not very positive, but, over time, the overall average consumption by the supplemented groups became similar to that by the control group. As this particularly occurred for the group supplemented with the highest concentration of EE, this difference may be due to the extract's flavor, which might have been too strong or overbearing. The values regarding drink intake were, however, consistent with those of the aforementioned works.

The relative weight of the heart in EE12 significantly increased when compared with the other groups, probably due to residual blood after cardiac puncture. Regarding hydropic changes, animals from the control group exhibited a generalized distribution of these changes. According to the literature, these represent acute and reversible, sublethal cell injuries [35], and in the absence of biochemical evidence of liver damage in these animals, the changes observed in the control group were not regarded as having pathological significance. In the present study, the animals supplemented with EE at the lowest concentration showed minor and zone-restricted hydropic changes when compared with the control group. This effect may be related to the anthocyanins present in the EE, because, in chronic ethanol-induced liver injury in male Wistar rats and CCl4-induced Kunming mice [36,37], these substances have been reported to have an hepatoprotective effect.

Our results regarding the renal function markers, particularly the urea levels in EE12, which were considerably greater than those of the control group, were in conflict with the literature, as anthocyanins are usually associated with a reduction in urea levels [38,39]. Our values are, however, in accordance with those of other studies regarding this parameter in FVB/n mice and do not raise suspicion of renal impairment [28,40]. As an increase in urea levels can occur in settings unrelated to renal disease, creatinine is seen as a more reliable marker of renal function [41]. Although there was a slight increase in creatinine levels, there was no statistically significant difference between the groups. The left kidney's relative weight was significantly higher in EE12 and EE24 compared with that of the control group, but, as no kidney lesions were registered by the histological analysis, this increase in the relative weight and biochemical parameters may be unrelated to renal function impairment.

It is also crucial to note that the EE supplementation was chronic, i.e., the animals consumed the extract in their drinking water for a lengthy period of time without interruption, which may not occur if it is incorporated into a product that is not the consumer's whole diet.

As mentioned above, elderberry extract supplementation did not cause hepatotoxicity at the biochemical level, as the hepatic transaminases (ALAT and ASAT) showed no significant statistical differences. These results are corroborated by another study that used an extract of *S. nigra* [42].

In terms of the comet assay results, our findings show that this extract did not appear to be genotoxic. Interestingly, when the Fpg enzyme was added, there were statistically significant differences between the control group and the two highest concentrations. As such, this leads us to believe that this extract, although not antigenotoxic, may protect against oxidative DNA damage. These findings are consistent with the literature [13,43], which indicates that *S. nigra* has antimutagenic and antigenotoxic effects, mainly due to its antioxidant properties. Olejnik et al. (2016) reported that, in a non-transformed, non-tumorigenic colon cell line (NCM460), an *S. nigra* extract was capable of protecting these cells against oxidative stress's detrimental effects to DNA [44]. In another study, Ferreira-Santos and collaborators employed an *S. nigra* aqueous extract for treating human colon carcinoma cells (RKO) and observed that this extract at concentrations of 200 and 400 µg/mL neither protected RKO cells from the oxidative activity induced by $H_2O_2$ nor triggered DNA repair activity [45]. The authors only observed a reduction in DNA damage of around 20% when the cells were treated for 24 h. The same was not observed when the cells were treated for 48 h; however, the results were not shown to be significantly different from those of the control group at both times [45]. Our findings regarding $NSS_{Fpg}$ indicate that this extract, particularly the two highest concentrations (EE24 and EE48), had a strong capacity for repairing oxidative DNA damage.

The group treated with a higher concentration of extract (EE48) also exhibited increased SOD activity, the primary intrinsic antioxidant enzyme, in the hepatic and renal tissue, confirming the stimulation of antioxidant activity previously reported for anthocyanins in the extracts of *S. nigra* [46,47]. However, CAT activity decreased in the same group only at the hepatic level. This discrepancy in activities recorded for SOD and CAT has been previously reported between tissues for the effect of anthocyanins [48] and in the same tissue attributed to different forms of modulation [49] that need further studies to understand the antioxidant mechanisms of these compounds. Moreover, the remaining parameters of oxidative stress did not vary significantly, which confirms the safety of the use of this extract at the tested doses.

Although synthetic food colorants are frequently used in the food industry, consuming them continuously could be toxic and have several side effects [50]. For this reason, natural origin colorants have gradually taken their place, adding potential health benefits as well [51]. Given their numerous reported medicinal properties, including antioxidant activity and anti-carcinogenic and anti-inflammatory properties, these natural dyes, particularly anthocyanins, have a significant potential to replace synthetic colorants in the food industry [52].

## 5. Conclusions

As a result of the decreased liver lesions and the potent defense against oxidative DNA damage, our research shows that this anthocyanin-rich elderberry extract has a favorable toxicological profile. The two highest concentrations (EE24 and EE48) appeared to perform exactly as intended, which was to create a natural colorant that had no negative effects on the health of individuals who consumed it and might even have some positive effects. In the future, further research may be needed to confirm our findings.

**Supplementary Materials:** The following supporting information can be downloaded at: https://www.mdpi.com/article/10.3390/app122311928/s1, Table S1. Hematological parameters evaluated (mean ± SE).

**Author Contributions:** Methodology and writing—original draft: T.A. and T.F.; Conceptualization: T.A., T.F., R.M.G.d.C., M.M.S.M.B., R.M., L.B. and P.A.O.; Supervision and project administration: R.M.G.d.C., M.M.S.M.B., R.M., L.B. and P.A.O.; Sample preparation and HPLC: M.I.D. and L.B.; Conducted the experiments with live animals and preparation of the drink: T.A., T.F. and R.S.-R.; Participated in the sacrifice of the animals: T.A., T.F., R.S.-R., M.J.P. and P.A.O.; Participated in biochemical analysis and microhematocrit: M.J.N. and M.J.P.; Performed the comet assay: T.A., J.F. and I.G.; Prepared the histology samples, histological analysis, and image capture: T.A., T.F. and M.d.L.P.; Participated in oxidative stress: F.T., D.F., L.F. and C.V.; Statistical analysis: T.A., T.F., R.M.G.d.C., M.M.S.M.B., R.M. and P.A.O. All authors have read and agreed to the published version of the manuscript.

**Funding:** This work is supported by National Funds by FCT—Portuguese Foundation for Science and Technology—under the project UIDB/04033/2020. This work was supported by the projects UIDB/CVT/00772/2020+UIDP/CVT/00772/2020 and LA/P/0059/2020 funded by the Portuguese Foundation for Science and Technology (FCT). Thanks are due to CESAM and FCT/MCTES for the financial support to UIDP/50017/2020+UIDB/50017/2020 through national funds. The authors acknowledge the financial support from the Portuguese Foundation for Science and Technology (FCT) through a Doctoral Grant (2020.04789.BD, Tiago Ferreira). The authors are grateful to the Foundation for Science and Technology (FCT, Portugal) for financial support through national funds FCT/MCTES (PIDDAC) to CIMO (UIDB/00690/2020 and UIDP/00690/2020) and SusTEC (LA/P/0007/2021); and L. Barros and M.I. Dias thank FCT, P.I., through the institutional scientific employment program contract, for their contracts (CEEC Institutional). This work was supported by LA/P/0045/2020 (ALiCE), UIDB/00511/2020 and UIDP/00511/2020 (LEPABE), funded by nationals funds through FCT/MCTES (PIDDAC); 2SMART (NORTE-01-0145-FEDER-000054), supported by Norte Portugal Regional Operational Programme (NORTE 2020), under the PORTUGAL 2020 Partnership Agreement, through the European Regional Development Fund (ERDF).

**Institutional Review Board Statement:** This study was conducted according to the guidelines of the Portuguese Competent Authority ("DGAV—Direção Geral de Alimentação e Veterinária", approval no. 10/2013) and reviewed by an Ethics Review Body ("ORBEA—Órgão Responsável pelo Bem-Estar e Ética Animal" under reference 0421/000/000/2014).

**Informed Consent Statement:** Not applicable.

**Data Availability Statement:** Data is contained within the article.

**Conflicts of Interest:** The authors declare no conflict of interest.

# References


1.  Schmitzer, V.; Veberic, R.; Stampar, F. European Elderberry (*Sambucus Nigra* L.) and American Elderberry (*Sambucus Canadensis* L.): Botanical, Chemical and Health Properties of Flowers, Berries and Their Products. In *Berries: Properties, Consumption and Nutrition*; Nova Biomedical: Waltham, MA, USA, 2012; pp. 127–148.
2.  Salvador, Â.C.; Silvestre, A.J.D.; Rocha, M.S. *Sambucus Nigra* L.: A Potential Source of Healthpromoting Components. *Front. Nat. Prod. Chem.* **2016**, *2*, 343–392.
3.  Młynarczyk, K.; Walkowiak-Tomczak, D.; Łysiak, G.P. Bioactive Properties of *Sambucus Nigra* L. as a Functional Ingredient for Food and Pharmaceutical Industry. *J. Funct. Foods* **2018**, *40*, 377–390. [CrossRef] [PubMed]
4.  Ağalar, H.G. Elderberry (*Sambucus Nigra* L.). In *Nonvitamin and Nonmineral Nutritional Supplements*; Elsevier: Amsterdam, The Netherlands, 2019; pp. 211–215, ISBN 978-0-12-812491-8.
5.  Atkinson, M.D.; Atkinson, E. *Sambucus Nigra* L. *J. Ecol.* **2002**, *90*, 895–923. [CrossRef]
6.  Veberic, R.; Jakopic, J.; Stampar, F.; Schmitzer, V. European Elderberry (*Sambucus Nigra* L.) Rich in Sugars, Organic Acids, Anthocyanins and Selected Polyphenols. *Food Chem.* **2009**, *114*, 511–515. [CrossRef]
7.  Sidor, A.; Gramza-Michałowska, A. Advanced Research on the Antioxidant and Health Benefit of Elderberry (*Sambucus Nigra*) in Food—A Review. *J. Funct. Foods* **2015**, *18*, 941–958. [CrossRef]
8.  Charlebois, D.; Byers, P.L.; Finn, C.E.; Thomas, A.L. Elderberry: Botany, Horticulture, Potential. *Horticultural Reviews*; Janick, J., Ed.; John Wiley & Sons, Inc.: Hoboken, NJ, USA, 2010; Volume 37, pp. 213–280, ISBN 978-0-470-54367-2.
9.  Barros, L.; Dueñas, M.; Carvalho, A.M.; Ferreira, I.C.F.R.; Santos-Buelga, C. Characterization of Phenolic Compounds in Flowers of Wild Medicinal Plants from Northeastern Portugal. *Food Chem. Toxicol.* **2012**, *50*, 1576–1582. [CrossRef]

10. Vlachojannis, C.; Zimmermann, B.F.; Chrubasik-Hausmann, S. Quantification of Anthocyanins in Elderberry and Chokeberry Dietary Supplements. *Phytother. Res.* **2015**, *29*, 561–565. [CrossRef]
11. Cejpek, K.; Maloušková, I.; Konečný, M.; Velíšek, J. Antioxidant Activity in Variously Prepared Elderberry Foods and Supplements. *Czech J. Food Sci.* **2009**, *27*, S45–S48. [CrossRef]
12. Małgorzata, D.; Aleksandra, P.; Monika, T.; Ireneusz, K. A New Black Elderberry Dye Enriched in Antioxidants Designed for Healthy Sweets Production. *Antioxidants* **2019**, *8*, 257. [CrossRef]
13. Ferreira, S.S.; Silva, A.M.; Nunes, F.M. *Sambucus Nigra* L. Fruits and Flowers: Chemical Composition and Related Bioactivities. *Food Rev. Int.* **2020**, *38*, 1–29. [CrossRef]
14. Khoo, H.E.; Azlan, A.; Tang, S.T.; Lim, S.M. Anthocyanidins and Anthocyanins: Colored Pigments as Food, Pharmaceutical Ingredients, and the Potential Health Benefits. *Food Nutr. Res.* **2017**, *61*, 1361779. [CrossRef]
15. Mateus, N.; de Freitas, V. Anthocyanins as Food Colorants. In *Anthocyanins: Biosynthesis, Functions, and Applications*; Winefield, C., Davies, K., Gould, K., Eds.; Springer: New York, NY, USA, 2009; pp. 284–304, ISBN 978-0-387-77335-3.
16. Saxena, S.; Raja, A.S.M. Natural Dyes: Sources, Chemistry, Application and Sustainability Issues. In *Roadmap to Sustainable Textiles and Clothing: Eco-Friendly Raw Materials, Technologies, and Processing Methods*; Muthu, S.S., Ed.; Textile Science and Clothing Technology; Springer: Singapore, 2014; pp. 37–80, ISBN 978-981-287-065-0.
17. Da Silva, R.F.R.; Barreira, J.C.M.; Heleno, S.A.; Barros, L.; Calhelha, R.C.; Ferreira, I.C.F.R. Anthocyanin Profile of Elderberry Juice: A Natural-Based Bioactive Colouring Ingredient with Potential Food Application. *Molecules* **2019**, *24*, 2359. [CrossRef] [PubMed]
18. Ding, M.; Feng, R.; Wang, S.Y.; Bowman, L.; Lu, Y.; Qian, Y.; Castranova, V.; Jiang, B.-H.; Shi, X. Cyanidin-3-Glucoside, a Natural Product Derived from Blackberry, Exhibits Chemopreventive and Chemotherapeutic Activity. *J. Biol. Chem.* **2006**, *281*, 17359–17368. [CrossRef] [PubMed]
19. Cho, E.; Chung, E.Y.; Jang, H.-Y.; Hong, O.-Y.; Chae, H.S.; Jeong, Y.-J.; Kim, S.-Y.; Kim, B.-S.; Yoo, D.J.; Kim, J.-S.; et al. Anti-Cancer Effect of Cyanidin-3-Glucoside from Mulberry via Caspase-3 Cleavage and DNA Fragmentation in Vitro and in Vivo. *Anticancer Agents Med. Chem.* **2017**, *17*. [CrossRef] [PubMed]
20. Ma, X.; Ning, S. Cyanidin-3-glucoside Attenuates the Angiogenesis of Breast Cancer via Inhibiting STAT3/VEGF Pathway. *Phytother. Res.* **2019**, *33*, 81–89. [CrossRef] [PubMed]
21. You, Y.; Han, X.; Guo, J.; Guo, Y.; Yin, M.; Liu, G.; Huang, W.; Zhan, J. Cyanidin-3-Glucoside Attenuates High-Fat and High-Fructose Diet-Induced Obesity by Promoting the Thermogenic Capacity of Brown Adipose Tissue. *J. Funct. Foods* **2018**, *41*, 62–71. [CrossRef]
22. Bastos, C.; Barros, L.; Dueñas, M.; Calhelha, R.C.; Queiroz, M.J.R.P.; Santos-Buelga, C.; Ferreira, I.C.F.R. Chemical Characterisation and Bioactive Properties of *Prunus Avium* L.: The Widely Studied Fruits and the Unexplored Stems. *Food Chem.* **2015**, *173*, 1045–1053. [CrossRef] [PubMed]
23. Oliveira, M.; Nascimento-Gonçalves, E.; Silva, J.; Oliveira, P.A.; Ferreira, R.; Antunes, L.; Arantes-Rodrigues, R.; Faustino-Rocha, A.I. Implementation of Human Endpoints in a Urinary Bladder Carcinogenesis Study in Rats. *In Vivo* **2017**, *31*. [CrossRef]
24. Forbes, D.; Blom, H.J.M.; Kostomitsopoulos, N.; Moore, G.; Perretta, G. *FELASA EUROGUIDE. On the Accomodation and Care of Animals Used for Experimental and Other Scientific Purposes*; The Royal Society of Medicine Press Ltd.: London, UK, 2007; ISBN 978-1-85315-751-6.
25. Collins, A.R. The Comet Assay for DNA Damage and Repair: Principles, Applications, and Limitations. *Mol. Biotechnol.* **2004**, *26*, 249–261. [CrossRef]
26. Marques, A.; Ferreira, J.; Abreu, H.; Pereira, R.; Pinto, D.; Silva, A.; Gaivão, I.; Pacheco, M. Comparative Genoprotection Ability of Wild-Harvested vs. Aqua-Cultured *Ulva Rigida* Coupled with Phytochemical Profiling. *Eur. J. Phycol.* **2021**, *56*, 105–118. [CrossRef]
27. Martins, T.; Oliveira, P.A.; Pires, M.J.; Neuparth, M.J.; Lanzarin, G.; Félix, L.; Venâncio, C.; Pinto, M.d.L.; Ferreira, J.; Gaivão, I.; et al. Effect of a Sub-Chronic Oral Exposure of Broccoli (Brassica Oleracea L. Var. Italica) By-Products Flour on the Physiological Parameters of FVB/N Mice: A Pilot Study. *Foods* **2022**, *11*, 120. [CrossRef]
28. Ribeiro-Silva, C.M.; Faustino-Rocha, A.I.; Gil da Costa, R.M.; Medeiros, R.; Pires, M.J.; Gaivão, I.; Gama, A.; Neuparth, M.J.; Barbosa, J.V.; Peixoto, F.; et al. Pulegone and Eugenol Oral Supplementation in Laboratory Animals: Results from Acute and Chronic Studies. *Biomedicines* **2022**, *10*, 2595. [CrossRef] [PubMed]
29. Martins, T.; Matos, A.F.; Soares, J.; Leite, R.; Pires, M.J.; Ferreira, T.; Medeiros-Fonseca, B.; Rosa, E.; Oliveira, P.A.; Antunes, L.M. Comparison of Gelatin Flavors for Oral Dosing of C57BL/6J and FVB/N Mice. *J. Am. Assoc. Lab. Anim. Sci.* **2022**, *61*, 89–95. [CrossRef] [PubMed]
30. Rodrigues, P.; Ferreira, T.; Nascimento-Gonçalves, E.; Seixas, F.; Gil da Costa, R.M.; Martins, T.; Neuparth, M.J.; Pires, M.J.; Lanzarin, G.; Félix, L.; et al. Dietary Supplementation with Chestnut (Castanea Sativa) Reduces Abdominal Adiposity in FVB/n Mice: A Preliminary Study. *Biomedicines* **2020**, *8*, 75. [CrossRef] [PubMed]
31. Sachse, B.; Meinl, W.; Glatt, H.; Monien, B.H. Ethanol and 4-Methylpyrazole Increase DNA Adduct Formation of Furfuryl Alcohol in FVB/N Wild-Type Mice and in Mice Expressing Human Sulfotransferases 1A1/1A2. *Carcinogenesis* **2016**, *37*, 314–319. [CrossRef] [PubMed]
32. Porras-Mija, I.; Chirinos, R.; García-Ríos, D.; Aguilar-Galvez, A.; Huaman-Alvino, C.; Pedreschi, R.; Campos, D. Physico-Chemical Characterization, Metabolomic Profile and *in Vitro* Antioxidant, Antihypertensive, Antiobesity and Antidiabetic Properties of Andean Elderberry (*Sambucus Nigra* Subsp. Peruviana). *J. Berry Res.* **2020**, *10*, 193–208. [CrossRef]

33. Chrubasik, C.; Maier, T.; Dawid, C.; Torda, T.; Schieber, A.; Hofmann, T.; Chrubasik, S. An Observational Study and Quantification of the Actives in a Supplement with *Sambucus Nigra* and *Asparagus Officinalis* Used for Weight Reduction. *Phytother. Res.* **2008**, *22*, 913–918. [CrossRef]

34. Wu, T.; Qi, X.; Liu, Y.; Guo, J.; Zhu, R.; Chen, W.; Zheng, X.; Yu, T. Dietary Supplementation with Purified Mulberry (Morus Australis Poir) Anthocyanins Suppresses Body Weight Gain in High-Fat Diet Fed C57BL/6 Mice. *Food Chem.* **2013**, *141*, 482–487. [CrossRef] [PubMed]

35. Wallig, M.A.; Janovitz, E.B. Chapter 4—Morphologic Manifestations of Toxic Cell Injury. In *Haschek and Rousseaux's Handbook of Toxicologic Pathology*, 3rd ed.; Haschek, W.M., Rousseaux, C.G., Wallig, M.A., Eds.; Academic Press: Boston, MA, USA, 2013; pp. 77–105, ISBN 978-0-12-415759-0.

36. Hou, Z.; Qin, P.; Ren, G. Effect of Anthocyanin-Rich Extract from Black Rice ( *Oryza Sativa* L. *Japonica* ) on Chronically Alcohol-Induced Liver Damage in Rats. *J. Agric. Food Chem.* **2010**, *58*, 3191–3196. [CrossRef]

37. Wang, L.; Zhao, Y.; Zhou, Q.; Luo, C.-L.; Deng, A.-P.; Zhang, Z.-C.; Zhang, J.-L. Characterization and Hepatoprotective Activity of Anthocyanins from Purple Sweet Potato (*Ipomoea Batatas* L. Cultivar Eshu No. 8). *J. Food Drug Anal.* **2017**, *25*, 607–618. [CrossRef] [PubMed]

38. Curtis, P.J.; Kroon, P.A.; Hollands, W.J.; Walls, R.; Jenkins, G.; Kay, C.D.; Cassidy, A. Cardiovascular Disease Risk Biomarkers and Liver and Kidney Function Are Not Altered in Postmenopausal Women after Ingesting an Elderberry Extract Rich in Anthocyanins for 12 Weeks. *J. Nutr.* **2009**, *139*, 2266–2271. [CrossRef]

39. Salvador, Â.; Król, E.; Lemos, V.; Santos, S.; Bento, F.; Costa, C.; Almeida, A.; Szczepankiewicz, D.; Kulczyński, B.; Krejpcio, Z.; et al. Effect of Elderberry (*Sambucus Nigra* L.) Extract Supplementation in STZ-Induced Diabetic Rats Fed with a High-Fat Diet. *Int. J. Mol. Sci.* **2016**, *18*, 13. [CrossRef]

40. Medeiros-Fonseca, B.; Mestre, V.F.; Colaço, B.; Pires, M.J.; Martins, T.; Gil da Costa, R.M.; Neuparth, M.J.; Medeiros, R.; Moutinho, M.S.S.; Dias, M.I.; et al. *Laurus Nobilis* (Laurel) Aqueous Leaf Extract's Toxicological and Anti-Tumor Activities in HPV16-Transgenic Mice. *Food Funct.* **2018**, *9*, 4419–4428. [CrossRef]

41. Gounden, V.; Bhatt, H.; Jialal, I. Renal Function Tests. In *StatPearls*; StatPearls Publishing: Treasure Island, FL, USA, 2021.

42. Opris, R.; Tatomir, C.; Olteanu, D.; Moldovan, R.; Moldovan, B.; David, L.; Nagy, A.; Decea, N.; Kiss, M.L.; Filip, G.A. The Effect of *Sambucus Nigra* L. Extract and Phytosinthesized Gold Nanoparticles on Diabetic Rats. *Colloids Surf. B Biointerfaces* **2017**, *150*, 192–200. [CrossRef] [PubMed]

43. Gonçalves, S.; Gaivão, I. Natural Ingredients Common in the Trás-Os-Montes Region (Portugal) for Use in the Cosmetic Industry: A Review about Chemical Composition and Antigenotoxic Properties. *Molecules* **2021**, *26*, 5255. [CrossRef] [PubMed]

44. Olejnik, A.; Olkowicz, M.; Kowalska, K.; Rychlik, J.; Dembczyński, R.; Myszka, K.; Juzwa, W.; Białas, W.; Moyer, M.P. Gastrointestinal Digested *Sambucus Nigra* L. Fruit Extract Protects *In Vitro* Cultured Human Colon Cells against Oxidative Stress. *Food Chem.* **2016**, *197*, 648–657. [CrossRef] [PubMed]

45. Ferreira-Santos, P.; Badim, H.; Salvador, Â.C.; Silvestre, A.J.D.; Santos, S.A.O.; Rocha, S.M.; Sousa, A.M.; Pereira, M.O.; Wilson, C.P.; Rocha, C.M.R.; et al. Chemical Characterization of *Sambucus Nigra* L. Flowers Aqueous Extract and Its Biological Implications. *Biomolecules* **2021**, *11*, 1222. [CrossRef]

46. Domínguez, R.; Zhang, L.; Rocchetti, G.; Lucini, L.; Pateiro, M.; Munekata, P.E.S.; Lorenzo, J.M. Elderberry (Sambucus Nigra L.) as Potential Source of Antioxidants. Characterization, Optimization of Extraction Parameters and Bioactive Properties. *Food Chem.* **2020**, *330*, 127266. [CrossRef]

47. Silva, P.; Ferreira, S.; Nunes, F.M. Elderberry (*Sambucus Nigra* L.) by-Products a Source of Anthocyanins and Antioxidant Polyphenols. *Ind. Crop. Prod.* **2017**, *95*, 227–234. [CrossRef]

48. Hassimotto, N.M.; Lajolo, F.M. Antioxidant Status in Rats after Long-Term Intake of Anthocyanins and Ellagitannins from Blackberries: Anthocyanin Intake Improves Antioxidant Status in Rats. *J. Sci. Food Agric.* **2011**, *91*, 523–531. [CrossRef]

49. Fernández-Pachón, M.S.; Berná, G.; Otaolaurruchi, E.; Troncoso, A.M.; Martín, F.; García-Parrilla, M.C. Changes in Antioxidant Endogenous Enzymes (Activity and Gene Expression Levels) after Repeated Red Wine Intake. *J. Agric. Food Chem.* **2009**, *57*, 6578–6583. [CrossRef]

50. Amchova, P.; Kotolova, H.; Ruda-Kucerova, J. Health Safety Issues of Synthetic Food Colorants. *Regul. Toxicol. Pharmacol.* **2015**, *73*, 914–922. [CrossRef] [PubMed]

51. Rodriguez-Amaya, D.B. Update on Natural Food Pigments—A Mini-Review on Carotenoids, Anthocyanins, and Betalains. *Food Res. Int.* **2019**, *124*, 200–205. [CrossRef] [PubMed]

52. Roy, S.; Rhim, J.-W. Anthocyanin Food Colorant and Its Application in PH-Responsive Color Change Indicator Films. *Crit. Rev. Food Sci. Nutr.* **2021**, *61*, 2297–2325. [CrossRef] [PubMed]

*Article*

# Inhibitory Effect of Coumarins and Isocoumarins Isolated from the Stems and Branches of *Acer mono* Maxim. against *Escherichia coli β*-Glucuronidase

Nguyen Viet Phong [1,2], Byung Sun Min [3], Seo Young Yang [4,*] and Jeong Ah Kim [1,2,*]

1   Vessel-Organ Interaction Research Center, VOICE (MRC), College of Pharmacy, Kyungpook National University, Daegu 41566, Korea
2   BK21 FOUR Community-Based Intelligent Novel Drug Discovery Education Unit, College of Pharmacy and Research Institute of Pharmaceutical Sciences, Kyungpook National University, Daegu 41566, Korea
3   Drug Research and Development Center, College of Pharmacy, Daegu Catholic University, Gyeongsan-si 38430, Korea
4   Department of Pharmaceutical Engineering, Sangji University, Wonju 26339, Korea
*   Correspondence: syyang@sangji.ac.kr (S.Y.Y.); jkim6923@knu.ac.kr (J.A.K.); Tel.: +82-33-738-7921 (S.Y.Y.); +82-53-950-8574 (J.A.K.)

**Abstract:** We isolated eight known secondary metabolites, including two isocoumarins and six coumarins, from the stems and branches of *Acer mono* Maxim. Their structures were confirmed using nuclear magnetic resonance spectroscopy and by comparing the data to published reports. The inhibitory effects of all compounds (**1–8**) on *Escherichia coli β*-glucuronidase were evaluated for the first time using in vitro assays. 3-(3,4-Dihydroxyphenyl)-8-hydroxyisocoumarin (**1**) displayed an inhibitory effect against *β*-glucuronidase (IC$_{50}$ = 58.83 ± 1.36 μM). According to the findings of kinetic studies, compound **1** could function as a non-competitive inhibitor. Molecular docking indicated that compound **1** binds to the allosteric binding site of *β*-glucuronidase, and the results corroborated those from kinetic studies. Furthermore, molecular dynamics simulations of compound **1** were performed to identify the behavioral and dynamic properties of the protein–ligand complex. Our results reveal that compound **1** could be a lead metabolite for designing new *β*-glucuronidase inhibitors.

**Keywords:** *Acer mono* Maxim.; isocoumarins; *β*-glucuronidase; non-competitive inhibitor; allosteric binding site; molecular dynamics

**Citation:** Phong, N.V.; Min, B.S.; Yang, S.Y.; Kim, J.A. Inhibitory Effect of Coumarins and Isocoumarins Isolated from the Stems and Branches of *Acer mono* Maxim. against *Escherichia coli β*-Glucuronidase. *Appl. Sci.* **2022**, *12*, 10685. https://doi.org/10.3390/app122010685

Academic Editors: Luca Mazzoni, Maria Teresa Ariza Fernández and Franco Capocasa

Received: 4 October 2022
Accepted: 19 October 2022
Published: 21 October 2022

**Publisher's Note:** MDPI stays neutral with regard to jurisdictional claims in published maps and institutional affiliations.

**Copyright:** © 2022 by the authors. Licensee MDPI, Basel, Switzerland. This article is an open access article distributed under the terms and conditions of the Creative Commons Attribution (CC BY) license (https://creativecommons.org/licenses/by/4.0/).

## 1. Introduction

*β*-Glucuronidase is a well-known enzyme that hydrolyzes conjugated compounds, including *β*-glucuronic acid, into their derivatives and free glucuronic acid [1]. *β*-Glucuronidase is frequently found in unicellular microorganisms, such as *Escherichia coli* and *Peptostreptococcus* species, and multicellular organisms, including plants and mammals [2,3]. In particular, this enzyme can be detected in several human body organs, such as the kidney, liver, lung, and digestive system [4]. In 2010, Wallace et al. first reported the crystal structure of *E. coli β*-glucuronidase [5]. The *β*-glucuronidase asymmetric unit (139 kDa) comprises two monomers (597 residues) and is organized into the following three areas: the N-terminal contains 180 residues and is similar to the carbohydrate-binding domain of the glycoside hydrolase 2 members; the C-terminal contains residue 274 to residue 603 and comprises an αβ loop; and the region between terminals N and C contains an Ig-like *β*-sandwich domain, as is the case with other members of the glycoside hydrolase 2 family [5,6]. Inhibiting *β*-glucuronidase can be beneficial for preventing and treating various diseases [7]. *β*-Glucuronidase is generated in the synovial fluid under inflammatory conditions, such as rheumatoid arthritis [8]. Moreover, an increased risk of colon cancer has been associated with the role of *β*-glucuronidase in the disease, as well as enhanced intestinal enzyme levels [9]. Increased levels of $\beta$-glucuronidase in the blood, owing to liver injury, can lead to liver cancer [7,10]. Thus, the discovery and development of $\beta$-glucuronidase inhibitors may be valuable in reducing these carcinogenic risks.

*Acer* L., commonly known as maple, is a diverse genus in the Aceraceae family, which comprises over 125 species with several infraspecific taxa, and is widely distributed in northern temperate regions worldwide [11]. Many *Acer* species provide products of economic value, including furniture, lumber, horticultural plants, and herbal medicines, especially gamma-linolenic acid, a dietary supplement that can help treat cancer and cardiovascular diseases [12]. *A. mono* Maxim., a deciduous tree of the *Acer* genus, is commonly found in Korea, Japan, and Northeast China [13,14]. Over the years, this plant has attracted considerable attention from both chemists and pharmacists, given its noteworthy traditional uses and pharmacological activities. In Korea, the leaves of *A. mono* Maxim. have long been employed as a material for hemostasis, whereas the roots have been used in traditional Korean medicine to treat arthralgia and cataclasis [13–15]. The sap of *A. mono* Maxim. has been used to treat gout, neuralgia, urinary hesitancy, constipation, and other gastroenteric conditions [15,16]. *A. mono* Maxim. reportedly possesses various pharmacological properties, including hepatoprotective, antioxidant, and osteoporotic inhibitory effects [13,14]. Phytochemical constituents of *A. mono* Maxim. include stilbenes, flavonoids, and their derivatives [14]. Among them, 5-*O*-methyl-(*E*)-resveratrol 3-*O*-$\beta$-D-glucopyranoside and 5-*O*-methyl-(*E*)-resveratrol 3-*O*-$\beta$-D-apiofuranosyl-(1→6)-$\beta$-D-glucopyranoside, two stilbene glycosides isolated from the leaves of *A. mono* Maxim., could reduce the levels of DPPH radicals at the concentrations of 100 µM, thereby exhibiting significant free-radical scavenging effects, with $IC_{50}$ values of 103.6 and 80.5 µM, respectively [13].

As part of our ongoing investigation of the secondary metabolites and respective biological effects from herbal and medicinal plants in Korea, we isolated and structurally elucidated eight compounds, including two isocoumarins and six coumarins, from the stems and branches of *A. mono* Maxim. in the present study. To the best of our knowledge, this is the first report regarding the isolation of isocoumarins from the *Acer* genus. In vitro assays were performed to determine the $\beta$-glucuronidase inhibitory activity of all compounds isolated from *A. mono* Maxim. In addition, kinetic analysis studies, molecular docking, and molecular dynamics (MD) simulations were performed to comprehensively clarify the inhibition mode, critical amino acid interactions, and the protein–ligand binding mechanism of active compound **1** with $\beta$-glucuronidase proteins.

## 2. Results

### 2.1. Isolated Compounds from the Stems and Branches of A. mono Maxim.

Methanol residue from the stems and branches of *A. mono* Maxim. was suspended in distilled water and partitioned with *n*-hexane, $CH_2Cl_2$, and EtOAc to obtain four extracts. The $CH_2Cl_2$ extract (77.8 g) and water layer (170 g) were separated by repeated column chromatography (CC) on silica gel, RP-18 gel, and Diaion resins, followed by preparative high-performance liquid chromatography (HPLC) to isolate and purify two isocoumarins (**1** and **2**) and six coumarins (**3**−**8**) (Figure 1). Based on an in-depth analysis of nuclear magnetic resonance (NMR) results and comparison with corresponding data from previous reports, the structures of all the compounds isolated were elucidated as 3-(3,4-dihydroxyphenyl)-8-hydroxyisocoumarin (**1**) [17], hydrangenol (**2**) [18], scopoletin (**3**) [19], isoscopoletin (**4**) [19], isofraxidin (**5**) [20], fraxidin 8-*O*-$\beta$-D-glucopyranoside (**6**) [21], aquillochin (**7**) [22], and cleomiscosin D (**8**) [23].

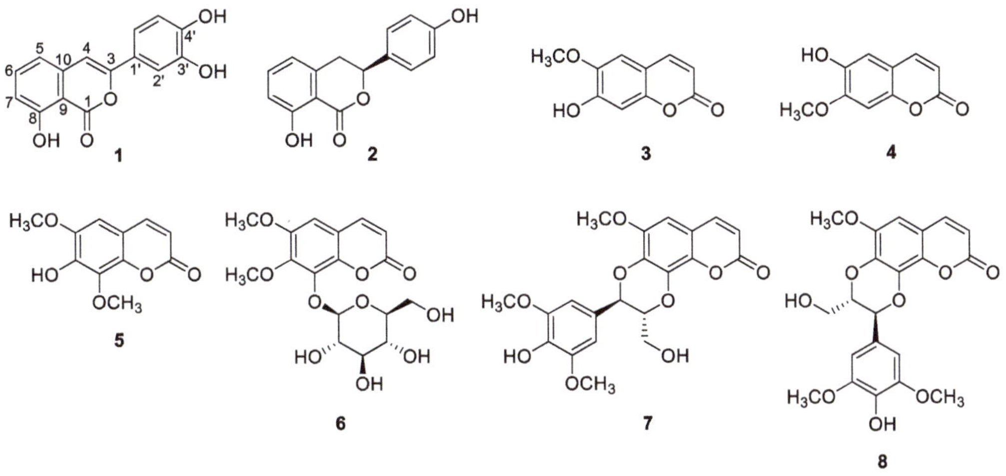

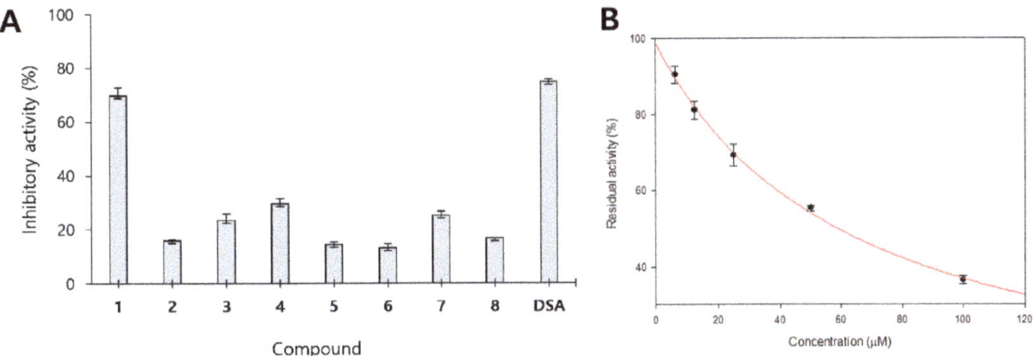

**Figure 1.** Chemical structures of isolated compounds from *Acer mono* Maxim. (**1−8**).

### 2.2. Inhibitory Activity of Isolated Compounds against β-Glucuronidase

All isolated compounds (**1−8**) were examined for their ability to inhibit β-glucuronidase using D-saccharic acid 1,4-lactone (DSA), a well-known inhibitor of β-glucuronidase, as a positive control [24]. The results are displayed in Figure 2A and Table 1, which revealed that 3-(3,4-dihydroxyphenyl)-8-hydroxyisocoumarin (**1**) exhibited β-glucuronidase inhibitory activity, with 69.85% of inhibition at 100 μM and an $IC_{50}$ value of 58.83 ± 1.36 μM. Compound **2** with the absence of the C-3−C-4 double bond in the lactone ring did not exhibit β-glucuronidase inhibitory activity (15.89% of inhibition at 100 μM). Coumarin and its derivatives (**3−8**) failed to exhibit inhibitory activity against β-glucuronidase ($IC_{50}$ >100 μM). These results suggested that the different positions of an oxygen atom and carbonyl group in the isocoumarin structure, compared with those of coumarin, could play an important role in β-glucuronidase inhibition.

**Figure 2.** Inhibitory activity of isolated compounds **1−8** at 100 μM (**A**) and determination of $IC_{50}$ value by compound **1** on β-glucuronidase (**B**).

**Table 1.** Inhibition of compounds **1–8** against $\beta$-glucuronidase.

| Compounds | Inhibition against $\beta$-Glucuronidase | |
|:---:|:---:|:---:|
| | Inhibition% (100 μM) | IC$_{50}$ (μM) [1] |
| **1** | 69.85 ± 2.93 | 58.83 ± 1.36 |
| **2** | 15.89 ± 0.55 | >100 |
| **3** | 23.46 ± 2.47 | >100 |
| **4** | 29.61 ± 1.83 | >100 |
| **5** | 14.34 ± 1.10 | >100 |
| **6** | 13.17 ± 1.46 | >100 |
| **7** | 25.21 ± 1.46 | >100 |
| **8** | 16.73 ± 0.09 | >100 |
| DSA [2] | 74.70 ± 0.95 | 24.56 ± 1.15 |

[1] The values (μM) represent 50% inhibition of $\beta$-glucuronidase. Results are presented as the mean ± standard error of triplicate experiments. [2] Positive control.

### 2.3. Enzyme Kinetics of Compound **1** against $\beta$-Glucuronidase

To determine the type of inhibition and the inhibitory constant $K_i$ of active compound **1**, enzyme kinetics were conducted at different concentrations of the substrate 4-nitrophenyl $\beta$-D-glucuronide (PNPG) and inhibitor [7]. In the Lineweaver−Burk plot, the x-axis is the reciprocal of the substrate concentration, or 1/(S), and the y-axis is the reciprocal of the reaction velocity (1/V). The non-competitive or competitive inhibition mode is indicated by the family of straight lines that intersect at the same point on the 1/(S) or 1/V axis, respectively, whereas mixed inhibition is represented by the straight lines of the inhibitor that intersect at the xy region. As depicted in Figure 3A, the Lineweaver−Burk plot revealed intersecting lines on the 1/(S) axis, indicating that compound **1** inhibited $\beta$-glucuronidase in a non-competitive inhibition mode. In addition, a Dixon plot was used to determine the $K_i$ value between the inhibitor and the enzyme, where the intersection value on the x-axis implies $K_i$. The $K_i$ is the concentration of an inhibitor needed to decrease the maximum rate of the reaction by 50% [25]. As presented in Figure 3B, the $K_i$ value of **1** was calculated to be 40.9 μM.

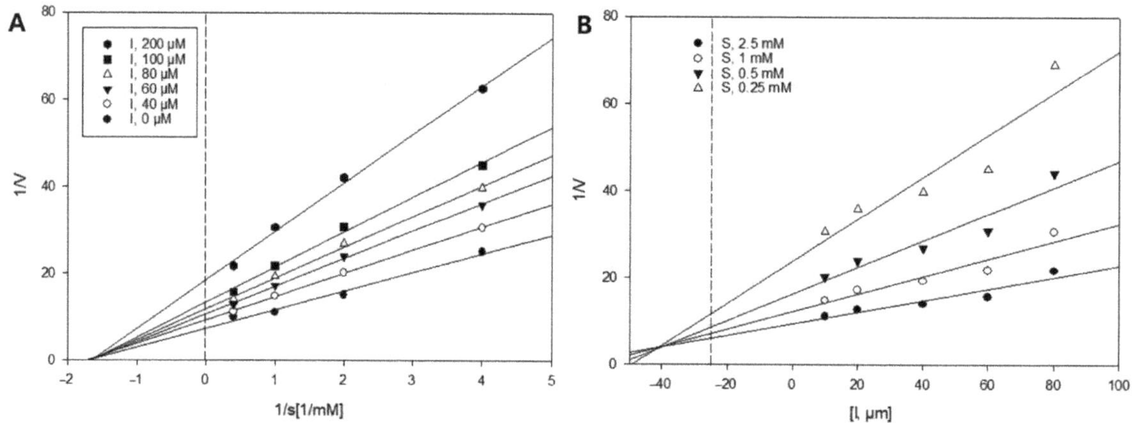


**Figure 3.** Lineweaver−Burk plot (**A**) and Dixon plot (**B**) analyses using active compound **1**.

### 2.4. Molecular Docking Studies

Molecular docking simulations were performed using AutoDock 4.2 to predict how compound **1** behaves in the binding site of the target protein and to clarify the fundamental biochemical processes of **1** with the $\beta$-glucuronidase enzyme. The results were analyzed and displayed using PyMOL and BIOVIA Discovery Studio (Figure 4). According to the kinetic analysis results, compound **1** displayed a non-competitive inhibition mode, suggesting that **1** could precisely bind to a specific region of $\beta$-glucuronidase, named the allosteric binding site, which differed from the active site (orange spheres) that binds to the substrate PNPG (red). The allosteric binding site of $\beta$-glucuronidase was predicted using the AlloSite 2.10 web server (green spheres) [26] and protein allosteric sites server PASSer2.0 (rainbow spheres) [27] (Figure 4A). To optimize the docking procedure, the substrate, PNPG, was docked as a native ligand into $\beta$-glucuronidase (PDB ID: 6LEL) (Figure 4B).

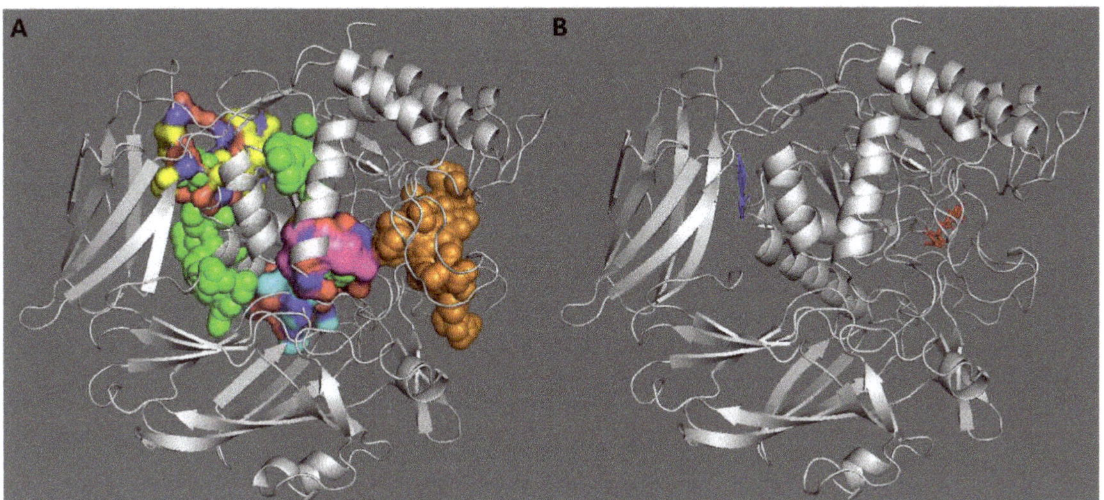

**Figure 4.** Predicted binding sites ((**A**); orange spheres: active site, green spheres: possible allosteric binding site predicted by AllositePro, and rainbow spheres: possible allosteric binding site predicted by PASSer2.0) and docking pose results of the substrate and compound **1** with the $\beta$-glucuronidase enzyme ((**B**); substrate: red; compound **1**: blue).

The docking results revealed that compound **1** could bind to the allosteric binding site of $\beta$-glucuronidase with a binding energy value of $-8.35$ kcal/mol (Figure 5 and Table 2). The carbonyl group and oxygen atom of **1** establish hydrogen bonding interactions with the amino acid residues, Thr191 and Thr438, whereas the hydroxy groups of **1** interact with Arg272, Leu435, and Asp436. Two benzene rings of **1** bind with residues Lys400 and Pro403 through $\pi-$alkyl interactions and Thr188 via $\pi-\sigma$ interactions. The lactone ring displayed $\pi-\sigma$ interactions with Val190, $\pi-$cation interactions with Arg439, and $\pi-$lone pair interactions with Val189. The other residues, including Gly270, Asn401, and Pro437, from different sites of the $\beta$-glucuronidase enzyme, bound to compound **1** via van der Waals interactions.

**Table 2.** Docking energies and binding site interactions of compound **1** with the $\beta$-glucuronidase enzyme.

| Comp. | Binding Energy (kcal/mol) | Hydrogen Bonds | van der Waals Interactions | Hydrophobic Interactions | Electrostatic Interactions | Others |
|---|---|---|---|---|---|---|
| **1** | $-8.35$ | Thr191<br>Arg272<br>Leu435<br>Thr438 | Gly270<br>Asn401<br>Pro437 | Thr188 ($\pi-\sigma$)<br>Val190 ($\pi-\sigma$)<br>Lys400 ($\pi-$alkyl)<br>Pro403 ($\pi-$alkyl) | Arg439 ($\pi-$cation) | Val189 ($\pi-$lone pair) |

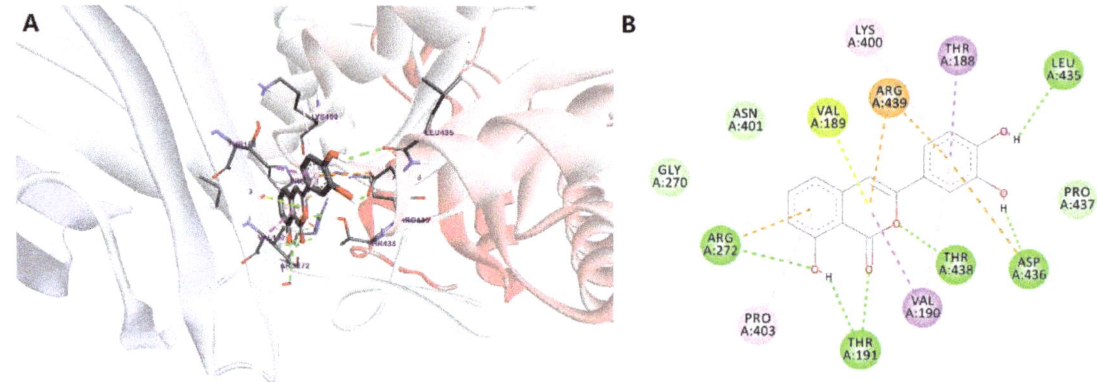

**Figure 5. The** 3D docking poses (**A**) and 2D interaction diagrams (**B**) of $\beta$-glucuronidase inhibition mediated by compound **1**. Green: hydrogen bonds, violet and pink: hydrophobic, orange: electrostatic, and light green: van der Waals interactions.

## 2.5. Molecular Dynamics Simulation of $\beta$-Glucuronidase Inhibition

To examine the structural stability and its variations of the **1**-6LEL complex, MD was performed after docking computations with a period of 100 ns, using the Desmond package (Schrödinger 2020-1, New York, NY, USA).

The general conditions of the simulation and its stabilization are obtained by root-mean-square deviation (RMSD) analysis [28]. The higher stability of the protein–ligand complex was demonstrated by a lower RMSD value, whereas decreased stability was indicated by an increased RMSD value [28]. The RMSD value of the **1**-6LEL complex increased rapidly during the initial equilibration fluctuation for 5 ns, slowly increasing from 5 to 38 ns, after which the RMSD was maintained between 2.2 and 2.4 Å, until the simulation was completed (Figure 6A). A maximum RMSD of 2.6 Å was observed for the target protein of $\beta$-glucuronidase, indicating that the **1**-6LEL complex maintained stability throughout the MD period. The flexibility and fluctuation of each residue in $\beta$-glucuronidase over the 100 ns simulation were represented by the root-mean-square fluctuation (RMSF) value used to predict the ligand binding-induced structural alterations in the protein structure [29]. Higher RMSF values represent greater flexibility in MD simulations [30]. The RMSF plot of the complex between compound **1** and $\beta$-glucuronidase is displayed in Figure 6B, where the peaks represent the $\beta$-glucuronidase regions that fluctuated the most during the 100 ns simulation. The RMSF values of amino acid residues in the allosteric site of $\beta$-glucuronidase were less than 1.3 Å. In contrast, the RMSF values of residues in the active site of $\beta$-glucuronidase fluctuate slightly more, between 2.0 and 3.5 Å, suggesting that the protein structure in ligand-bound conformations is stable during MD simulation. Hence, compound **1** may function as a non-competitive inhibitor of $\beta$-glucuronidase, which is consistent with the results of the kinetic study.

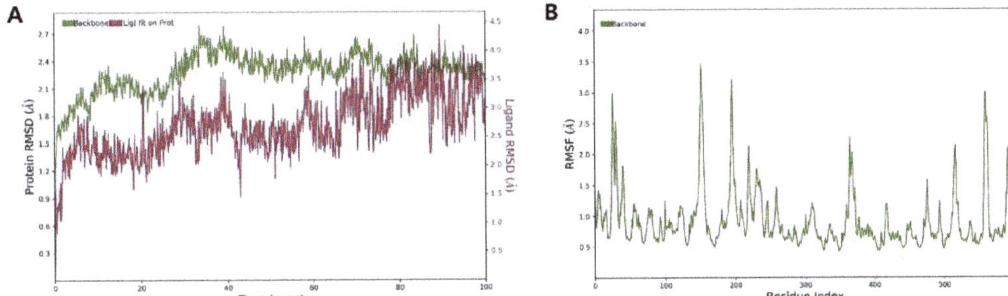

**Figure 6.** Molecular dynamics simulation of active compound **1** and β-glucuronidase (PDB ID: 6LEL) complex: RMSD ((**A**), protein RMSD: green line and RMSD of **1**: red line) and RMSF (**B**).

Figure 7 presents the protein–ligand contact diagram between compound **1** and β-glucuronidase. The hydroxy groups of **1** established hydrophobic contact with Leu435, polar interactions with Thr191 and His192, a positive charge with Lys400, and a negative charge with Asp436; these residues of the β-glucuronidase allosteric site contributed to the binding interactions of 24, 17, 13, 28, and 98%, respectively. In addition, the lactone ring of **1** was linked with Thr438 via polar interactions; Arg272 and Arg439 through π−cation interactions, accounting for 12 and 41% of the contribution, respectively. The MD results revealed the crucial role of the lactone ring in the β-glucuronidase inhibitory activity mediated by active compound **1**.

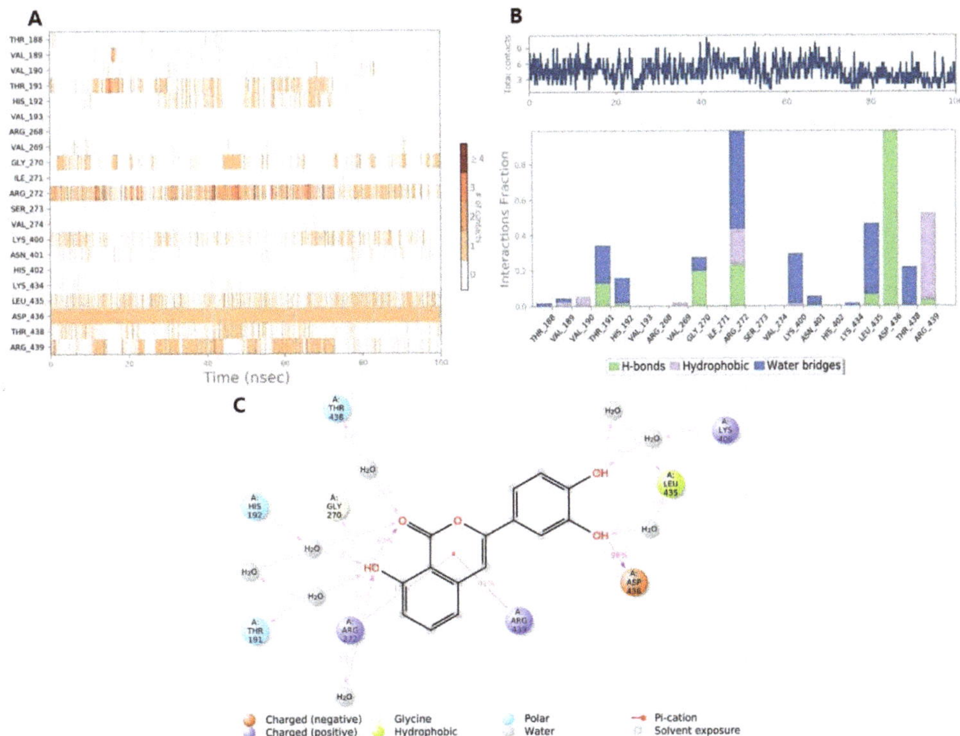



**Figure 7.** Protein–ligand contact analysis between compound **1** and β-glucuronidase complex ((**A**), timeline and (**B**), bar chart presentations) and 2D interaction diagram (**C**).

The ligand torsion plot that characterizes the conformational evolution of each rotatable bond (RB) in compound 1 during the 100 ns simulation trajectory is displayed in Figure 8A, where the 2D chemical structure of compound **1** with colored RB is followed by a same-colored dial and bar plots. A total of four RBs were observed in compound **1**, with the potential values of 3.36 kcal/mol and 7.16 kcal/mol for hydroxy groups and the linkage between the phenol ring and lactone ring, respectively.

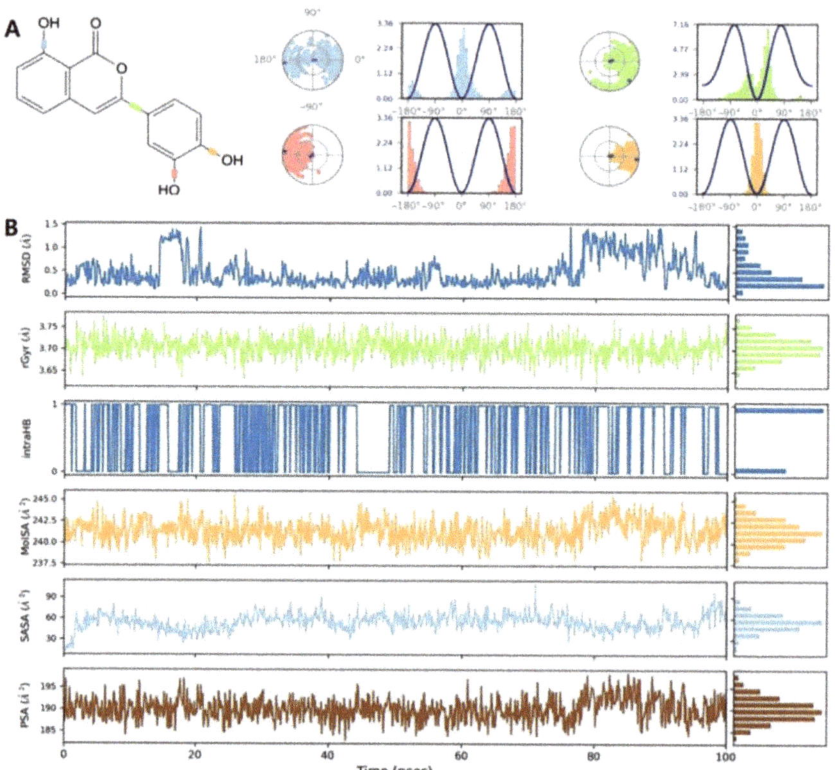

**Figure 8.** Ligand torsion profile (**A**) and ligand properties trajectory (**B**) of the 1-6LEL complex throughout the simulation trajectory (0 to 100 ns).

Considering ligand properties, we evaluated the ligand RMSD, the radius of gyration (rGyr), intramolecular hydrogen bonds (intraHB), molecular surface area (MolSA), solvent-accessible surface area (SASA), and polar surface area (PSA) (Figure 8B). The RMSD of compound **1**, with respect to the reference conformation, ranged from 0.2 to 1.5 Å, and its balance was approximated at 0.45 Å. rGyr was analyzed to examine the stability of compound **1** in the allosteric binding site of β-glucuronidase during the 100 ns simulation. The 1-6LEL complex exhibited an average rGyr value of 3.70 Å. No significant fluctuations were observed in rGyr, suggesting that the 1-6LEL complex exhibited steady behavior. IntraHB refers to the number of internal hydrogen bonds within the ligand. The constant intraHB value for ligand **1** indicated the consistency of **1** during the simulation process. The MolSA value was calculated using a 1.4 Å probe radius and was equivalent to the van der Waals surface area. The MolSA value for ligand **1** was slightly altered, from 237.5 to 245.0 Å, throughout the 100 ns MD simulation. The SASA value represents the surface area of a molecule that can be accessed by a water molecule. The SASA value of the 1-6LEL complex significantly increased from 25 to 60 Å until 4 ns of the MD simulation, followed by gradual stabilization at 56 Å. PSA is the SASA of a molecule, provided only by oxygen and nitrogen atoms. The PSA of compound **1** fluctuated at a consistent rate throughout the 100 ns MD simulation, while the PSA value ranged between 185 and 195 Å.

## 3. Discussion

All the secondary metabolites were isolated from *A. mono* Maxim. (**1**−**8**) and evaluated to determine their $\beta$-glucuronidase inhibitory activity. Compound **1** exhibited a $\beta$-glucuronidase inhibitory effect, with an $IC_{50}$ value of 58.83 µM, whereas the remaining compounds demonstrated no inhibitory activity against $\beta$-glucuronidase. The results suggest that the different positions of an oxygen atom and carbonyl group in the lactone ring of the isocoumarin structure, as well as the presence of a double bond between C-3 and C-4, could contribute to $\beta$-glucuronidase inhibition. Kinetic analysis indicated that active compound **1** displayed non-competitive inhibition. Thus, molecular docking studies were employed to determine the binding position and critical amino acid interactions of compound **1** with the allosteric binding site of the $\beta$-glucuronidase protein. The docking results indicated that compound **1** could tightly bind to the allosteric binding site of $\beta$-glucuronidase, with a binding energy of −8.35 kcal/mol. The lactone ring of compound **1** displays hydrogen bonds with Thr191 and Thr438, $\pi-\sigma$ interactions with Val190, $\pi-$cation interactions with Arg439, and $\pi-$lone pair interactions with Val189, which might explain their inhibitory activity against $\beta$-glucuronidase. In addition, the conformational stability of compound **1** complexed with the $\beta$-glucuronidase protein and its variations were investigated by performing MD simulation trajectories (100 ns). According to the MD results, all ligand properties fluctuated during the initial simulation period, gradually reaching equilibrium and a steady state as the simulation was completed. This indicated that compound **1** was stable in the allosteric binding site of $\beta$-glucuronidase. Based on the predicted pharmacokinetic properties (Supplementary Material), compound **1** exhibited no blood–brain barrier penetration and behaved as a safe drug candidate, given that it adhered to Lipinski's rule of five.

## 4. Materials and Methods

### 4.1. Experimental Procedures

$^{1}$H and $^{13}$C-NMR spectra were acquired on a Bruker Advance Digital 500 MHz instrument (Bruker, Karlsruhe, Germany). CC was conducted on silica gel 60 (230–400 mesh) and Cosmosil $C_{18}$ reversed phase gel (Nacalai Tesque, Kyoto, Japan). HPLC was conducted using a Waters HPLC system with a 2996 PDA detector (Waters, Milford, MA, USA). Thin-layer chromatography was conducted on pre-coated glass plates (silica gel 60 $F_{254}$ and RP-18 $F_{254s}$; Merck, Darmstadt, Germany). Plates were checked under ultraviolet radiation (254 and 365 nm), followed by spraying with sulfuric acid 10% and heating at 100–110 °C.

### 4.2. Chemicals and Reagents

HPLC solvents were supplied from Fisher Scientific Korea Ltd. (Seoul, Korea). *E. coli* $\beta$-glucuronidase enzyme (EC 3.2.1.31, G7396) and DSA (S0375) were provided by Sigma-Aldrich (St. Louis, MO, USA). PNPG (N0618) was provided by Tokyo Chemical Industry Co., Ltd. (Tokyo, Japan). All chemicals used in the experiments were provided by Duksan Pure Chemicals Inc. (Ansan, Gyeonggi, Korea).

### 4.3. Plant Material

Stems and branches of *A. mono* Maxim. were collected from the Medicinal Herb Garden of Daegu Catholic University in August 2021 and identified by Professor Byung Sun Min at the College of Pharmacy, Daegu Catholic University, Korea. A voucher specimen of *A. mono* Maxim. (23A-AM) was deposited at the Laboratory of Pharmacognosy, College of Pharmacy, Kyungpook National University, Korea.

*4.4. Extraction and Isolation*

Dried stems and branches of *A. mono* Maxim. (5.4 kg) were cut into small pieces and extracted with methanol (4 × 20 L) under reflux. The MeOH solvent was evaporated under reduced pressure to obtain the MeOH residue (350.0 g), which was subsequently suspended in $H_2O$ (3 L) and then partitioned with *n*-hexane, $CH_2Cl_2$, and EtOAc to obtain the *n*-hexane extract (35.2 g), $CH_2Cl_2$ extract (77.8 g), EtOAc extract (62.4 g), and water layer after removing the chemical solvents.

The $CH_2Cl_2$ extract was separated by CC on silica gel using a stepwise eluent of *n*-hexane–acetone (100:1–1:100, *v/v*) and $CH_2Cl_2$–MeOH (30:1–1:1, *v/v*) to yield seven fractions, 1A–G. Fraction 1D (8.1 g) was chromatographed on silica gel CC, eluted using a gradient mixture of $CH_2Cl_2$–acetone (100:1–1:100, *v/v*) and $CH_2Cl_2$–MeOH (20:1–1:1, *v/v*), to yield six subfractions, 1D1–6. Subfraction 1D2 (49.7 mg) was separated by RP-18 CC, using a mixture of acetone and $H_2O$ (1.5:1, *v/v*) as the eluent to obtain compound **3** (8.9 mg). Compounds **4** (5.8 mg) and **5** (2.0 mg) were purified from subfractions 1D3 (140.1 mg) and 1D4 (92.2 mg), respectively, by RP-18 CC elution with MeOH–$H_2O$ (1:1.5, *v/v*). Compounds **1** (4.9 mg) and **2** (8.9 mg) were isolated from subfraction 1D6 (223.2 mg) by HPLC using an isocratic mixture of MeOH and distilled $H_2O$ (40:60, *v/v*). Fraction 1E (9.5 g) was separated by silica gel CC, using $CH_2Cl_2$–acetone (gradient 100:1–1:100, *v/v*) and then $CH_2Cl_2$–MeOH (gradient 30:1–1:1, *v/v*) as the eluent to yield eight subfractions, 1E1−8. From subfraction 1E7 (239.2 mg), compounds **7** (44.5 mg) and **8** (17.4 mg) were isolated by HPLC, using 52% MeOH in distilled $H_2O$ as the eluent.

The water layer was chromatographed by Diaion CC using MeOH–$H_2O$ (gradient 0:1−1:0, *v/v*) as the eluent to yield three fractions, 2A−C. Fraction 2B (43.7 g) was chromatographed on silica gel and eluted with $CH_2Cl_2$–MeOH (gradient 100:1–1:100, *v/v*), followed by RP-18 CC using a MeOH–$H_2O$ mixture (1.5:1, *v/v*) as the mobile phase to obtain four fractions, 2B1−4. Compound **6** (7.2 mg) was isolated from fraction 2B1 (3.8 g) by HPLC, using an isocratic mixture of MeOH and $H_2O$ (50:50, *v/v*) as the eluent.

3-(3,4-Dihydroxyphenyl)-8-hydroxyisocoumarin (**1**), yellow needles, $^1$H-NMR (500 MHz, $CD_3OD$-$d_4$): $\delta_H$ 7.56 (1H, t, *J* = 7.9 Hz, H-6), 7.46 (1H, d, *J* = 2.0 Hz, H-2′), 7.32 (1H, d, *J* = 7.6 Hz, H-5), 7.11 (1H, dd, *J* = 8.2, 2.0 Hz, H-6′), 6.87 (1H, d, *J* = 8.1 Hz, H-7), 6.80 (1H, d, *J* = 7.6 Hz, H-5′), 6.46 (1H, s, H-4); $^{13}$C-NMR (125 MHz, $CD_3OD$-$d_4$): $\delta_C$ 167.9 (C-1), 144.1 (C-3), 108.6 (C-4), 116.6 (C-5), 137.8 (C-6), 116.4 (C-7), 158.3 (C-8), 109.9 (C-9), 143.8 (C-10), 126.9 (C-1′), 111.7 (C-2′), 146.5 (C-3′), 147.5 (C-4′), 117.7 (C-5′), 124.2 (C-6′).

*4.5. β-Glucuronidase Inhibition Assay*

The β-glucuronidase inhibition assay was evaluated as previously described [7].

*4.6. β-Glucuronidase Kinetics Assay*

The β-glucuronidase kinetics assay was performed as previously described [7].

*4.7. Molecular Docking*

Docking simulations were conducted using AutoDock 4.2, following our previously described protocol [7]. The crystallographic structure of β-glucuronidase was retrieved from the RCSB PDB website (PDB ID: 6LEL) [31]. The allosteric site of β-glucuronidase was predicted and generated using the AllositePro method provided by the AlloSite 2.10 web server and PASSer2.0 protein allosteric sites server [26,27].

*4.8. Molecular Dynamics Simulation*

MD simulations were performed using the Desmond package (Schrödinger 2020-1, New York, NY, USA). The protein–ligand complex was prepared in a 10.0 × 10.0 × 10.0 Å orthorhombic box (simple point-charge solvation model) [32]. Next, a 0.15 M NaCl solution and counter-ions were added to the system for neutralization. The solvated system was energy-minimized, and its position was restrained with the OPLS3e force field. The minimized system was implemented in an NPT ensemble at 300 K and 1 atm. Finally, the

MD simulation was conducted to run for 100 ns, and 1000 frames were generated, with a recording interval of 100 ps.

*4.9. Statistics*

All results are expressed as the mean $\pm$ standard error (SEM) of the three independent experiments. Statistical significance was analyzed using one-way analysis of variance (ANOVA) and Duncan's test ($p$-value < 0.05).

## 5. Conclusions

In the present study, we performed a chemical investigation of the stems and branches of *A. mono* Maxim., which resulted in the purification and structural elucidation of eight known compounds (two isocoumarins and six coumarins). To the best of our knowledge, our study is the first report that isolated isocoumarins from an *Acer* species. In addition, we, for the first time, determined the inhibitory activity of the isolated molecules against $\beta$-glucuronidase. Our results revealed that 3-(3,4-dihydroxyphenyl)-8-hydroxyisocoumarin (**1**) inhibited $\beta$-glucuronidase activity. The results of the kinetic analysis were consistent with the molecular docking studies, suggesting that compound **1** could bind to the $\beta$-glucuronidase allosteric site. This result was supported by MD studies up to 100 ns, which confirmed the binding stability of the protein–ligand complex in the trajectory analysis. These findings imply that the complex of $\beta$-glucuronidase and compound **1** is quite stable in biological systems. Moreover, the pharmacokinetic properties of compound **1** were analyzed, and the results suggested that compound **1** could be a promising drug candidate, given that no violations of the drug-likeness rules were observed.

**Supplementary Materials:** The following supporting information can be downloaded at: https://www.mdpi.com/article/10.3390/app122010685/s1, Figure S1. Bioavailability radar of compound **1**; Table S1. Drug-likeness properties of compound **1** [33–37].

**Author Contributions:** Conceptualization and methodology, S.Y.Y.; investigation, and data curation, N.V.P.; formal analysis, writing—original draft preparation, N.V.P. and S.Y.Y.; writing—review and editing, B.S.M. and J.A.K.; supervision, J.A.K. All authors have read and agreed to the published version of the manuscript.

**Funding:** This research was supported by the National Research Foundation of Korea (NRF) grant funded by the Korean government (MSIT) (No. NRF-2020R1A5A2017323 and NRF-2022R1C1C1004636).

**Institutional Review Board Statement:** Not applicable.

**Informed Consent Statement:** Not applicable.

**Data Availability Statement:** Not applicable.

**Conflicts of Interest:** The authors declare no conflict of interest.

## References

1. Bano, B.; Arshia; Khan, K.M.; Kanwal; Fatima, B.; Taha, M.; Ismail, N.H.; Wadood, A.; Ghufran, M.; Perveen, S. Synthesis, in vitro $\beta$-glucuronidase inhibitory potential and molecular docking studies of quinolines. *Eur. J. Med. Chem.* **2017**, *139*, 849–864. [CrossRef] [PubMed]
2. Taha, M.; Baharudin, M.S.; Ismail, N.H.; Selvaraj, M.; Salar, U.; Alkadi, K.A.A.; Khan, K.M. Synthesis and in silico studies of novel sulfonamides having oxadiazole ring: As $\beta$-glucuronidase inhibitors. *Bioorg. Chem.* **2017**, *71*, 86–96. [CrossRef] [PubMed]
3. Kaliannan, K.; Donnell, S.O.; Murphy, K.; Stanton, C.; Kang, C.; Wang, B.; Li, X.-Y.; Bhan, A.K.; Kang, J.X. Decreased tissue omega-6/omega-3 fatty acid ratio prevents chemotherapy-induced gastrointestinal toxicity associated with alterations of gut microbiome. *Int. J. Mol. Sci.* **2022**, *23*, 5332. [CrossRef] [PubMed]
4. Jain, S.; Drendel, W.B.; Chen, Z.-w.; Mathews, F.S.; Sly, W.S.; Grubb, J.H. Structure of human $\beta$-glucuronidase reveals candidate lysosomal targeting and active-site motifs. *Nat. Struct. Mol. Biol.* **1996**, *3*, 375–381. [CrossRef]
5. Wallace, B.D.; Wang, H.; Lane, K.T.; Scott, J.E.; Orans, J.; Koo, J.S.; Venkatesh, M.; Jobin, C.; Yeh, L.-A.; Mani, S.; et al. Alleviating cancer drug toxicity by inhibiting a bacterial enzyme. *Science* **2010**, *330*, 831–835. [CrossRef]
6. Jacobson, R.H.; Zhang, X.J.; DuBose, R.F.; Matthews, B.W. Three-dimensional structure of $\beta$-galactosidase from *E. coli*. *Nature* **1994**, *369*, 761–766. [CrossRef]

7. Phong, N.V.; Zhao, Y.; Min, B.S.; Yang, S.Y.; Kim, J.A. Inhibitory activity of bioactive phloroglucinols from the rhizomes of *Dryopteris crassirhizoma* on *Escherichia coli* β-glucuronidase: Kinetic analysis and molecular docking studies. *Metabolites* **2022**, *12*, 938. [CrossRef]

8. Weissmann, G.; Zurier, R.B.; Spieler, P.J.; Goldstein, I.M. Mechanisms of lysosomal enzyme release from leukocytes exposed to immune complexes and other particles. *J. Exp. Med.* **1971**, *134*, 149–165. [CrossRef]

9. Goldin, B.R.; Gorbach, S.L. The relationship between diet and rat fecal bacterial enzymes implicated in colon cancer. *J. Natl. Cancer Inst.* **1976**, *57*, 371–375. [CrossRef]

10. George, J. Elevated serum β-glucuronidase reflects hepatic lysosomal fragility following toxic liver injury in rats. *Biochem. Cell Biol.* **2008**, *86*, 235–243. [CrossRef]

11. Yang, J.; Tao, L.; Yang, J.; Wu, C.; Sun, W. Integrated conservation of *Acer yangbiense*: A case study for conservation methods of plant species with extremely small populations. *Plants People Planet* **2022**, in press. [CrossRef]

12. Bi, W.; Gao, Y.; Shen, J.; He, C.; Liu, H.; Peng, Y.; Zhang, C.; Xiao, P. Traditional uses, phytochemistry, and pharmacology of the genus *Acer* (maple): A review. *J. Ethnopharmacol.* **2016**, *189*, 31–60. [CrossRef] [PubMed]

13. Yang, H.; Lee, M.K.; Kim, Y.C. Protective activities of stilbene glycosides from *Acer mono* leaves against $H_2O_2$-induced oxidative damage in primary cultured rat hepatocytes. *J. Agric. Food. Chem.* **2005**, *53*, 4182–4186. [CrossRef] [PubMed]

14. Yang, H.; Sung, S.H.; Kim, Y.C. Two new hepatoprotective stilbene glycosides from *Acer mono* leaves. *J. Nat. Prod.* **2005**, *68*, 101–103. [CrossRef] [PubMed]

15. Bae, K.H. *Medicinal Plants of Korea*; Kyo-Hak Publishing Co.: Seoul, Korea, 2000.

16. Park, C.-H.; Son, H.-U.; Son, M.; Lee, S.-H. Protective effect of *Acer mono* Max. sap on water immersion restraint stress-induced gastric ulceration. *Exp. Ther. Med.* **2011**, *2*, 843–848. [CrossRef]

17. Adam, K.P.; Becker, H. Phenanthrenes and other phenolics from in vitro cultures of *Marchantia polymorpha*. *Phytochemistry* **1993**, *35*, 139–143. [CrossRef]

18. Myung, D.-B.; Han, H.-S.; Shin, J.-S.; Park, J.Y.; Hwang, H.J.; Kim, H.J.; Ahn, H.S.; Lee, S.H.; Lee, K.-T. Hydrangenol isolated from the leaves of *Hydrangea serrata* attenuates wrinkle formation and repairs skin moisture in uvb-irradiated hairless mice. *Nutrients* **2019**, *11*, 2354. [CrossRef]

19. Kim, B.G.; Lee, Y.; Hur, H.-G.; Lim, Y.; Ahn, J.-H. Production of three *O*-methhylated esculetins with *Escherichia coli* expressing *O*-methyltransferase from poplar. *Biosci. Biotechnol. Biochem.* **2006**, *70*, 1269–1272. [CrossRef]

20. Hu, H.-B.; Zheng, X.-D.; Jian, Y.-F.; Liu, J.-X.; Zhu, J.-H. Constituents of the root of *Anemone tomentosa*. *Arch. Pharm. Res.* **2011**, *34*, 1097. [CrossRef]

21. Garcez, F.R.; Garcez, W.S.; Martins, M.; Cruz, A.C. A bioactive lactone from *Nectandra gardneri*. *Planta Med.* **1999**, *65*, 775. [CrossRef]

22. Jin, W.; Thuong, P.T.; Su, N.D.; Min, B.S.; Son, K.H.; Chang, H.W.; Kim, H.P.; Kang, S.S.; Sok, D.E.; Bae, K. Antioxidant activity of cleomiscosins A and C isolated from *Acer okamotoanum*. *Arch. Pharm. Res.* **2007**, *30*, 275–281. [CrossRef] [PubMed]

23. Kumar, S.; Ray, A.B.; Konno, C.; Oshima, Y.; Hikino, H. Cleomiscosin D, a coumarino-lignan from seeds of *Cleome viscosa*. *Phytochemistry* **1988**, *27*, 636–638. [CrossRef]

24. Karunairatnam, M.C.; Levvy, G.A. The inhibition of β-glucuronidase by saccharic acid and the role of the enzyme in glucuronide synthesis. *Biochem. J.* **1949**, *44*, 599–604. [CrossRef] [PubMed]

25. Yung-Chi, C.; Prusoff, W.H. Relationship between the inhibition constant ($K_i$) and the concentration of inhibitor which causes 50 per cent inhibition ($I_{50}$) of an enzymatic reaction. *Biochem. Pharmacol.* **1973**, *22*, 3099–3108. [CrossRef]

26. Huang, W.; Lu, S.; Huang, Z.; Liu, X.; Mou, L.; Luo, Y.; Zhao, Y.; Liu, Y.; Chen, Z.; Hou, T.; et al. Allosite: A method for predicting allosteric sites. *Bioinformatics* **2013**, *29*, 2357–2359. [CrossRef]

27. Tian, H.; Jiang, X.; Tao, P. PASSer: Prediction of allosteric sites server. *Mach. Learn. Sci. Technol.* **2021**, *2*, 035015. [CrossRef] [PubMed]

28. Singh, A.; Kumar, P.; Sarvagalla, S.; Bharadwaj, T.; Nayak, N.; Coumar, M.S.; Giri, R.; Garg, N. Functional inhibition of c-Myc using novel inhibitors identified through "hot spot" targeting. *J. Biol. Chem.* **2022**, *298*, 101898. [CrossRef] [PubMed]

29. Dong, Y.-w.; Liao, M.-l.; Meng, X.-l.; Somero, G.N. Structural flexibility and protein adaptation to temperature: Molecular dynamics analysis of malate dehydrogenases of marine molluscs. *Proc. Natl. Acad. Sci. USA* **2018**, *115*, 1274–1279. [CrossRef]

30. Zhao, Y.; Zeng, C.; Massiah, M.A. Molecular dynamics simulation reveals insights into the mechanism of unfolding by the A130T/V mutations within the MID1 zinc-binding Bbox1 domain. *PLoS ONE* **2015**, *10*, e0124377. [CrossRef]

31. Lin, H.-Y.; Chen, C.-Y.; Lin, T.-C.; Yeh, L.-F.; Hsieh, W.-C.; Gao, S.; Burnouf, P.-A.; Chen, B.-M.; Hsieh, T.-J.; Dashnyam, P.; et al. Entropy-driven binding of gut bacterial β-glucuronidase inhibitors ameliorates irinotecan-induced toxicity. *Commun. Biol.* **2021**, *4*, 280. [CrossRef]

32. Berendsen, H.J.C.; Grigera, J.R.; Straatsma, T.P. The missing term in effective pair potentials. *J. Phys. Chem.* **1987**, *91*, 6269–6271. [CrossRef]

33. Zrieq, R.; Ahmad, I.; Snoussi, M.; Noumi, E.; Iriti, M.; Algahtani, F.D.; Patel, H.; Saeed, M.; Tasleem, M.; Sulaiman, S.; et al. Tomatidine and patchouli alcohol as inhibitors of SARS-CoV-2 enzymes (3CLpro, PLpro and NSP15) by molecular docking and molecular dynamics simulations. *Int. J. Mol. Sci.* **2021**, *22*, 10693. [CrossRef]

34. Lipinski, C.A.; Lombardo, F.; Dominy, B.W.; Feeney, P.J. Experimental and computational approaches to estimate solubility and permeability in drug discovery and development settings. *Adv. Drug. Deliv. Rev.* **2012**, *64*, 4–17. [CrossRef]

35. Daina, A.; Michielin, O.; Zoete, V. SwissADME: A free web tool to evaluate pharmacokinetics, drug-likeness and medicinal chemistry friendliness of small molecules. *Sci. Rep.* **2017**, *7*, 42717. [CrossRef] [PubMed]
36. Feng, Z.; Cao, J.; Zhang, Q.; Lin, L. The drug likeness analysis of anti-inflammatory clerodane diterpenoids. *Chin. Med.* **2020**, *15*, 126. [CrossRef]
37. Daina, A.; Zoete, V. A BOILED-Egg to predict gastrointestinal absorption and brain penetration of small molecules. *ChemMedChem* **2016**, *11*, 1117–1121. [CrossRef]

Review

# Volatiles in Berries: Biosynthesis, Composition, Bioavailability, and Health Benefits

Inah Gu, Luke Howard and Sun-Ok Lee *

Department of Food Science, University of Arkansas, Fayetteville, AR 72704, USA
* Correspondence: sunok@uark.edu; Tel.: +1-479-575-6921

**Abstract:** Volatile compounds in fruits are responsible for their aroma. Among fruits, berries contain many volatile compounds, mainly esters, alcohols, terpenoids, aldehydes, ketones, and lactones. Studies for volatile compounds in berries have increased extensively as the consumption of berry products rapidly increased. In this paper, we reviewed biosynthesis and profiles of volatiles in some berries (strawberry, blueberry, raspberry, blackberry, and cranberry) and their bioavailability and health benefits, including anti-inflammatory, anti-cancer, anti-obesity, and anti-diabetic effects in vitro and in vivo. Each berry had different major volatiles, but monoterpene had an important role in all berries as aroma-active components. Volatile compounds were nonpolar and hydrophobic and rapidly absorbed and eliminated from our body after administration. Among them, monoterpenes, including linalool, limonene, and geraniol, showed many health benefits against inflammation, cancer, obesity, and diabetes in vitro and in vivo. More research on the health benefits of volatile compounds from berries and their bioavailability would be needed to confirm the bioactivities of berry volatiles.

**Keywords:** berry volatiles; biosynthesis; chemical composition; bioavailability; health benefits

**Citation:** Gu, I.; Howard, L.; Lee, S.-O. Volatiles in Berries: Biosynthesis, Composition, Bioavailability, and Health Benefits. *Appl. Sci.* **2022**, *12*, 10238. https://doi.org/10.3390/app122010238

Academic Editors: Luca Mazzoni, Maria Teresa Ariza Fernández and Franco Capocasa

Received: 30 August 2022
Accepted: 5 October 2022
Published: 12 October 2022

**Publisher's Note:** MDPI stays neutral with regard to jurisdictional claims in published maps and institutional affiliations.

**Copyright:** © 2022 by the authors. Licensee MDPI, Basel, Switzerland. This article is an open access article distributed under the terms and conditions of the Creative Commons Attribution (CC BY) license (https://creativecommons.org/licenses/by/4.0/).

## 1. Introduction

Berries, one of the most common fruits in the human diet (strawberry, blueberry, red raspberry, black raspberry, blackberry, and cranberry in the United States), are rich in minerals, vitamins, dietary fibers, and especially polyphenols and volatiles [1–3]. Volatile compounds in berries are responsible for the unique aroma of berries [4]. Fruit volatile compounds are mainly comprised of diverse chemicals, including esters, alcohols, terpenoids, aldehydes, ketones, and lactones [5]. The volatile composition of berries is complex and different by many factors, including the cultivar, ripeness, pre- and post-harvest storage conditions, fruit samples, temperature, and experimental conditions [6–11]. Blueberries (*Vaccinium ashei*) showed linalool increasing and α-terpineol and β-caryophyllene decreasing during the maturation of the blueberry [6]. Full-red harvested strawberries contained more volatile compounds than ¾-red harvested strawberries, regardless of the storage duration [12]. Raspberries (raw, frozen, or frozen for a year) were examined to compare the long-term frozen storage. The changes in volatile composition during long-term frozen storage were negligible except for an increase in α-ionone and caryophyllene [8].

Volatile compounds are small and light molecules (below 250–300 Da) with low polarity and high vapor pressure [13]. Plants synthesize and release volatile compounds to communicate and interact with environments, compensating for the immobility of plants [14]. Volatile compounds play an important role in pollination by attracting pollinators, protecting from pathogens and herbivores, and even communicating with inter- and intra-plants [15]. Volatiles can be divided into primary compounds and secondary compounds. Primary compounds are synthesized during maturation by anabolic or catabolic pathways of the plant [16]. Secondary volatile compounds are produced from tissue disruption by autoxidation or enzyme catalyzing reactions [17–19].

A mixture of many different volatile compounds makes a unique aroma. Although a lot of compounds were found as volatile compounds in fruits, only a few compounds have been identified as aroma compounds of fruit flavor based on their quantitative abundance and olfactory thresholds [20]. With the increasing consumption of berries and berry products such as fruits, juice, puree, jams, and other berry ingredients, there have been many studies conducted about identifying aroma volatile compounds of berries for developing consumer acceptability [21] and the studies about the health beneficial effect of berries, especially berry polyphenols. Berry polyphenols are composed of flavonoids, phenolic acids, tannins, stilbenes, lignans, and others and have been shown to possess many health effects, such as antioxidant, anti-inflammatory, and anti-cancer activities [22–24]. Unlike berry polyphenols, although there have been extensive analyses of volatile berry composition, there is still a very limited number of studies on the bioavailability and health benefits of berry volatiles. Recently, volatile compounds in plants have been reported to have health-promoting activities such as anti-inflammatory effects [25,26]. Many review articles mainly focused on the composition of berry volatiles and affecting factors such as different locations, ripeness, cultivar/genotypes, harvest and storage conditions, etc., but there is a lack of information about the bioavailability and biological activities of berry volatiles in our body. In this article, the biosynthesis of volatiles in plants, the chemical composition of some berries commonly consumed in the U.S. (blackberry, blueberry, cranberry, raspberry, and strawberry), bioavailability, and the health benefits of volatile compounds that are rich in berries were reviewed.

## 2. Biosynthesis of Volatiles in Plants

Plant volatiles can be grouped into terpenoids, phenylpropanoids/benzenoids, and fatty acid derivatives based on chemical structure and biochemical synthesis [14,27,28]. Terpenes/terpenoids are the most abundant and diverse family of secondary plant metabolites and essential oils, including more than 30,000 compounds [29,30]. Terpenes are classified by the amount of C5 isoprene units in the structure: isoprene (C5), monoterpenes (C10), sesquiterpenes (C15), diterpenes (C20), and so on, with polyterpenes (C5n where n can be ~30,000) [30]. Monoterpenes and sesquiterpenes are the most abundant terpenes found in essential oils [31]. Although they have diverse chemical structures, they all share common biosynthesis pathways, and they are synthesized in all parts of the plants, such as the leaves, fruits, flowers, stems, and roots [15]. Since terpenes/terpenoids are the major class of berry volatiles, we focused more on the biosynthesis of terpenes, such as monoterpenes and sesquiterpenes, in this review.

### 2.1. Biosynthesis of Terpenes

All terpenes are synthesized from two universal precursors (C5), isopentenyl diphosphate (IPP) and dimethylallyl pyrophosphate (DMAPP) [32]. A series of condensation reactions of IPP and DMAPP produce prenyl diphosphates, which are precursors of terpenes. Condensation of IPP and DMAPP produces geranyl diphosphate (GPP, C10), the precursor of monoterpenes, by catalyzation of geranyl diphosphate synthase (GDS). Two IPPs and a DMAPP are condensed to the precursor of sesquiterpenes, farnesyl diphosphate (FPP, C15), by farnesyl diphosphate synthase (FDS) [33]. Three IPPs and a DMAPP also produce geranylgeranyl pyrophosphate (GGPP, C20), the precursor of diterpenes. Geranyl-farnesyl diphosphate (GFPP, C25) is the recently discovered precursor of sesterterpenes [33]. These prenyl precursors of terpenes are converted to terpenes (iso-, mono-, sesquiterpenes, and so on) by terpene synthases (TPS). Monoterpenes (C10) are one of the major groups of terpenes in essential oils and berries [34]. During the conversion of prenyl precursors to monoterpenes, many different catalytic reactions, including hydroxylation, oxidation, reduction, acetylation, methylation, glycosylation, isomerization, conjugation, and others, occur to modify the structure of monoterpenes and produce many different compounds of terpenes [30,35,36].

### 2.2. MVA and MEP Pathways

The two precursors of all terpenes, IPP and DMAPP, are generated from two different pathways in different subcellular compartments: the mevalonic acid (MVA) pathway in the cytosol and the 2-methylerythritol 4-phosphate (MEP) pathway (1-deoxy-xylulose-5-phosphate (DOXP) pathway) in plastids (Figure 1) [37]. The MVA pathway generates IPP from acetyl-CoA, while the MEP pathway produces IPP and DMAPP from pyruvate and glyceraldehyde-3-phosphate (GA-3P) [38,39]. In the MVA pathway, there are six enzymatic reactions to generate IPP. Three acetyl-CoA are sequentially condensed, reduced to mevalonate (MVA), then form IPP through two phosphorylations and decarboxylation by mevalonate kinase, phosphomevalonate kinase, and mevalonate diphosphate decarboxylase, respectively [40,41]. Produced IPP in the MVA pathway further forms its allylic isomer, DMAPP, by isopentenyl diphosphate isomerase [42]. In the MEP pathway, seven enzymatic actions are involved in generating IPP and DMAPP [43]. Condensation of pyruvate and GA-3P generates DOXP, and DOXP synthesizes MEP by DOXP reductoisomerase (DXR). Further transformations form 1-hydroxy-2-methyl-2-(E)-butenyl 4-diphosphate (HMBPP) and (E)-4-hydroxy-3-methlbut-2-enyl diphosphate reductase (HDR) catalyzes the conversion of HMBPP to IPP and DMAPP [44]. MVA pathway mainly synthesizes sesquiterpenes, which account for about 28% of total flower terpenes, while the MEP pathway synthesizes more monoterpenes and diterpenes, accounting for about 53% and 1% of total flower terpenes, respectively [45,46]. However, metabolic crosstalk exists between the MVA and MEP pathways, especially from plastids to cytosol [47–49].

### 2.3. Biosynthesis of Other Plant Volatiles

#### 2.3.1. Phenylpropanoids/Benzenoids

Phenylpropanoids/benzenoids are synthesized from the shikimate pathway [31]. The shikimate pathway starts with the condensation of phosphoenolpyruvate and erythrose 4-phosphate [51]. Chorismic acid is formed by the elimination of ring alcohol from shikimic acid, and this forms the phenylpropionic acid skeleton [29]. Cinnamic acid and p-hydroxycinnamic acid undergo many enzymatic reactions to produce volatile compounds, including eugenol and benzyl benzoate [52].

#### 2.3.2. Volatile Fatty Acid Derivatives

Fatty acid-derived volatiles are synthesized via the lipoxygenase pathway [53]. C18 fatty acids, including linoleic acid and linolenic acid, are catalyzed by lipoxygenases and generate 9-hydroperoxy and 13-hydroperoxy derivatives of fatty acids [54]. These two intermediates turn into fatty acid derivatives, including methyl jasmonate and green leaf volatiles. The lipoxygenase pathway also can synthesize oxylipins, isoprene, carotenoid derivatives, indoles, phenolics, methyl salicylate, and aromatic volatile organic compounds [55,56].

### 2.4. Application of Plant Volatile Biosynthesis

Volatile compounds that are emitted from plants have an important role in many different functions, such as pollinator attraction, direct and indirect defenses against herbivores, insects, and microorganisms, and communication between and within plants [27,57,58]. In addition, natural volatile compounds such as methyl jasmonate, allyl isothiocyanate, and tea tree oil have been used for modulating volatile biosynthesis and controlling the pre- and post-harvest quality of berries [9,59]. Those volatile compounds also suppressed the decay in strawberries and blackberries stored at 10 °C [60]. Sangiorgio et al. found a positive correlation between *Lactobacillus*, *Paenibacillus* spp., and norisoprenoids and a negative correlation between Enterobacteriaceae and monoterpenes [61]. From these results, the raspberry microbiome can be selectively chosen for the overall better quality of fruit, including its aroma, shelf-life, and safety [61]. Although there is still more research on the mechanisms required, accumulated results in metabolomic and genomic approaches can be used for making advances in fruit ripeness, quality, and consumer acceptability [62–64].

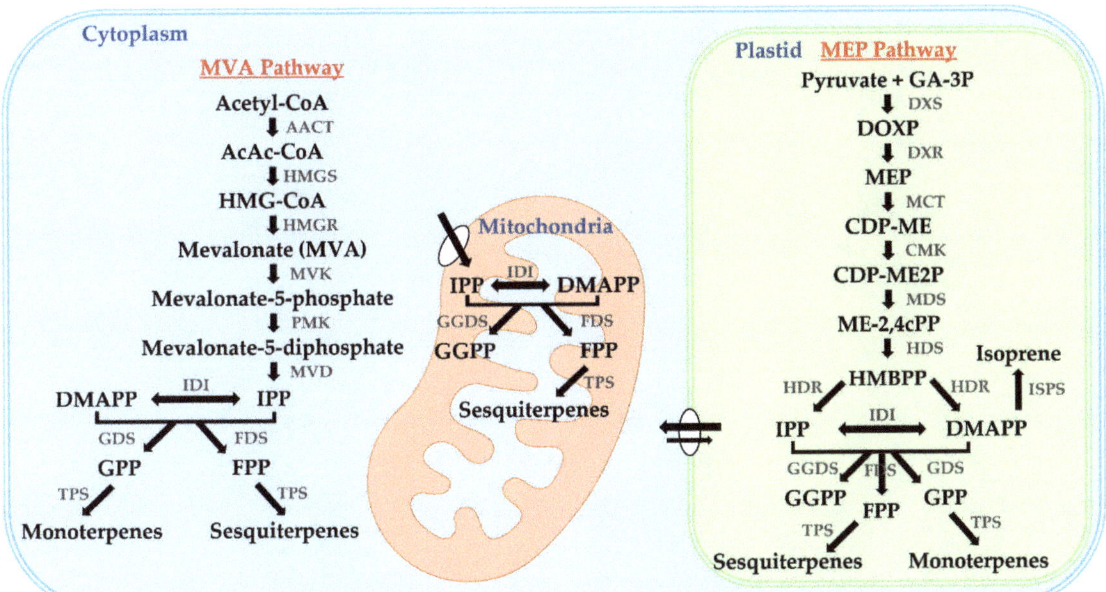

**Figure 1.** Biosynthetic pathways of volatile terpenes in plants (adopted and modified from Nagegowda [50]). AACT = acetoacetyl-CoA thiolase; AcAc-CoA = acetoacetyl-CoA; CDP-ME = 4-(cytidine 50-diphospho)-2-C-methyl-D-erythritol; CDP-ME2P = 4-(cytidine 50-diphospho)-2-C-methyl-D-erythritol phosphate; CMK = CDP-ME kinase; DMAPP = dimethylallyldiphosphate; DOXP = 1-deoxy-D-xylulose 5-phosphate; DXR = DOXP re-ductoisomerase; DXS = DOXP synthase; FDS = farnesyl diphosphate synthase; FPP = farnesyl diphosphate; GA-3P = glyceraldehyde-3-phosphate; GDS = geranyl diphosphate synthase; GGDS = geranyl geranyl diphosphate synthase; GGPP = geranyl geranyl di-phosphate; GPP = geranyldiphosphate; HDR = (E)-4-hydroxy-3-methylbut-2-enyl di-phosphate reductase; HDS = (E)-4-hydroxy-3-methylbut-2-enyl diphosphate synthase; HMBPP = (E)-4-hydroxy-3-methylbut-2-enyl diphosphate; HMG-CoA = 3-hydroxy-3-methylglutaryl-CoA; HMGR = HMG-CoA reductase; HMGS = HMG-CoA synthase; IDI = isopentenyl diphosphateisomerase; IPP = isopentenyl diphosphate; ISPS = isoprene synthase; MCT = 2-C-methyl-D-erythritol 4-phosphate cytidylyltransferase; MDS = 2-C-methyl-D-erythritol 2,4-cyclodiphosphate synthase; ME-2,4cPP = 2-C-methyl-D-erythritol 2,4-cyclodiphosphate; MEP = 2-C-methyl-D-erythritol 4-phosphate; MVD = mevalonate diphosphate decar-boxylase; MVK = mevalonate ki-nase; PMK = phosphomevalonate kinase; TPS=terpene synthase. Names of the enzymes are in gray.

## 3. The Chemical Composition of Volatile Compounds in Berries

The major volatile compounds identified from five common berries in the U.S. (strawberry, blueberry, raspberry, blackberry, and cranberry) were summarized (Table 1).

**Table 1.** The major volatile compounds in berries.

| Compound | Strawberry | Blueberry | Raspberry | Blackberry | Cranberry |
|---|---|---|---|---|---|
| *Esters* | | | | | |
| Methyl butanoate | [65,66] | | | | |
| Methyl hexanoate | [65–67] | | | | |
| Ethyl acetate | [65] | [68] | [69] | [70] | |
| Hexyl acetate | [65] | | | | |
| Ethyl butanoate | [66,69] | | | [71] | |

**Table 1.** *Cont.*

| Compound | Strawberry | Blueberry | Raspberry | Blackberry | Cranberry |
|---|---|---|---|---|---|
| Ethyl hexanoate | [66,72] | | | [73] | |
| Ethyl 2-methylbutanoate | [67] | [74] | | [71,73] | [75] |
| Ethyl 2-methylpropanoate | [67] | | [76] | | |
| 3-methylbutylacetate | [67] | | | | |
| Hexyl butanoate | [69] | | | | |
| Ethyl propanoate | | [74] | | | |
| Methyl 2-methylbutanoate | | [74] | | | |
| Methyl 3-methylbutanoate | | [74] | | | |
| Ethyl 3-methylbutanoate | | [74] | | | |
| (Z)-3-hexyl acetate | | [74] | | | |
| (E)-2-hexyl acetate | | [74] | | | |
| Geranyl acetate | | [74] | | | |
| 3-cis-hexenyl formate | | | | | [77] |
| Ethyl benzoate | | | | | [69] |
| *Ketones* | | | | | |
| 2-heptanone | [67] | [74,78] | [79] | [73] | |
| 2,3-butanedione | [67] | | | [80] | |
| 1-octen-3-one | | [74] | | | |
| 2-nonanone | | [74] | | | |
| 6-methyl-5-hepten-2-one | | [74] | | | |
| Raspberry ketone | | | [76] | | |
| β-damascenone | | | [76] | [71] | |
| 3-hydroxy-2-butanone | | | | [80] | |
| 2-undecanone | | | | [73] | |
| Isophorone | | | | [69] | |
| *Terpenes* | | | | | |
| Limonone | [65] | [74] | | [69,73] | |
| α-terpinene | [65] | | | | |
| Linalool | [65,66] | [69,74,78] | [67,76] | [71,73] | [69] |
| Nerolidol | [66] | | | | |
| Myrtenol | [69] | | [69] | [69,73] | |
| Geraniol | | [68,74] | [76,79] | [71] | |
| Citronellol | | [74,81] | | | [69] |
| α-terpineol | | [69,74,81] | [69,76] | [70] | [69,77,82,83] |
| Nerol | | [74] | [76] | | |
| Eucalyptol (1,8-cineolo) | | [74,78] | | | [69,83] |
| Dihydrolinalool oxide | | [74] | | | |
| δ-elemene | | [78] | | | |
| (E)-caryophyllene | | [78] | | | |
| Caryophyllene oxide | | [78] | | | |
| Linalool oxide | | [69] | | | [83] |
| β-ionone | | [69] | [76,79] | [73] | [75] |
| α-ionone | | | [69,76,79] | [73] | |
| β-pinene | | | [76] | | |
| Terpinen-4-ol | | | [79] | [69] | |
| α-pinene | | | | [73] | |
| p-cymene | | | | [73] | |
| Sabinene | | | | [73] | |
| *Acids* | | | | | |
| Butanoic acid | [69] | | | [69] | [77] |
| Hexanoic acid | | [78] | [69] | [69,73] | |
| Octanoic acid | | [78] | | | |
| Nonanoic acid | | [78] | | | |
| Decanoic acid | | [78] | | | |
| 3-methylbutanoic acid | | | [69] | [73] | |
| 2-methylbutanoic acid | | | | [73,80] | [69,75] |
| Acetic acid | | | | [73] | |

**Table 1.** *Cont.*

| Compound | Strawberry | Blueberry | Raspberry | Blackberry | Cranberry |
|---|---|---|---|---|---|
| *Alcohols* | | | | | |
| Cis-3-hexen-1-ol | [65] | | | | |
| (E)-2-hexenol | | [68] | | | |
| (Z)-3-hexenol | | [68] | [76] | | |
| 2-phenylethanol | | [81] | | | |
| (Z)-3-hexenol | | [74] | | | |
| 2-heptanol | | [74] | | [73] | |
| Phenylethyl alcohol | | [69] | | | |
| 2-ethylhexanol | | [69] | [69] | [69] | |
| 1-octanol | | | [76] | [73,80] | |
| (Z)-hexenol | | | [79] | | |
| Ethanol | | | | [73,80] | |
| 1-hexanol | | | | [73,80] | |
| p-cymen-8-ol | | | | [73] | |
| Nopol | | | | [73] | |
| 4-methyl-1-pentanol | | | | [69] | |
| 4-penten-2-ol | | | | | [77] |
| Benzyl alcohol | | | | | [77,82,84] |
| *Aldehydes* | | | | | |
| Hexanal | [65,85] | [68,74,78] | [76,79] | [73] | [75] |
| Trans-2-hexenal | [65] | | | [73,80] | |
| Cis-3-hexenal | [67] | | | | |
| (E)-2-hexenal | | [68,74,78] | [76,79] | | [75] |
| Vanilllin | | [81] | [69] | | |
| (Z)-3-hexenal | | [74,78] | | | |
| (E,Z)-2,6-nonadienal | | [74] | | | |
| (E,E)-2,4-nonedienal | | [74] | | | |
| Pentanal | | [74] | | | [75] |
| Octanal | | [74] | | | [75] |
| (E,E)-2,4-hexadienal | | [74] | | | |
| Decanal | | [74] | | | |
| Hexadienal | | [78] | | | |
| Heptanal | | [78] | [76] | | |
| Benzaldehyde | | | [76] | | [77] |
| cuminaldehyde | | | [69] | | |
| Methylbutanal | | | | [80] | |
| Methional | | | | [71] | |
| Trans,cis-2,6-nonadienal | | | | [71] | |
| 3-methylbutanal | | | | [73] | |
| 2-methylbutanal | | | | [73] | |
| (E)-2-heptenal | | | | | [75] |
| (E)-2-octenal | | | | | [75] |
| (E)-2-nonenal | | | | | [75] |
| Trans-2-decanal | | | | | [83] |
| 2-octanal | | | | | [83] |
| *Surfurs* | | | | | |
| methanethiol | [66] | | | | |
| *Norisoprenoids* | | | | | |
| β-damascenone | | [78] | | | |
| Cis-1,5-octadien-3-one | | | | [71] | |
| *Furanones* | | | | | |
| Furaneol | [65,66] | [74] | [76] | [71] | |
| Mesifurane | [65–67,69] | | | | |

**Table 1.** *Cont.*

| Compound | Strawberry | Blueberry | Raspberry | Blackberry | Cranberry |
|---|---|---|---|---|---|
| *Hydrocarbons* | | | | | |
| Octane | | [78] | | | |
| Ethyl benzene | | [78] | | | |
| p-xylene | | [78] | | | |
| Mxylene | | [78] | | | |
| B-ocimene | | | [76] | | |
| *Lactones* | | | | | |
| γ-decalactone | [65] | | | | |
| Butyrolactone | | [78] | | | |
| δ-octalactone | | | [79] | | |
| δ-decalactone | | | [79] | | |

### 3.1. Strawberry

Strawberries (*Fragaria* spp.) are the most consumed berry fruit for their sweet taste and unique aroma [86]. The consumption of strawberry products, such as jams, juices, and puree, has significantly increased [87,88]. Strawberries are rich in volatile compounds responsible for the strawberry flavor and aroma [89,90]. Volatile compounds in strawberries have been extensively studied, and more than 360 volatile compounds have been identified [91]. These compounds include esters, which were qualitatively and quantitatively dominant, terpenes, furanones, sulfur, lactones, alcohols, and carbonyls. In the study of Lu et al., a total of 42 volatiles were detected, with 19 esters, 10 alcohols, and 6 terpenes being the most abundant in the strawberry samples analyzed [65].

Esters are the most abundant and major aroma volatiles affecting the aroma of strawberries [92]. Among 19 ester compounds, methyl butanoate, methyl hexanoate, ethyl acetate, and hexyl acetate were the major ester compounds [65]. Terpenes are important compounds to the flavor and possess many preferable aroma profiles [93]. Among six terpenes, limonene and α-terpinene were the main compounds in strawberries [65]. Although strawberries contained many alcohol compounds, alcohols did not affect the strawberry flavor notably. In addition, c-decalactone (peach-like aroma) contributes significantly to the strawberry flavor. Hexanal, trans-2-hexenal, and cis-3-hexen-1-ol are responsible for the green, unripe aroma of the strawberry. The furanones 2,5-dimethyl-4-hydroxy-3(2H)-furanone (furaneol) and 2,5-dimethyl-4-methoxy-3(2H)-furanone (mesifurane), which have the sweet, floral, and fruity aroma, are major furans found in strawberries [65].

Summarizing some strawberry volatile studies, volatile components that are consistently considered important aroma compounds include ethyl butanoate, ethyl hexanoate, methyl butanoate, ethyl 3-methylbutanoate, fureneol, and linalool [66,85,94]. Additional compounds that have been reported include 2-heptanone, mesifurane, cis-3-hexenal, ethyl 2-methylpropanoate, 2,3-butanedione, 3-methylbutyl acetate, methyl hexanoate, and ethyl 2-methylbutanoate [67]. Eight different strawberry varieties (Sabrosa, Albion, Sweet Ann, Festival, Fortuna, Ventana, Camarosa, and Rubygem) were examined to investigate the volatile composition [72]. With headspace solid-phase micro-extraction gas chromatography-mass spectrometry (HS-SPME/GC-MS), esters (31 volatiles), especially ethyl hexanoate were detected as major compounds [72]. The most detected compounds were esters (31 compounds), which give fruity and floral characteristics to strawberries [72]. The SPME/GC-MS is one of the most common and effective methods for volatile analysis. SPME fiber samples the volatile compounds from air and thermally desorbs the sample in the injection port of a GC system [95,96]. In Gu et al., 55 volatiles were found in the strawberry extract, with monoterpene being the predominant volatiles (43% of total volatile concentration) [69]. It was followed by acids, esters, furan, aldehydes, alcohol, ketones, and alkylbenzene. Predominant compounds in strawberry extract were myrtenol, butanoic acid, mesifuran, ethyl butanoate, and hexyl butanoate.

*3.2. Blueberry*

Blueberries (*Vaccinium* spp.) are the second most popular berry fruit in the U.S. after strawberries [97]. The blueberry market has increased 10–20% annually over the last 5 years [98]. The increased blueberry consumption is due to its well-known health benefits and flavor [99]. Highbush blueberries mainly contained ethyl acetate, (E)-2-hexenal, (E)-2-hexenol, hexanal, (Z)-3-hexenol, linalool, and geraniol [68]. Others, including citronellol, α-terpineol, 2-phenylethanol, and vanillin, were also considered to have the highbush blueberry aroma [81]. A total of 38 aroma volatiles were detected from southern highbush blueberry [74]. There were nine aldehydes, eight esters, seven terpenes, five ketones, two alcohols, two acids, two sulfurs, and three miscellaneous compounds. Aldehydes were the most abundant chemical group within southern highbush blueberry and a major volatile compound group to the blueberry aroma. Hexanal, (Z)-3-hexenal, (E)-2-hexenal, (E,Z)-2,6-nonadienal, and (E,E)-2,4-nonedienal had "fresh green", "grassy", and "fruity" aroma characteristics, whereas pentanal, octanal, (E,E)-2,4-hexadienal, and decanal had "fatty" and "citrus" [74]. Esters were the second most abundant in southern highbush blueberry, and they included ethyl propanoate, methyl 2-methylbutanoate, methyl 3-methylbutanoate, ethyl 2-methylbutanoate, ethyl 3-methylbutanoate, (Z)-3-hexyl acetate, (E)-2-hexyl acetate, and geranyl acetate, with "green", "sweet", "fruity", "apple", "banana", "pear", and "floral" aromatic characteristics. Unlike strawberries, apples, and bananas, where esters are the main contributors to the aroma, fewer esters were found in highbush blueberry [66,100]. Terpenes including linalool, citronellol, nerol, and geraniol showed "sweet", "floral", "fruity", "citrus", and "berry-like", while 1,8-cineole, dihydrolinalool oxide, and α-terpineol had "woody", "herbaceous", and "piney" characteristics. Linalool was one of the major volatiles in southern highbush blueberry. Two alcohols, including (Z)-3-hexenol and 2-heptanol, were aroma active. Only three ketones, including 2-heptanone, 1-octen-3-one, and 2-nonanone with "fruity" "mushroom", "earthy", and "cheese-like", were found to have aroma activity. Furaneol was found in southern highbush blueberries. Furaneol had "sweet", "candy", and "caramel" characteristics. In another study, five *Vaccinium* cultivars ("Biloxi", "Brigitta Blue", "Centurion", "Chandler", and "Ozark Blue") were found to have 106 volatile organic compounds [78]. Esters, 25 compounds, were the major compounds and were followed by 18 aldehydes, 16 alcohols, 14 monoterpenes, 7 ketones, 4 acids, 4 hydrocarbons, 3 sesquiterpenes, 1 lactone, and 1 norisoprenoid. Aldehydes, including (E)-2-hexenal, hexanal, (Z)-3-hexenal, hexadienal, or heptenal, are the most abundant, accounting for almost half of the *Vaccinium* volatile composition. Monoterpenes, including 1,8-cineole and linalool, were also important compounds in the volatile blueberry profile [78]. Even though esters had smaller content compared to others, esters had a unique aroma that characterizes the aroma of blueberry. Among seven ketones identified, 2-heptanone and 6-methyl-5-hepten-2-one were the ones with the highest contents. Octane, ethyl benzene, p-xylene, and mxylene were identified as hydrocarbons. Other volatile compounds in the blueberry aroma profile were hexanoic acid, octanoic acid, nonanoic acid, decanoic acid (acids), d-elemene, (E)-caryophyllene, and caryophyllene oxide (sesquiterpenes), b-damascenone (norisoprenoid), and butyrolactone (lactone). Dymerski et al. conducted identification of volatile blueberry compounds, and alcohol (51.8%), ester (32.8%), and carboxylic acid (6.9%) were mainly detected [101]. Forty-six blueberry volatiles were identified by Gu et al. by using GC-MS [69]. Monoterpenes accounted for 45% of total volatile concentration, with alcohols (17%), aldehydes (8%), C13 norisoprenoids and esters (each 7%), furans (5%), ketones (4%), and others. The major individual blueberry volatiles were linalool, linalool oxide, phenylethyl alcohol, 2-ethylhexanol, α-terpineol, and β-ionone.

*3.3. Raspberry*

In the Rosaceae family, raspberry (*Rubus spp.*) is a fruit with an attractive appearance and unique flavor [102,103]. There are red raspberry (*Rubus idaeus*) and black raspberry (*Rubus occidentalis*). Different cultivars and varieties of raspberries are grown worldwide, in Europe, North America, and Asia [88]. Raspberries have been reported to contain an aroma impact compound, which is a single compound that has an odor characteristic of raspberry [67]. This compound has been identified as 1-(phydroxyphenyl)-3-butanone and is referred to as raspberry ketone. Approximately 200 volatile compounds were detected in raspberries [76]. Raspberry ketone, α-ionone, β-ionone, linalool, (Z)-3-hexenol, geraniol, nerol, α-terpineol, furaneol, hexanal, β-ocimene, 1-octanol, β-pinene, β-damascenone, ethyl 2-methylpropanoate, (E)-2-hexenal, heptanal, and benzaldehyde have been identified as the raspberry aroma. Among them, α-ionone, β-ionone, geraniol, nerol, linalool, and raspberry ketone especially contributed to the red raspberry aroma. Monoterpene is an important class of fruit volatile organic compounds (VOCs) [104]. This class contains some of the most aroma-active compounds, such as citronellol, nerol, geraniol, α-terpineol, and linalool. The volatile composition of raspberries was identified with 30 compounds, including (Z)-hexenol, hexanal, (E)-2-hexenal, 2-heptanone, δ-octalactone, δ-decalactone, geraniol, α-ionone, β-ionone, and terpinen-4-ol [79]. The main volatile compounds in raspberries include monoterpenes (20%), acids (14%), alcohols (12%), esters (12%), aldehydes (8%), ketones (7%), C-13 norisopernoids (6%), hydrocarbons (6%), lactones (4%), sesquiterpenes (4%), furans (3%), sulfur (3%), and phenols (1%) [105]. Gu et al. identified 78 and 73 volatiles from black and red raspberry extracts, respectively [69]. The major chemical class was monoterpene (61% in black raspberries, 47% in red raspberries) in both raspberries. In black raspberries, (−)-myrtenol, linalool, α-terpineol, 2-ethylhexanol, cuminaldehyde, hexanoic acid, ethyl acetate, and (+)-myrtenol were the major compounds. In red raspberries, myrtenol, butanoic acid, linalool, eugenol, 3-methylbutanoic acid, α-ionone, and vanillin were found as major compounds.

*3.4. Blackberry*

Blackberries (*Rubus* spp.) are produced worldwide and consumed mostly as fresh but also as frozen, preserves, jelly, wine, dietary supplements, and others [106]. There have been studies identifying and analyzing the volatile composition of blackberries, but they are still limited. In D'Agostino et al., thirteen *Rubus ulmifolius* schott blackberries from different locations in Italy and Spain were used to identify the volatile composition by using SPME and GC-MS [80]. They identified a total of 74 volatiles from blackberry samples. Esters and aliphatic alcohols were the major classes, and methylbutanal, ethanol, 2,3-butanedione, trans-2-hexenal, 3-hydroxy-2-butanone, 1-hexanol, 1-octanol, and methylbutanoic acid were mainly found in all samples, which were 76.4% and 65.1% of volatile blackberry profiles from Italy and Spain, respectively. Wang et al. compared the aroma compositions of Chickasaw blackberries grown in Arkansas and Oregon [71]. A total of 84 compounds, including 19 esters, 18 terpenes and terpenoids, 15 alcohols, 13 aldehydes, 4 ketones, 4 acids, 4 lactones, 2 furans, 2 sulfur-containing compounds, 1 pyrazine, and 2 miscellaneous compounds, were identified. Even though they were the same cultivar, climate difference in the two regions strongly affected their blackberry aroma. The most attributing aromas of Chickasaw from Oregon were ethyl butanoate, linalool, methional, trans,cis-2,6-nonadienal, cis-1,5-octadien-3-one, and 2,5-dimethyl-4-hydroxy-3(2H)-furanone. However, the most potent aromas in Chickasaw from Arkansas were ethyl butanoate, linalool, methional, ethyl 2-methylbutanoate, β-damascenone, and geraniol. In sensory evaluations, Oregon samples were evaluated to have green, fruity, citrus, and watermelon aromas, while Arkansas samples were to have cinnamon, piney, floral, sweet, and caramel aromas [71]. Qian and Wang also investigated the volatile compositions of Marion and Thornless Evergreen blackberries by using GC-MS [73]. Acids (53.83%) and alcohols (24.25%) were the most abundant compounds in Marion, while alcohols (46.62%) were the most abundant in Thornless Evergreen. Thornless Evergreen blackberries showed much more amounts of volatiles (27.33 ppm) compared to Marion blackberries (8.62 ppm). The most abundant individual volatiles were acetic, hexanoic, decanoic, and 2/3-methylbutanoic acids, ethanol, and linalool for Marion, and 2-heptanol, octanol, α-pinene, hexanol, p-cymen-8-ol, and nopol for Thornless Evergreen. Based on odor activity values (OAVs), the most potent odorants were ethyl hexanoate, β-ionone, linalool, 2-heptanone, 2-undecanone, α-ionone, and hexanal for Marion, and ethyl hexanoate, 2-heptanone, ethyl 2-methylbutanoate, 2-heptanol, 3-methylbutanal, α-pinene, limonene, p-cymene, linalool, t-2-hexenal, myrtenol, hexanal, 2-methylbutanal, and sabinene [73]. Sixty-one volatiles were identified from volatile blackberry extract: 24 monoterpenes, 12 alcohols, 6 esters, 4 ketones, 4 C13 norisoprenoids, 3 furans, 2 acids, a lactone, and a phenolic [69]. Acids (57%) accounted for the highest concentration, followed by alcohols (18%), esters (10%), monoterpenes (10%), and others. Major individual compounds were butanoic acid, hexanoic acid, 4-methyl-1-pentanol, myrtenol, 2-ethylhexanol, isophorone, limonene, and 4-terpineol. Morin et al. identified a total of 80 volatiles in volatile extracts from three blackberry genotypes: Natchez and two University of Arkansas breeding lines, A2528T and A2587T [70]. Monoterpenes, alcohols, and esters were the predominant chemical classes in the three genotypes. As individual volatile compounds, ethyl acetate and α-terpineol were found to be the major volatiles in all three genotypes.

*3.5. Cranberry*

Cranberries (*Vaccinium* spp.) are native to North America, and production has been highly increased due to their well-known health benefits of cranberries [107]. The unique cranberry aroma is developed during ripening [75]. In 1981, Hirvi et al. detected 70 volatile compounds from European (*Vaccinium oxycoccus*, L.) and American cranberries (*Vaccinium macrocarpon*, Ait.) [82]. In this study, benzyl alcohol accounted for 29.2% and 21.6% in European and American cranberries, respectively. α-terpineol was 13% and 9.7% of total volatiles in European and American cranberries, respectively. Ruse et al. identified 21 volatiles from fresh cranberries [77]. Common volatile compounds detected from wild (*Vaccinium oxycoccus* L.) and different cultivars of cranberries (*Vaccinium macrocarpon* Ait., 'Early Black', 'Ben Lear', 'Steven', Bergman' and 'Pilgrim') were 4-penten-2-ol, 3-cis-hexenyl formate, benzaldehyde, α-1-terpineol, butyric acid, and benzyl alcohol. Zhu et al. analyzed the cranberry (*Vaccinium macrocarpon* Ait.) volatile composition of four cultivars ('Early Black', 'Howes', 'Searles', and 'McFarlin') by using GC-MS and GC-olfactometry (GC-O) [75]. A total of 33-36 volatiles were detected as odor-active compounds by GC-O. Hexanal, pentanal, (E)-2-heptenal, (E)-2-hexenal, (E)-2-octenal, (E)-2-nonenal, ethyl 2-methylbutyrate, β-ionone, 2-methylbutyric acid, and octanal were mainly contributing to the cranberry aroma. Khomych et al. detected 54 aromatic compounds in cranberry juice [84]. Twenty-three alcohols (41.2% of total concentration) were predominant in cranberry juice, followed by eight acids (40.7%), ten aldehydes (1.7%), five ketones (2.2%), five ethers (1.4%), three lactones, and each heterocyclic and unidentified compound (less than 1% each). Among alcohols, benzyl alcohol was the major volatile, accounting for 23.1% of total volatile concentration. Moore et al. detected 23 cranberry volatiles by using GC-MS [83]. In terms of total volatile concentration, Monoterpene (84%) was predominantly contained in cranberries, followed by aldehyde (8%) and alcohols (3%). The major volatile compounds were α-terpineol, linalool oxide, eucalyptol, trans-2-decanal, and 2-octanal. In cranberry volatile extract, 35 volatiles were found: 16 monoterpenes, 8 alcohols, 6 aldehydes, 2 esters, 2 ketones, and an acid [81]. Monoterpenes (60%) were predominant in total volatile concentration. α-terpineol, eucalyptol, 2-methylbutyric acid, ethyl benzoate, citronellol, and linalool were the major individual volatile compounds in cranberry.

## 4. Bioavailability of Berry Volatiles

The definition of bioavailability by the U.S. Food and Drug Administration (FDA) is "the rate and extent to which the active ingredient or active moiety is absorbed from a drug product and becomes available at the site of action" [108]. More proper meaning is the part of ingested compound reaching the systemic circulation and specific site where it is available in the body [109]. Investigating the bioavailability of a compound is important to find the clinical relevance of the health-promoting activities of the bioactive compound in our body [110]. Thus, it is necessary to study the absorption, distribution, metabolism, and excretion of bioactive compounds.

Bioavailability can be variable due to many different factors, including chemical structure, physical state, solubility, route of administration, and distribution via biotransformation and excretion [111,112]. Based on solubility, volatile compounds and essential oils are relatively more nonpolar hydrophobic compounds than polyphenols that are more polar hydrophilic nutraceuticals [113,114]. Regarding the route of administration, volatile compounds are more suitable for pulmonary administration through inhalation, while polyphenols are normally administrated orally [115]. Oral administration generally takes, on average, 30–90 min of action, while inhalation of gaseous compounds takes, on average, only 2–3 min [116]. The bioavailability of volatiles is largely affected by volatility, instability, and hydrophobicity [117].

Although there are many types of research conducted about the identification and quantification of berry volatiles, information is still lacking on the bioavailability of berry volatiles in animals and humans [118]. In this review, bioavailability studies of essential oils and herbal medicinal products that contain volatiles commonly present in berries were selected to estimate the bioavailability of volatile berry compounds. Unfortunately, the studies found are limited and mostly include animal models.

Most of the bioavailability studies of essential oils showed that the volatile compounds in essential oils are rapidly absorbed and eliminated after pulmonary, dermal, and oral administration [115]. The compounds were mostly metabolized and eliminated within an hour of elimination half-life through the kidney after phase-II conjugation or $CO_2$ exhalation.

In Igimi et al., the absorption, distribution, and excretion of d-limonene, a monoterpene in many essential oils but also found in black raspberries and blackberries, were investigated in rats [119]. $^{14}C$-labeled d-limonene was orally administered to 21 male Wistar rats. The maximum radioactivity was obtained 2 h after administration in blood and 1–2 h after administration in tissues. High radioactivity in the liver, kidney, and blood became not significant after 48 h. The excretion of d-limonene was 60% in urine, 5% in feces, and 2% in expired $CO_2$ in 48 h. About 25% of administered d-limonene was eliminated in bile in 24 h in bile duct cannulated rats. Biotransformation studies of (+)-limonene in humans showed that the main metabolites of biotransformation were perillic acid, dihydroperillic acid, and limonene-10-ol, and their glucuronides, perillyl alcohol, p-mentha-1,8-dien-carboxylic acid, cis- and trans-dihydroperillic acid, limonene, 1,2-diol and limonene-8,9-diol [120].

Dermal application of α-pinene on humans (ointment) and mice (bath) resulted in rapid absorption in plasma, reaching maximum plasma levels in 10 min of application [115]. Inhalation of α-pinene in humans resulted in 61% absorption of α-pinene [115]. However, only 4–6% of α-pinene was found to be absorbed in the blood. In α-pinene dermal and pulmonary administration studies [115], the half-life was short in the α-phase (5 min) and longer in the β-phase (26–38 min). α- and β-pinene in humans were metabolized to trans- and cis-verbenol, respectively, and they were further hydroxylated to diols. In rabbits, trans-verbenol was major, and myrtenol and myrtenic acid were minor metabolites of α-pinene, while cis-verbenol was the major metabolite of β-pinene [121]. In a recent open-label, single-arm study, ten male subjects consumed Mastiha oil (1 mL) containing rich monoterpenes, and blood samples were collected at 0–24 h after Mastiha oil administration [122]. Mastiha oil contained α-pinene (82.2%), myrcene (8.5%), and β-pinene (2.4%) as the major terpenes and also had linalool and limonene (0.8% each). In subjects' blood samples, the three major terpenes were detected. Myrcene reached its peak at 2.2 h (966.6 µg/L), and α-pinene and β-pinene reached their peaks at 3.8 (914.8 µg/L) and 3.6 h (18 µg/L), respectively [122].

Linalool, one of the major berry volatile compounds, was metabolized to dihydrolinalool, tetrahydrolinalool, and 8-hydroxylinalool, then further oxidized to 8-carboxylinalool by cytochrome P450 (CYP). Metabolites derived by CYP formed glucuronide conjugates [123]. Oral administration of α-terpineol to rats resulted in the metabolization of alpha-terpineol to p-menthane-1,2,8-triol. The major biotransformation occurred in 1,2-double bound with allylic methyl oxidation and reduction [124].

## 5. Health Benefits of Berry Volatiles

Despite the sensory properties of fruits and vegetables, there are studies demonstrating that the role of aroma compounds is more than their odor impact [18,125]. Recently, volatile compounds in plants have been reported to have health-promoting activities, including anti-inflammatory [25,126], anti-cancer [26], anti-obesity, and anti-diabetic effects [127]. However, since there are not many studies on the health-promoting effects of berry volatiles, studies of volatile compounds from essential oils and other fruits and plants that berries commonly contain were used to review the potential health benefits of volatile compounds in berries (Figure 2).

**Figure 2.** Potential bioactivities of berry volatiles. COX-2 = cyclooxygenase-2; ERK = extracellular signal-regulated kinase; IL-1β = interleukin-1β; IL-6 = interleukin-6; JNK = c-Jun N-terminal kinase; MAPK = mitogen-activated protein kinase; NF-κB = nuclear factor kappa B; NO = nitric oxide; Nrf2 = nuclear factor erythroid 2-related factor 2; PGE₂ = prostaglandin 2; TNF-α = tumor necrosis factor-α; ↓ = decrease.

### 5.1. Inflammation

Infection, inflammation, or any cellular damage/stimuli are detected by macrophages and dendritic cells through pattern recognition receptors (PRRs) with pathogen-associated molecular patterns and danger-associated molecular patterns [128,129]. Toll-like receptors and intracellular nucleotide-binding domain leucine-rich-repeat-containing receptors recognize these stimuli and stimulate signal transductions, mitogen-activated protein kinases (MAPKs) [130–132]. MAPK signal transduction pathways include extracellular signal-regulated kinase (ERK), c-Jun N-terminal kinase (JNK), and p38 and regulate downstream protein kinases and transcription factors [133]. Stimulated signal transductions activate pro-inflammatory transcription factors, such as nuclear factor kappa-B (NF-κB) and nuclear factor erythroid 2-related factor 2 (Nrf2) [132]. NF-κB regulates the production of pro-inflammatory cytokines and chemokines, including interleukin (IL)-1ß, IL-6, nitric oxide (NO), prostaglandin (PGE) 2, and tumor necrosis factor (TNF)-α [134,135]. NF-κB is activated by IκB kinase (IKK) phosphorylating NF-κB-inhibitory protein (IκBα), and translocating into the nucleus, promoting transcription of pro-inflammatory mediators [136]. Nrf2 is another transcription factor related to oxidative damage and inflammation [137]. During the cascade of inflammatory responses, reactive oxidative stress (ROS) increases oxidative stress on cells, leading to the autophagy of cells [138,139]. However, recent studies found that volatile compounds in plants, especially terpenes, mitigate inflammation by suppressing many different inflammatory processes [140,141]. In this section, the effects of volatile compounds rich in berries against inflammation in different steps of inflammatory processes were summarized (Table 2).

**Table 2.** The effect of volatile compounds rich in berries on inflammation models.

| Volatile Compound | Inflammation Model | Effect | References |
|---|---|---|---|
| Limonene | Carrageenan-induced mice subcutaneous air pouch mice model | IFN-$\gamma$, IL-1$\beta$, NO, and TNF-$\alpha$ production$\downarrow$ | [142] |
| Limonene | LPS-induced acute lung injury mice model | NF-$\kappa$B and MAPK activation$\downarrow$ (I$\kappa$B$\alpha$, NF-$\kappa$B p65, ERK, JNK, and p38 MAPK phosphorylation$\downarrow$) | [143] |
| D-limonene | Doxorubicin-induced rat model | COX-2, iNOS, NO, PGE$_2$, and TNF-$\alpha$ production$\downarrow$ NF-$\kappa$B activation$\downarrow$ | [144] |
| D-limonene | Ulcerative colitis rat model | COX-2, iNOS, PGE$_2\downarrow$ | [145] |
| Limonene, myrcene | IL-1$\beta$-induced human chondrocyte model | JNK and p38 phosphorylation$\downarrow$ NF-$\kappa$B activation$\downarrow$ | [146] |
| Rheosmin | LPS-induced RAW264.7 cells | COX-2, iNOS, NO, and PGE$_2$ production$\downarrow$ | [147] |
| $\alpha$-terpineol | LPS-induced RAW264.7 cells | NO production$\downarrow$ | [83] |
| $\gamma$-terpinene | Carrageenan-induced peritonitis mice model | IL-1$\beta$ and TNF-$\alpha$ production$\downarrow$ | [146] |
| Terpinen-4-ol | DSS-induced colitis mice model | NF-$\kappa$B activation$\downarrow$ | [148] |
| Terpinen-4-ol | LPS-induced acute lung injury mice model | IL-1$\beta$ and TNF-$\alpha$ production$\downarrow$ I$\kappa$B$\alpha$ and NF-$\kappa$B p65 phosphorylation$\downarrow$ | [149] |
| Linalool | Aged triple transgenic Alzheimer's mice model | COX-2, IL-1$\beta$, and iNOS production$\downarrow$ P38 MAPK production$\downarrow$ | [150] |
| Linalool | Endotoxin-induced mice model | IFN-$\gamma$, IL-1$\beta$, IL-18, NO, and TNF-$\alpha$ production$\downarrow$ TLR4 expression$\downarrow$ NF-$\kappa$B activation$\downarrow$ | [151] |
| Linalool | Ovalbumin-induced pulmonary inflammation mice model | iNOS and MCP-1 production$\downarrow$ MAPK and NF-$\kappa$B activation$\downarrow$ | [152] |
| Linalool | Cigarette smoke-induced acute lung inflammation mice model | IL-1$\beta$, IL-6, IL-8, MCP-1, and TNF-$\alpha$ production$\downarrow$ NF-$\kappa$B activation$\downarrow$ | [153] |
| Linalool | LPS-induced RAW264.7 cells | IL-6 and TNF-$\alpha$ production$\downarrow$ | [154] |
| Linalool | *Pasteurella multocida*-induced lung inflammation mice model | IL-6 and TNF-$\alpha$ production$\downarrow$ Nrf2 nuclear translocation$\uparrow$ | [155] |
| Linalool | LPS-induced BV2 microglia cells | IL-1$\beta$, NO, PGE$_2$, and TNF-$\alpha$ production$\downarrow$ NF-$\kappa$B activation$\downarrow$ Nrf2 nuclear translocation$\uparrow$ HO-1 expression$\uparrow$ | [156] |
| $\alpha$-pinene | LPS-induced mouse peritoneal macrophages | COX-2, IL-6, iNOS, NO, and TNF-$\alpha$ production$\downarrow$ MAPK and NF-$\kappa$B activation$\downarrow$ | [157] |
| $\alpha$-pinene, 1,8-cineole | H$_2$O$_2$-stimulated U373-MG cells (human astrocytoma cell line) | ROS formation$\downarrow$ | [158] |
| Berry volatile extracts | LPS-induced RAW264.7 cells | COX-2$\downarrow$, IL-6$\downarrow$, NO$\downarrow$, PGE$_2\downarrow$, TNF-$\alpha\downarrow$ I$\kappa$B$\alpha$ and NF-$\kappa$B p65 phosphorylation$\downarrow$ | [69] |

COX-2 = cyclooxygenase-2; ERK = extracellular signal-regulated kinase; HO-1 = heme oxygenase-1; IFN-$\gamma$ = interferon-$\gamma$; I$\kappa$B$\alpha$ = I$\kappa$B kinase phosphorylating NF-$\kappa$B-inhibitory protein; IL-18 = interleukin-18; IL-1$\beta$ = interleukin-1$\beta$; IL-6 = interleukin-6; iNOS = inducible nitric oxide synthase; JNK = c-Jun N-terminal kinase; MAPK = mitogen-activated protein kinase; MCP-1 = monocyte chemoattractant protein-1; NF-$\kappa$B = nuclear factor kappa B; NO = nitric oxide; Nrf2 = nuclear factor erythroid 2-related factor 2; PGE$_2$ = prostaglandin 2; ROS = reactive oxidative stress; TLR4 = toll-like receptor 4; TNF-$\alpha$ = tumor necrosis factor-$\alpha$; $\uparrow$ = increase; $\downarrow$ = decrease.

### 5.1.1. Modulation of Pro-Inflammatory Mediators

Many volatile compounds in plants showed anti-inflammatory effects by reducing the level of pro-inflammatory mediators and cytokines, such as NO, $PGE_2$, cyclooxygenase (COX-2), TNF-$\alpha$, and interleukins [69,83,142,145–147,150–154,159,160]. Amorim et al. indicated that the essential oils obtained from Citrus species and limonene demonstrated a significant anti-inflammatory effect by reducing cytokine production, including NO, TNF-$\alpha$, and IL-1$\beta$ [142]. The essential oils of citrus fruit peel contain abundant monoterpenes such as limonene, geranial, $\beta$-pinene, and $\gamma$-terpinene, which are one of the major volatile compounds in berries [159]. In Rehman et al., d-limonene at 5% and 10% doses mixed in a diet given to rats for 20 days effectively reduced doxorubicin-induced COX-2, inducible nitric oxide synthase (iNOS), and NO [160]. In an ulcerative colitis rat model, rats (n = 8/group) fed d-limonene (50 and 100 mg/kg) for 7 days showed anti-inflammatory effects by suppressing the level of iNOS, COX-2, and $PGE_2$ [145]. In lipopolysaccharide (LPS)-induced RAW264.7 cells, 125–1000 µg/mL of rheosmin (raspberry ketone) isolated from pine needles exerted an anti-inflammatory activity with reduced NO, $PGE_2$, iNOS, and COX-2 production [147]. In the LPS-induced murine macrophage RAW264.7 cell model, $\alpha$-terpineol treatment (1.16 µg/mL) before and after LPS stimulation showed significant inhibition on the level of NO [83]. In a mouse model of carrageenan-induced peritonitis, oral administration of $\gamma$-terpinene 1 h before intraperitoneal carrageenan injection significantly attenuated the TNF-$\alpha$ and IL-1$\beta$ production [146]. In a triple transgenic Alzheimer's mouse model, oral administration of 25 mg/kg linalool, every 48 h for 3 months, markedly decreased the production of iNOS, COX-2, and IL-1$\beta$ [150]. Mice administered 2.6 and 5.2 mg/kg linalool before injecting endotoxin remarkably suppressed the nitrate/nitrite, IL-1$\beta$, IL-18, TNF-$\alpha$, and interferon (IFN)-$\gamma$ production [151]. Oral linalool administration (15 and 30 mg/kg) also lowered iNOS levels in lung tissues in mice with allergic asthma [152]. Intraperitoneal injection of 10, 20, and 40 mg/kg linalool two hours before cigarette smoke exposure for five days ameliorated the lung inflammation by suppressing the level of pro-inflammatory cytokines (TNF-$\alpha$, IL-6, IL-1$\beta$, IL-8, and monocyte chemoattractant protein (MCP)-1) [153]. Linalool (40–120 µg/mL) attenuated the LPS-induced inflammation on RAW264.7 cells with the suppressed level of TNF-$\alpha$ and IL-6 [154]. In the LPS-stimulated mouse macrophage RAW264.7 cell model, 1 h pretreatment of volatile extracts (50-fold dilution) from blackberry, black raspberry, blueberry, cranberry, red raspberry, and strawberry significantly reduced production in NO, $PGE_2$, and COX-2 [69]. Blackberry, blueberry, cranberry, and volatile strawberry extracts also effectively suppressed LPS-induced TNF-$\alpha$ and IL-6 production.

### 5.1.2. Regulation of Inflammatory Transcription Factors and Signal Transduction

NF-$\kappa$B is a crucial target for anti-inflammation since it is one of the main transcription factors regulating pro-inflammatory mediators [161]. Linalool significantly reduced the NF-$\kappa$B activation in mice with endotoxin injection [151], cigarette smoke-induced acute lung inflammation [153], and airway allergic inflammation [152]. Limonene (25–75 mg/kg) intraperitoneal injection 1 h before LPS administration down-regulated the phosphorylation of I$\kappa$B$\alpha$, NF-$\kappa$B p65, p38 MAPK, JNK, and ERK in LPS-induced acute lung injury mice [143]. In Rehman et al., d-limonene also suppressed NF-$\kappa$B activation in a doxorubicin-stimulated inflammation rat model [160]. Limonene and myrcene attenuated the IL-1$\beta$-induced inflammation by suppressing NF-$\kappa$B and JNK activation in human chondrocytes [144]. Terpinen-4-ol inhibited NF-$\kappa$B in the dextran sulfate sodium (DSS)-increased experimental colitis in mice [148]. Terpinen-4-ol also attenuated the LPS-stimulated I$\kappa$B$\alpha$ and NF-$\kappa$B p65 phosphorylation in acute lung injury mice [149]. In Wu et al., linalool increased nuclear translocation of Nrf2 in mice with pneumonia infected by *Pasteurella multocida* [155]. Linalool (162–648 µM), one of the major berry volatiles, reduced LPS-stimulated inflammation on BV2 microglia cells through Nrf2/HO-1 signaling pathway [156]. $\alpha$-Pinene in coniferous trees and rosemary oils suppressed the MAPK and NF-$\kappa$B activation in LPS-induced macrophages [157]. In LPS-induced RAW264.7 murine macrophage cells, volatile extract (50-fold dilution) from blackberry, black raspberry, blueberry, and cranberry significantly suppressed the NF-κB activation by down-regulating phosphorylation of NF-κB p65 and IκBα [69].

### 5.1.3. Attenuation of Oxidative Stress and Autophagy

A disparity between the production and elimination of ROS causes oxidative stress. Excessively produced ROS can damage tissues, increasing inflammatory responses and leading to cell death, such as necrosis and apoptosis [162]. There have been many examinations of the antioxidant activities of volatile compounds in plants against oxidative stress in vitro. α-terpinene, γ-terpinene, and linalool showed antioxidant activities in 2,2′-azino-bis-3-ethylbenzthiazoline-6-sulphonic acid (ABTS), chelating power, 2,2-diphenyl-1-picrylhydrazyl (DPPH) and oxygen radical absorbance capacity (ORAC) assays [163]. α-pinene, 1,8-cineole, and d-limonene remarkably ameliorated the formation of ROS in $H_2O_2$-stimulated oxidative stress [158].

### 5.2. Cancer

Kim et al. demonstrated that geraniol inhibits human prostate cancer cell PC-3 proliferation in in vitro and in vivo xenograft mice models [164]. Geraniol at 0.5 and 1 mM significantly suppressed the cell growth of PC-3 by increasing cell cycle arrest and apoptosis. Balb/C nude mice inoculated with PC-3 cells took intratumoral geraniol injection daily for 38 days at 0, 12, 60, or 300 mg/kg. The mice treated with 60 or 300 mg/kg geraniol showed significantly decreased tumor volume and weight. Injection of geraniol at 20 mg/kg also sensitized the chemotherapeutic agent Docetaxel (2 mg/kg) in the xenograft mice model. In Lee et al., geraniol inhibited prostate cancer growth by a down-regulating E2F8 transcription factor and inducing G2/M phase cell cycle arrest [165]. In gastric adenocarcinoma AGS cells, geraniol showed cytotoxicity by inhibiting the JNK/ERK signaling pathway [166]. Lavender essential oil, its active compounds linalool and linalyl acetate [167], and ethyl acetate fraction of Ajwa dates [168] also inhibited PC-3 cell proliferation by increasing apoptosis and cell cycle arrest. Linalool also reduced tumor growth in PC-3 xenograft mice [167] and the 22Rv1 xenograft mice model [169]. Ethyl acetate exerted an anti-proliferative activity on human breast cancer MCF7 and SKBR3 cells [170] and human cervical cancer HeLa cells [171]. α-terpineol [172] and linalool [173] also showed strong cytotoxicity on HeLa cells with apoptosis and cell cycle arrest. In human acute myeloid leukemia U937 cells [173], human oral cancer cells [174,175], and lung adenocarcinoma A549 cells [176], linalool significantly suppressed the cell growth. D-limonene [177] and d-limonene-rich blood orange volatile oils [178] inhibited the proliferation of lung cancer A549 and H1299 cells and human colon adenocarcinoma cells SW480 and HT-29 cells, respectively. Limonene (9 μM) significantly reduced the proliferation of human bladder cancer cell T24 after 24 h, showing induced apoptosis with increased G2/M cell cycle arrest and apoptotic markers (Bax, and cleaved caspase-3, 8, and 9) [179]. In Yu et al., d-limonene induced apoptosis and autophagy-related genes in a lung cancer model [177]. In an acetic acid-induced gastric ulcer rat model, 7-day oral administration of (-)-myrtenol at 50–100 mg/kg increased the healing of the ulcer [180]. Myrcene640 μM) significantly decreased the proliferation of SCC9 oral cancer cells after 24 h [181]. Myrcene also showed increased apoptosis with the concentration of 5–20 μM and significantly suppressed the migration of SCC9 cells at 10 μM myrcene treatment. The effects of volatile compounds rich in berries on cancer models were summarized in Table 3.

**Table 3.** The effect of volatile compounds rich in berries on cancer models.

| Volatile Compound | Cancer Model | Effect | References |
|---|---|---|---|
| Geraniol | Human prostate cancer PC-3 cells, in vitro and in vivo xenograft mice model | Cell proliferation↓<br>Cell cycle arrest and apoptosis↑<br>Tumor volume and weight↓<br>Docetaxel sensitization↑ | [164] |

**Table 3.** *Cont.*

| Volatile Compound | Cancer Model | Effect | References |
|---|---|---|---|
| Geraniol | Human prostate cancer PC-3 cells | E2F8 transcription factor↓<br>G2/M phase cell cycle arrest↑ | [165] |
| Geraniol | Gastric adenocarcinoma AGS cells | ERK, JNK, p38 MAPK activation↓<br>Apoptosis↑ | [166] |
| Linalool, linalyl acetate | Human prostate cancer PC-3 and DU145 cells, PC-3 cell-transplanted xenograft mice model | Apoptosis and G2/M phase cell cycle arrest↑<br>Tumor growth↓ | [167] |
| Linalool | Human prostate cancer 22Rv1 cells | Cell proliferation↓<br>Apoptosis↑ | [169] |
| Linalool | Human leukemia U937 cells and human cervical adenocarcinoma HeLa cells | Apoptosis and cell cycle arrest↑ | [173] |
| Linalool | Human oral cancer OECM1 and KB cells | Cell proliferation↓<br>Apoptosis and sub-G1 phase cell cycle arrest↑ | [174]<br>[175] |
| Linalool, 1,8-cineole | Human lung adenocarcinoma A549 cells | Cell proliferation↓<br>Cell cycle arrest↑<br>No apoptosis | [176] |
| Ethyl acetate | Human prostate cancer PC-3 cells | Apoptosis and S phase cell cycle arrest↑<br>Oxidative stress↑<br>Mitochondrial membrane potential (MMP)↓ | [168] |
| Ethyl acetate | Human breast cancer MCF7 and SKBR3 cells | Sub G1 phase cell cycle arrest↑<br>ROS production↑<br>MMP↓ | [170] |
| Ethyl acetate | Human cervical cancer HeLa cells | Apoptosis and G2/M phase cell cycle arrest↑ | [171] |
| α-terpineol | Human cervical cancer HeLa cells | Apoptosis and G1 phase cell cycle arrest↑ | [172] |
| D-limonene | Lung cancer A549 and H1299 cells | Tumor growth↓<br>Apoptosis and autophagy-related gene expression↑ | [177] |
| Limonene | Human bladder cancer T24 cells | Cell proliferation↓<br>Apoptosis and G2/M cell cycle arrest↑<br>Bax, cleaved caspase-3, 8, and 9 expression↑<br>Bcl-2 expression↓ | [179] |
| Myrcene | Oral cancer SCC9 cells | Apoptosis↑<br>Cell migration↓ | [181] |

ERK = extracellular signal-regulated kinase; JNK = c-Jun N-terminal kinase; MAPK = mitogen-activated protein kinase; ROS = reactive oxidative stress; ↑ = increase; ↓ = decrease.

### 5.3. Obesity

A high-fat diet for obese humans and animals can increase endothelial dysfunction. It can lead to many other severe cardiovascular diseases and metabolic disorders. In Wang et al., geraniol was examined for the effect on endothelial function in high-fat diet (HFD)-fed mice [182]. Forty mice were fed HFD for 8 weeks, while 20 mice had a normal diet. Then, HFD-fed mice were randomly assigned to intraperitoneal geraniol treatment (20 mice) or vehicle treatment (20 mice) group for 6 weeks. As a result, geraniol protected and improved HFD-induced endothelial dysfunction in HFD-fed mice by reducing aortic NADPH oxidases and ROS production. In Sousa et al., α-terpineol enantiomers were examined for their effect on the biological markers in HFD-induced obese rats [183]. Six weeks of daily α-terpineol supplementation (50–100 mg/kg of diet) suppressed pro-inflammatory cytokines (TNF-α and IL-1β), serum thiobarbituric acid reactive substances (TBARS), and recovered insulin sensibility. In Li et al., Microcapsules of d-limonene-rich sweet orange essential oil (SOEO) were orally administered to HFD-induced obese rats for 15 days [184]. SOEO microcapsules decreased the body weight in obese rats by protecting the gut barrier, increasing *Bifidobacterium*, and reducing low-grade inflammation. While white adipocytes store the excessive energy in triglyceride forms, brown adipocytes burn the calories in heat form by non-shivering thermogenesis [185]. Lone and Yun showed that limonene increased 3T3-L1 adipocytes browning through the activation of the β3-adenergenic receptor and ERK signaling pathway [185]. In Ayala-Ruiz et al., male Wister rats (*n* = 6 per group) were administered with control, high-fat-sucrose diet (HFSD) and 0.6 mL of corn oil, HFSD with 1,8-cineole (0.88 mg/kg), limonene (0.43 mg/kg), α-terpineol (0.32 mg/kg), or the mixture of three terpenes per gavage for 15 weeks [186]. Rats fed with HFSD with terpenes significantly reduced weight gain compared to the ones with only HFSD. In addition, all terpenes suppressed the fat deposition, serum glucose levels, and triacylglycerol levels. The effects of volatile compounds rich in berries against obesity were summarized (Table 4).

**Table 4.** The effect of volatile compounds rich in berries on obesity models.

| Volatile Compound | Obesity Model | Effect | References |
|---|---|---|---|
| Geraniol | High-fat diet (HFD)-fed mice | Aortic NADPH oxidases, ROS production↓ | [182] |
| α-terpineol | HFD-induced obese rats | IL-1β and TNF-α↓<br>Serum TBARS↓<br>Insulin sensibility↑ | [183] |
| D-limonene-rich sweet orange essential oil | HFD-induced obese rats | Body weight↓<br>Relative abundance of *Bifidobacterium*↑ | [184] |
| Limonene | Mouse preadipocytes 3T3-L1 | Adipocyte browning↑ | [185] |
| Limonene, α-terpineol, 1,8-cineole | High-fat-sucrose diet (HFSD)-fed rats | Body weight↓<br>Fat deposition↓<br>Serum glucose level↓<br>Triacylglycerol level↓ | [186] |

IL-1β = interleukin-1β; NADPH = nicotinamide adenine dinucleotide phosphate; ROS = reactive oxidative stress; TBARS = thiobarbituric acid reactive substances; TNF-α = tumor necrosis factor-α; ↑ = increase; ↓ = decrease.

### 5.4. Diabetes

In Bacanlı et al., streptozotocin (STZ) (60 mg/kg) was injected into Wistar rats to induce type 1 diabetes [187]. Diabetic rats were orally treated with d-limonene (50 mg/kg body weight) for 28 days. D-limonene treatment significantly reduced DNA damage and induced the level of antioxidant enzymes (catalase, superoxide dismutase, and total glutathione). D-limonene also altered hepatic enzyme and lipid profile, suggesting the potential of d-limonene being protective against diabetes in the liver and kidney in rats. D-limonene showed potential antihyperglycemic activities [188,189] and reduced lipid peroxidation with increased antioxidant activity [190]. In El-Bassossy et al., geraniol (150 mg/kg) was orally treated in STZ-induced obese rats for 7 weeks [191]. Geraniol significantly reduced systolic cardiac function related to diabetes by alleviating oxidative stress. Geraniol treatment also reduced GLUT 2 transporter [192], hyperglycemia [193], diabetic nephropathy [194], and improved impaired vascular reactivity [195] in STZ-induced diabetic rats. Linalool also exerted a reduction in fasting blood glucose level, insulin resistance, glycation oxidative stress [196], and nephropathic changes in kidneys [197] on STZ-induced diabetic rats. Xuemei et al. investigated the effect of myrtenol on STZ-induced gestational diabetes mellitus (GDM) in rats [198]. GDM is diabetes that occurs only during pregnancy. Twenty-five mg/kg of STZ was injected into the pregnant rats to induce GDM. Myrtenol (50 mg/kg) was orally administered for 2 weeks. Myrtenol oral administration helped decrease blood glucose levels and pro-inflammatory markers. It also increased high-density lipoprotein (HDL) and antioxidant status in diabetic pregnant rats. The effects of volatile compounds rich in berries against diabetes were summarized in Table 5.

**Table 5.** The effect of volatile compounds rich in berries on diabete models.

| Volatile Compound | Diabetes Model | Effect | References |
|---|---|---|---|
| D-limonene | Streptozotocin-induced diabetic rat model | DNA damage↓<br>Antioxidant enzyme activities↑ | [187] |
| D-limonene | Streptozotocin-induced diabetic rat model | Antihyperglycemic activities↑ | [188] |
| Limonene, linalool | Streptozotocin-induced diabetic rat model | Blood glucose level↓<br>Antioxidant enzyme activities↑ | [189] |
| D-limonene | Streptozotocin-induced diabetic rat model | Lipid peroxidation↓<br>Antioxidant activity↑ | [190] |
| Geraniol | Streptozotocin-induced diabetic rat model | Oxidative stress↓ | [191] |
| Geraniol | Streptozotocin-induced diabetic rat model | GLUT2 expression↓<br>Kidney glucose release↓ | [192] |
| Geraniol | Streptozotocin-induced diabetic rat model | Insulin resistance↓<br>Plasma glucose level↓ | [193] |
| Geraniol | Streptozotocin-induced diabetic rat model | Redox balance↑<br>Lipid peroxidation↓ | [194] |
| Geraniol | Streptozotocin-induced diabetic rat model | Vasoconstriction↓ | [195] |
| Linalool | Streptozotocin-induced diabetic rat model | NF-κB and TGF-$\beta_1$ expression↓ | [197] |
| Myrtenol | Streptozotocin-induced gestational diabetic pregnant rat model | Blood glucose level↓<br>Pro-inflammatory markers↓<br>HDL and antioxidant activity↑ | [198] |

HDL = high-density lipoprotein; NF-κB = nuclear factor kappa B; TGF-$\beta_1$ = Transforming growth factor beta-1; ↑ = increase; ↓ = decrease.

## 6. Conclusions

As berry consumption through the fruit and products, including berry-flavored water, juice, and others, have rapidly increased, many studies about monitoring and improving overall berry quality, including flavor, aroma, appearance, shelf-life, and safety, were conducted to target consumer acceptability. However, the beneficial health effects of berry volatiles have not been extensively studied. In this article, we looked into the biosynthesis of plant volatiles, volatile composition, and possible bioavailability and health benefits of some berry volatiles were reviewed. Major terpene volatiles were synthesized via MVA and MEP pathways. Major chemical classes in berries were esters, alcohols, terpenoids, aldehydes, ketones, and lactones. Berries had different profiles of volatiles, but monoterpene showed a crucial role in characterizing the unique berry aroma in all five berries. Volatile compounds were nonpolar and hydrophobic and rapidly absorbed and eliminated from our body after administration. Among them, monoterpenes, including linalool, limonene, and geraniol, showed many health benefits associated with inflammation, cancer, obesity, and diabetes in vitro and in vivo, suggesting potential health beneficial effects of berry volatiles. More research on animal and human models of the health benefits of berry volatiles and bioavailability would be needed to confirm their bioactivities.

**Author Contributions:** Conceptualization, L.H. and S.-O.L.; writing—original draft preparation, I.G.; writing—review and editing, S.-O.L., L.H. and I.G.; supervision, S.-O.L. and L.H.; funding acquisition, L.H. and S.-O.L. All authors have read and agreed to the published version of the manuscript.

**Funding:** The work was supported by the Arkansas Biosciences Institute.

**Institutional Review Board Statement:** Not applicable.

**Informed Consent Statement:** Not applicable.

**Data Availability Statement:** Not applicable.

**Conflicts of Interest:** The authors declare no conflict of interest.

# References


1. Kader, A.A. Flavor Quality of Fruits and Vegetables. *J. Sci. Food Agric.* **2008**, *88*, 1863–1868. [CrossRef]
2. Du, X.; Qian, M. Flavor chemistry of small fruits: Blackberry, raspberry, and blueberry. In *Flavor and Health Benefits of Small Fruits*; Qian, M.C., Rimando, A.M., Eds.; ACS Symposium Series; American Chemical Society: Washington, DC, USA, 2010; Volume 1035, pp. 27–43.
3. Afrin, S.; Giampieri, F.; Gasparrini, M.; Forbes-Hernandez, T.Y.; Varela-López, A.; Quiles, J.L.; Mezzetti, B.; Battino, M. Chemopreventive and Therapeutic Effects of Edible Berries: A Focus on Colon Cancer Prevention and Treatment. *Molecules* **2016**, *21*, 169. [CrossRef] [PubMed]
4. Goff, S.A.; Klee, H.J. Plant Volatile Compounds: Sensory Cues for Health and Nutritional Value? *Science* **2006**, *311*, 815–819. [CrossRef] [PubMed]
5. Forney, C.F.; Song, J. Flavors and Aromas: Chemistry and Biological Functions. In *Fruit and Vegetable Phytochemicals*; John Wiley & Sons, Ltd.: Hoboken, NJ, USA, 2017; pp. 515–540. ISBN 978-1-119-15804-2.
6. Horvat, R.J.; Schlotzhauer, W.S.; Chortyk, O.T.; Nottingham, S.F.; Payne, J.A. Comparison of Volatile Compounds from Rabbiteye Blueberry (*Vaccinium Ashei*) and Deerberry (*V. Stamineum*) during Maturation. *J. Essent. Oil Res.* **1996**, *8*, 645–648. [CrossRef]
7. Ibáñez, E.; López-Sebastián, S.; Ramos, E.; Tabera, J.; Reglero, G. Analysis of Volatile Fruit Components by Headspace Solid-Phase Microextraction. *Food Chem.* **1998**, *63*, 281–286. [CrossRef]
8. de Ancos, B.; Ibañez, E.; Reglero, G.; Cano, M.P. Frozen storage effects on anthocyanins and volatile compounds of raspberry fruit. *J. Agric. Food Chem.* **2000**, *48*, 873–879. [CrossRef]
9. Wang, C.Y. Maintaining Postharvest Quality of Raspberries with Natural Volatile Compounds. *Int. J. Food Sci. Technol.* **2003**, *38*, 869–875. [CrossRef]
10. Duan, W.; Sun, P.; Chen, L.; Gao, S.; Shao, W.; Li, J. Comparative Analysis of Fruit Volatiles and Related Gene Expression between the Wild Strawberry *Fragaria pentaphylla* and cultivated *Fragaria × ananassa*. *Eur. Food Res. Technol.* **2018**, *244*, 57–72. [CrossRef]
11. Yuan, F.; Cheng, K.; Gao, J.; Pan, S. Characterization of Cultivar Differences of Blueberry Wines Using GC-QTOF-MS and Metabolic Profiling Methods. *Molecules* **2018**, *23*, 2376. [CrossRef] [PubMed]
12. Li, H.; Brouwer, B.; Oud, N.; Verdonk, J.C.; Tikunov, Y.; Woltering, E.; Schouten, R.; Pereira da Silva, F. Sensory, GC-MS and PTR-ToF-MS Profiling of Strawberries Varying in Maturity at Harvest with Subsequent Cold Storage. *Postharvest Biol. Technol.* **2021**, *182*, 111719. [CrossRef]
13. Pichersky, E.; Noel, J.P.; Dudareva, N. Biosynthesis of plant volatiles: Nature's diversity and ingenuity. *Science* **2006**, *311*, 808–811. [CrossRef] [PubMed]
14. Dudareva, N.; Negre, F.; Nagegowda, D.A.; Orlova, I. Plant Volatiles: Recent Advances and Future Perspectives. *Crit. Rev. Plant Sci.* **2006**, *25*, 417–440. [CrossRef]
15. Dudareva, N.; Klempien, A.; Muhlemann, J.K.; Kaplan, I. Biosynthesis, Function and Metabolic Engineering of Plant Volatile Organic Compounds. *New Phytol.* **2013**, *198*, 16–32. [CrossRef] [PubMed]
16. Song, J.; Forney, C.F. Flavour Volatile Production and Regulation in Fruit. *Can. J. Plant Sci.* **2008**, *88*, 537–550. [CrossRef]
17. Bood, K.G.; Zabetakis, I. The Biosynthesis of Strawberry Flavor (II): Biosynthetic and Molecular Biology Studies. *J. Food Sci.* **2002**, *67*, 2–8. [CrossRef]
18. Siegmund, B. Biogenesis of aroma compounds: Flavour formation in fruits and vegetables. In *Flavour Development, Analysis and Perception in Food and Beverages*; Woodhead Publishing: Cambridge, UK, 2015; pp. 127–149.
19. Giuggioli, N.R.; Briano, R.; Baudino, C.; Peano, C. Effects of Packaging and Storage Conditions on Quality and Volatile Compounds of Raspberry Fruits. *CyTA-J. Food* **2015**, *13*, 512–521. [CrossRef]
20. Saftner, R.; Polashock, J.; Ehlenfeldt, M.; Vinyard, B. Instrumental and Sensory Quality Characteristics of Blueberry Fruit from Twelve Cultivars. *Postharvest Biol. Technol.* **2008**, *49*, 19–26. [CrossRef]
21. Dymerski, T.; Namieśnik, J.; Leontowicz, H.; Leontowicz, M.; Vearasilp, K.; Martinez-Ayala, A.L.; González-Aguilar, G.A.; Robles-Sánchez, M.; Gorinstein, S. Chemistry and biological properties of berry volatiles by two-dimensional chromatography, fluorescence and Fourier transform infrared spectroscopy techniques. *Food Res. Int.* **2016**, *83*, 74–86. [CrossRef]
22. Bowen-Forbes, C.S.; Zhang, Y.; Nair, M.G. Anthocyanin content, antioxidant, anti-inflammatory and anticancer properties of blackberry and raspberry fruits. *J. Food Compos. Anal.* **2010**, *23*, 554–560. [CrossRef]
23. Gasparrini, M.; Forbes-Hernandez, T.Y.; Giampieri, F.; Afrin, S.; Alvarez-Suarez, J.M.; Mazzoni, L.; Mezzetti, B.; Quiles, J.L.; Battino, M. Anti-Inflammatory Effect of Strawberry Extract against LPS-Induced Stress in RAW 264.7 Macrophages. *Food Chem. Toxicol.* **2017**, *102*, 1–10. [CrossRef]

24. Seeram, N.P.; Adams, L.S.; Zhang, Y.; Lee, R.; Sand, D.; Scheuller, H.S.; Heber, D. Blackberry, black raspberry, blueberry, cranberry, red raspberry, and strawberry extracts inhibit growth and stimulate apoptosis of human cancer cells in vitro. *J. Agric. Food Chem.* **2006**, *54*, 9329–9339. [CrossRef] [PubMed]
25. Gomes, B.S.; Neto, B.P.S.; Lopes, E.M.; Cunha, F.V.M.; Araújo, A.R.; Wanderley, C.W.S.; Wong, D.V.T.; Júnior, R.C.P.L.; Ribeiro, R.A.; Sousa, D.P.; et al. Anti-Inflammatory Effect of the Monoterpene Myrtenol Is Dependent on the Direct Modulation of Neutrophil Migration and Oxidative Stress. *Chem.-Biol. Interact.* **2017**, *273*, 73–81. [CrossRef] [PubMed]
26. Aguiar, J.; Gonçalves, J.L.; Alves, V.L.; Câmara, J.S. Relationship between Volatile Composition and Bioactive Potential of Vegetables and Fruits of Regular Consumption—An Integrative Approach. *Molecules* **2021**, *26*, 3653. [CrossRef] [PubMed]
27. Pichersky, E.; Gershenzon, J. The Formation and Function of Plant Volatiles: Perfumes for Pollinator Attraction and Defense. *Curr. Opin. Plant Biol.* **2002**, *5*, 237–243. [CrossRef]
28. Muhlemann, J.K.; Klempien, A.; Dudareva, N. Floral Volatiles: From Biosynthesis to Function. *Plant Cell Environ.* **2014**, *37*, 1936–1949. [CrossRef]
29. Dewick, P.M. The Biosynthesis of C5–C25 Terpenoid Compounds. *Nat. Prod. Rep.* **2002**, *19*, 181–222. [CrossRef]
30. McGarvey, D.J.; Croteau, R. Terpenoid Metabolism. *Plant Cell* **1995**, *7*, 1015–1026. [CrossRef]
31. Zuzarte, M.; Salgueiro, L. Essential Oils Chemistry. In *Bioactive Essential Oils and Cancer*; de Sousa, D.P., Ed.; Springer International Publishing: Cham, Switzerland, 2015; pp. 19–61. ISBN 978-3-319-19144-7.
32. Croteau, R.; Kutchan, T.M.; Lewis, N.G. Natural Products (Secondary Metabolites). In *Biochemistry and Molecular Biology of Plants*; American Society of Plant Physiologists: Rockville, MD, USA, 2000.
33. Liu, Y.; Luo, S.-H.; Schmidt, A.; Wang, G.-D.; Sun, G.-L.; Grant, M.; Kuang, C.; Yang, M.-J.; Jing, S.-X.; Li, C.-H.; et al. A Geranylfarnesyl Diphosphate Synthase Provides the Precursor for Sesterterpenoid (C$_{25}$) Formation in the Glandular Trichomes of the Mint Species *Leucosceptrum canum*. *Plant Cell* **2016**, *28*, 804–822. [CrossRef]
34. Loza-Tavera, H. Monoterpenes in Essential Oils. In *Chemicals via Higher Plant Bioengineering*; Shahidi, F., Kolodziejczyk, P., Whitaker, J.R., Munguia, A.L., Fuller, G., Eds.; Advances in Experimental Medicine and Biology; Springer: Boston, MA, USA, 1999; pp. 49–62. ISBN 978-1-4615-4729-7.
35. Dudareva, N.; Pichersky, E.; Gershenzon, J. Biochemistry of Plant Volatiles. *Plant Physiol.* **2004**, *135*, 1893–1902. [CrossRef]
36. Degenhardt, J.; Köllner, T.G.; Gershenzon, J. Monoterpene and sesquiterpene synthases and the origin of terpene skeletal diversity in plants. *Phytochemistry* **2009**, *70*, 1621–1637. [CrossRef]
37. Pulido, P.; Perello, C.; Rodriguez-Concepcion, M. New Insights into Plant Isoprenoid Metabolism. *Mol. Plant* **2012**, *5*, 964–967. [CrossRef] [PubMed]
38. Newman, J.D.; Chappell, J. Isoprenoid Biosynthesis in Plants: Carbon Partitioning Within the Cytoplasmic Pathway. *Crit. Rev. Biochem. Mol. Biol.* **1999**, *34*, 95–106. [CrossRef] [PubMed]
39. Rohmer, M.; Knani, M.; Simonin, P.; Sutter, B.; Sahm, H. Isoprenoid Biosynthesis in Bacteria: A Novel Pathway for the Early Steps Leading to Isopentenyl Diphosphate. *Biochem. J.* **1993**, *295*, 517–524. [CrossRef]
40. Lange, B.M.; Wildung, M.R.; Stauber, E.J.; Sanchez, C.; Pouchnik, D.; Croteau, R. Probing Essential Oil Biosynthesis and Secretion by Functional Evaluation of Expressed Sequence Tags from Mint Glandular Trichomes. *Proc. Natl. Acad. Sci. USA* **2000**, *97*, 2934–2939. [CrossRef] [PubMed]
41. Tholl, D. Biosynthesis and Biological Functions of Terpenoids in Plants. In *Biotechnology of Isoprenoids*; Schrader, J., Bohlmann, J., Eds.; Advances in Biochemical Engineering/Biotechnology; Springer International Publishing: Cham, Switzerland, 2015; pp. 63–106. ISBN 978-3-319-20107-8.
42. Berthelot, K.; Estevez, Y.; Deffieux, A.; Peruch, F. Isopentenyl Diphosphate Isomerase: A Checkpoint to Isoprenoid Biosynthesis. *Biochimie* **2012**, *94*, 1621–1634. [CrossRef]
43. Rohdich, F.; Hecht, S.; Gärtner, K.; Adam, P.; Krieger, C.; Amslinger, S.; Arigoni, D.; Bacher, A.; Eisenreich, W. Studies on the Nonmevalonate Terpene Biosynthetic Pathway: Metabolic Role of IspH (LytB) Protein. *Proc. Natl. Acad. Sci. USA* **2002**, *99*, 1158–1163. [CrossRef]
44. Tritsch, D.; Hemmerlin, A.; Bach, T.J.; Rohmer, M. Plant Isoprenoid Biosynthesis via the MEP Pathway: In Vivo IPP/DMAPP Ratio Produced by (E)-4-Hydroxy-3-Methylbut-2-Enyl Diphosphate Reductase in Tobacco BY-2 Cell Cultures. *FEBS Lett.* **2010**, *584*, 129–134. [CrossRef]
45. Vranová, E.; Coman Schmid, D.; Gruissem, W. Network Analysis of the MVA and MEP Pathways for Isoprenoid Synthesis. *Annu. Rev. Plant Biol.* **2013**, *64*, 665–700. [CrossRef]
46. Abbas, F.; Ke, Y.; Yu, R.; Yue, Y.; Amanullah, S.; Jahangir, M.M.; Fan, Y. Volatile Terpenoids: Multiple Functions, Biosynthesis, Modulation and Manipulation by Genetic Engineering. *Planta* **2017**, *246*, 803–816. [CrossRef]
47. Hemmerlin, A.; Hoeffler, J.-F.; Meyer, O.; Tritsch, D.; Kagan, I.A.; Grosdemange-Billiard, C.; Rohmer, M.; Bach, T.J. Cross-Talk between the Cytosolic Mevalonate and the Plastidial Methylerythritol Phosphate Pathways in Tobacco Bright Yellow-2 Cells *. *J. Biol. Chem.* **2003**, *278*, 26666–26676. [CrossRef]
48. Laule, O.; Fürholz, A.; Chang, H.-S.; Zhu, T.; Wang, X.; Heifetz, P.B.; Gruissem, W.; Lange, M. Crosstalk between Cytosolic and Plastidial Pathways of Isoprenoid Biosynthesis in Arabidopsis Thaliana. *Proc. Natl. Acad. Sci. USA* **2003**, *100*, 6866–6871. [CrossRef] [PubMed]

49. Dudareva, N.; Andersson, S.; Orlova, I.; Gatto, N.; Reichelt, M.; Rhodes, D.; Boland, W.; Gershenzon, J. The Nonmevalonate Pathway Supports Both Monoterpene and Sesquiterpene Formation in Snapdragon Flowers. *Proc. Natl. Acad. Sci. USA* **2005**, *102*, 933–938. [CrossRef] [PubMed]

50. Nagegowda, D.A. Plant Volatile Terpenoid Metabolism: Biosynthetic Genes, Transcriptional Regulation and Subcellular Compartmentation. *FEBS Lett.* **2010**, *584*, 2965–2973. [CrossRef] [PubMed]

51. Weaver, L.M.; Herrmann, K.M. Dynamics of the Shikimate Pathway in Plants. *Trends Plant Sci.* **1997**, *2*, 346–351. [CrossRef]

52. Sangwan, N.S.; Farooqi, A.H.A.; Shabih, F.; Sangwan, R.S. Regulation of Essential Oil Production in Plants. *Plant Growth Regul.* **2001**, *34*, 3–21. [CrossRef]

53. Shrivastava, G.; Rogers, M.; Wszelaki, A.; Panthee, D.R.; Chen, F. Plant Volatiles-Based Insect Pest Management in Organic Farming. *Crit. Rev. Plant Sci.* **2010**, *29*, 123–133. [CrossRef]

54. Porta, H.; Rocha-Sosa, M. Plant Lipoxygenases. Physiological and Molecular Features. *Plant Physiol.* **2002**, *130*, 15–21. [CrossRef]

55. Tholl, D. Terpene Synthases and the Regulation, Diversity and Biological Roles of Terpene Metabolism. *Curr. Opin. Plant Biol.* **2006**, *9*, 297–304. [CrossRef]

56. Tholl, D.; Boland, W.; Hansel, A.; Loreto, F.; Röse, U.S.R.; Schnitzler, J.-P. Practical Approaches to Plant Volatile Analysis. *Plant J.* **2006**, *45*, 540–560. [CrossRef]

57. Gershenzon, J.; Croteau, R. Regulation of monoterpene biosynthesis in higher plants. In *Biochemistry of the Mevalonic Acid pathway to Terpenoids*; Springer: Boston, MA, USA, 1990; pp. 99–160.

58. Schenkel, D.; Maciá-Vicente, J.G.; Bissell, A.; Splivallo, R. Fungi Indirectly Affect Plant Root Architecture by Modulating Soil Volatile Organic Compounds. *Front. Microbiol.* **2018**, *9*, 1847. [CrossRef]

59. Wang, S.-Y.; Shi, X.-C.; Liu, F.-Q.; Laborda, P. Effects of Exogenous Methyl Jasmonate on Quality and Preservation of Postharvest Fruits: A Review. *Food Chem.* **2021**, *353*, 129482. [CrossRef] [PubMed]

60. Chanjirakul, K.; Wang, S.Y.; Wang, C.Y.; Siriphanich, J. Natural Volatile Treatments Increase Free-Radical Scavenging Capacity of Strawberries and Blackberries. *J. Sci. Food Agric.* **2007**, *87*, 1463–1472. [CrossRef]

61. Sangiorgio, D.; Cellini, A.; Spinelli, F.; Pastore, C.; Farneti, B.; Savioli, S.; Rodriguez-Estrada, M.T.; Donati, I. Contribution of Fruit Microbiome to Raspberry Volatile Organic Compounds Emission. *Postharvest Biol. Technol.* **2022**, *183*, 111742. [CrossRef]

62. Egea, M.B.; Bertolo, M.R.V.; de Oliveira Filho, J.G.; Lemes, A.C. A Narrative Review of the Current Knowledge on Fruit Active Aroma Using Gas Chromatography-Olfactometry (GC-O) Analysis. *Molecules* **2021**, *26*, 5181. [CrossRef] [PubMed]

63. Ferrão, L.F.V.; Johnson, T.S.; Benevenuto, J.; Edger, P.P.; Colquhoun, T.A.; Munoz, P.R. Genome-Wide Association of Volatiles Reveals Candidate Loci for Blueberry Flavor. *New Phytol.* **2020**, *226*, 1725–1737. [CrossRef] [PubMed]

64. Colantonio, V.; Ferrão, L.F.V.; Tieman, D.M.; Bliznyuk, N.; Sims, C.; Klee, H.J.; Munoz, P.; Resende, M.F.R. Metabolomic Selection for Enhanced Fruit Flavor. *Proc. Natl. Acad. Sci. USA* **2022**, *119*, e2115865119. [CrossRef] [PubMed]

65. Lu, H.; Ban, Z.; Wang, K.; Li, D.; Li, D.; Poverenov, E.; Li, L.; Luo, Z. Aroma Volatiles, Sensory and Chemical Attributes of Strawberry (*Fragaria* × *Ananassa* Duch.) Achenes and Receptacle. *Int. J. Food Sci. Technol.* **2017**, *52*, 2614–2622. [CrossRef]

66. Yan, J.; Ban, Z.; Lu, H.; Li, D.; Poverenov, E.; Luo, Z.; Li, L. The Aroma Volatile Repertoire in Strawberry Fruit: A Review. *J. Sci. Food Agric.* **2018**, *98*, 4395–4402. [CrossRef]

67. Forney, C.F. Horticultural and Other Factors Affecting Aroma Volatile Composition of Small Fruit. *HortTechnology* **2001**, *11*, 529–538. [CrossRef]

68. Du, X.; Plotto, A.; Song, M.; Olmstead, J.; Rouseff, R. Volatile Composition of Four Southern Highbush Blueberry Cultivars and Effect of Growing Location and Harvest Date. *J. Agric. Food Chem.* **2011**, *59*, 8347–8357. [CrossRef]

69. Gu, I.; Brownmiller, C.; Stebbins, N.B.; Mauromoustakos, A.; Howard, L.; Lee, S.-O. Berry Phenolic and Volatile Extracts Inhibit Pro-Inflammatory Cytokine Secretion in LPS-Stimulated RAW264.7 Cells through Suppression of NF-KB Signaling Pathway. *Antioxidants* **2020**, *9*, 871. [CrossRef] [PubMed]

70. Morin, P.; Howard, L.; Tipton, J.; Lavefve, L.; Brownmiller, C.; Lee, S.-O.; Gu, I.; Liyanage, R.; Lay, J.O. Blackberry Phenolic and Volatile Extracts Inhibit Cytokine Secretion in LPS-Inflamed RAW264.7 Cells. *J. Food Bioact.* **2021**, *16*, 34–47. [CrossRef]

71. Wang, Y.; Finn, C.; Qian, M.C. Impact of Growing Environment on Chickasaw Blackberry (*Rubus*, L.) Aroma Evaluated by Gas Chromatography Olfactometry Dilution Analysis. *J. Agric. Food Chem.* **2005**, *53*, 3563–3571. [CrossRef] [PubMed]

72. Oz, A.T.; Baktemur, G.; Kargi, S.P.; Kafkas, E. Volatile Compounds of Strawberry Varieties. *Chem. Nat. Compd.* **2016**, *52*, 507–509. [CrossRef]

73. Qian, M.C.; Wang, Y. Seasonal Variation of Volatile Composition and Odor Activity Value of 'Marion' (*Rubus* Spp. Hyb) and 'Thornless Evergreen' (*R. Laciniatus*, L.) Blackberries. *J. Food Sci.* **2005**, *70*, C13–C20. [CrossRef]

74. Du, X.; Rouseff, R. Aroma Active Volatiles in Four Southern Highbush Blueberry Cultivars Determined by Gas Chromatography–Olfactometry (GC-O) and Gas Chromatography–Mass Spectrometry (GC-MS). *J. Agric. Food Chem.* **2014**, *62*, 4537–4543. [CrossRef]

75. Zhu, J.; Chen, F.; Wang, L.; Niu, Y.; Chen, H.; Wang, H.; Xiao, Z. Characterization of the Key Aroma Volatile Compounds in Cranberry (Vaccinium Macrocarpon Ait.) Using Gas Chromatography–Olfactometry (GC-O) and Odor Activity Value (OAV). *J. Agric. Food Chem.* **2016**, *64*, 4990–4999. [CrossRef]

76. El Hadi, M.A.M.; Zhang, F.-J.; Wu, F.-F.; Zhou, C.-H.; Tao, J. Advances in Fruit Aroma Volatile Research. *Molecules* **2013**, *18*, 8200–8229. [CrossRef]

77. Ruse, K.; Sabovics, M.; Rakcejeva, T.; Dukalska, L.; Galoburda, R.; Berzina, L. The Effect of Drying Conditions on the Presence of Volatile Compounds in Cranberries. *Int. J. Agric. Biosyst. Eng.* **2012**, *6*, 163–169.

78. Farneti, B.; Khomenko, I.; Grisenti, M.; Ajelli, M.; Betta, E.; Algarra, A.A.; Cappellin, L.; Aprea, E.; Gasperi, F.; Biasioli, F.; et al. Exploring Blueberry Aroma Complexity by Chromatographic and Direct-Injection Spectrometric Techniques. *Front. Plant Sci.* **2017**, *8*, 617. [CrossRef]
79. Malowicki, S.M.M.; Martin, R.; Qian, M.C. Volatile Composition in Raspberry Cultivars Grown in the Pacific Northwest Determined by Stir Bar Sorptive Extraction—Gas Chromatography—Mass Spectrometry. *J. Agric. Food Chem.* **2008**, *56*, 4128–4133. [CrossRef] [PubMed]
80. D'Agostino, M.F.; Sanz, J.; Sanz, M.L.; Giuffrè, A.M.; Sicari, V.; Soria, A.C. Optimization of a Solid-Phase Microextraction Method for the Gas Chromatography–Mass Spectrometry Analysis of Blackberry (*Rubus Ulmifolius Schott*) Fruit Volatiles. *Food Chem.* **2015**, *178*, 10–17. [CrossRef] [PubMed]
81. Gilbert, J.L.; Schwieterman, M.L.; Colquhoun, T.A.; Clark, D.G.; Olmstead, J.W. Potential for Increasing Southern Highbush Blueberry Flavor Acceptance by Breeding for Major Volatile Components. *HortScience* **2013**, *48*, 835–843. [CrossRef]
82. Hirvi, T.; Honkanen, E.; Pyysalo, T. The Aroma of Cranberries. *Z. Für Lebensm. -Unters. Und Forsch.* **1981**, *172*, 365–367. [CrossRef]
83. Moore, K.; Howard, L.; Brownmiller, C.; Gu, I.; Lee, S.-O.; Mauromoustakos, A. Inhibitory Effects of Cranberry Polyphenol and Volatile Extracts on Nitric Oxide Production in LPS Activated RAW 264.7 Macrophages. *Food Funct.* **2019**, *10*, 7091–7102. [CrossRef]
84. Khomych, G.; Matsuk, Y.; Nakonechnaya, J.; Oliynyk, N.; Medved, L. Study of the Chemical Composition of Cranberry and the Use of Berries in Food Technology. *EEJET* **2017**, *6*, 59–65. [CrossRef]
85. Schwieterman, M.L.; Colquhoun, T.A.; Jaworski, E.A.; Bartoshuk, L.M.; Gilbert, J.L.; Tieman, D.M.; Odabasi, A.Z.; Moskowitz, H.R.; Folta, K.M.; Klee, H.J.; et al. Strawberry Flavor: Diverse Chemical Compositions, a Seasonal Influence, and Effects on Sensory Perception. *PLoS ONE* **2014**, *9*, e88446. [CrossRef]
86. Aharoni, A.; Giri, A.P.; Verstappen, F.W.A.; Bertea, C.M.; Sevenier, R.; Sun, Z.; Jongsma, M.A.; Schwab, W.; Bouwmeester, H.J. Gain and Loss of Fruit Flavor Compounds Produced by Wild and Cultivated Strawberry Species. *Plant Cell* **2004**, *16*, 3110–3131. [CrossRef]
87. Giuggioli, N.R.; Girgenti, V.; Baudino, C.; Peano, C. Influence of Modified Atmosphere Packaging Storage on Postharvest Quality and Aroma Compounds of Strawberry Fruits in a Short Distribution Chain. *J. Food Process. Preserv.* **2015**, *39*, 3154–3164. [CrossRef]
88. Baby, B.; Antony, P.; Vijayan, R. Antioxidant and Anticancer Properties of Berries. *Crit. Rev. Food Sci. Nutr.* **2018**, *58*, 2491–2507. [CrossRef]
89. Ménager, I.; Jost, M.; Aubert, C. Changes in Physicochemical Characteristics and Volatile Constituents of Strawberry (Cv. Cigaline) during Maturation. *J. Agric. Food Chem.* **2004**, *52*, 1248–1254. [CrossRef] [PubMed]
90. Kafkas, E.; Türemiş, N.; Bilgili, B.; Zarifikhosroshahi, M.; Burgut, A.; Kafkas, S. Aroma Profiles of Organically Grown 'Benicia' and 'Albion' Strawberries. *Acta Hortic.* **2017**, 703–708. [CrossRef]
91. Song, C.; Hong, X.; Zhao, S.; Liu, J.; Schulenburg, K.; Huang, F.-C.; Franz-Oberdorf, K.; Schwab, W. Glucosylation of 4-Hydroxy-2,5-Dimethyl-3(2H)-Furanone, the Key Strawberry Flavor Compound in Strawberry Fruit. *Plant Physiol.* **2016**, *171*, 139–151. [CrossRef] [PubMed]
92. Jetti, R.R.; Yang, E.; Kurnianta, A.; Finn, C.; Qian, M.C. Quantification of Selected Aroma-Active Compounds in Strawberries by Headspace Solid-Phase Microextraction Gas Chromatography and Correlation with Sensory Descriptive Analysis. *J. Food Sci.* **2007**, *72*, S487–S496. [CrossRef] [PubMed]
93. Hampel, D.; Mosandl, A.; Wüst, M. Biosynthesis of Mono- and Sesquiterpenes in Strawberry Fruits and Foliage: 2H Labeling Studies. *J. Agric. Food Chem.* **2006**, *54*, 1473–1478. [CrossRef]
94. Urrutia, M.; Rambla, J.L.; Alexiou, K.G.; Granell, A.; Monfort, A. Genetic Analysis of the Wild Strawberry (*Fragaria Vesca*) Volatile Composition. *Plant Physiol. Biochem.* **2017**, *121*, 99–117. [CrossRef]
95. Hook, G.L.; Kimm, G.L.; Hall, T.; Smith, P.A. Solid-Phase Microextraction (SPME) for Rapid Field Sampling and Analysis by Gas Chromatography-Mass Spectrometry (GC-MS). *TrAC Trends Anal. Chem.* **2002**, *21*, 534–543. [CrossRef]
96. Curran, A.M.; Rabin, S.I.; Prada, P.A.; Furton, K.G. Comparison of the Volatile Organic Compounds Present in Human Odor Using Spme-GC/MS. *J. Chem. Ecol.* **2005**, *31*, 1607–1619. [CrossRef]
97. Beaulieu, J.C.; Stein-Chisholm, R.E.; Lloyd, S.W.; Bett-Garber, K.L.; Grimm, C.C.; Watson, M.A.; Lea, J.M. Volatile, Anthocyanidin, Quality and Sensory Changes in Rabbiteye Blueberry from Whole Fruit through Pilot Plant Juice Processing. *J. Sci. Food Agric.* **2017**, *97*, 469–478. [CrossRef]
98. Zhu, N.; Zhu, Y.; Yu, N.; Wei, Y.; Zhang, J.; Hou, Y.; Sun, A. Evaluation of Microbial, Physicochemical Parameters and Flavor of Blueberry Juice after Microchip-Pulsed Electric Field. *Food Chem.* **2019**, *274*, 146–155. [CrossRef]
99. Bett-Garber, K.L.; Lea, J.M.; Watson, M.A.; Grimm, C.C.; Lloyd, S.W.; Beaulieu, J.C.; Stein-Chisholm, R.E.; Andrzejewski, B.P.; Marshall, D.A. Flavor of Fresh Blueberry Juice and the Comparison to Amount of Sugars, Acids, Anthocyanidins, and Physicochemical Measurements. *J. Food Sci.* **2015**, *80*, S818–S827. [CrossRef] [PubMed]
100. Eshghi, S.; Jamali, B.; Rowshan, V. Headspace Analysis of Aroma Composition and Quality Changes of Selva Strawberry (*Fragaria* x *Ananassa* Duch.), Fruits as Influenced by Salinity Stress and Application Timing of Nitric Oxide. *Anal. Chem. Lett.* **2014**, *4*, 178–189. [CrossRef]
101. Dymerski, T.; Namieśnik, J.; Vearasilp, K.; Arancibia-Avila, P.; Toledo, F.; Weisz, M.; Katrich, E.; Gorinstein, S. Comprehensive Two-Dimensional Gas Chromatography and Three-Dimensional Fluorometry for Detection of Volatile and Bioactive Substances in Some Berries. *Talanta* **2015**, *134*, 460–467. [CrossRef] [PubMed]

102. Klesk, K.; Qian, M.; Martin, R.R. Aroma Extract Dilution Analysis of Cv. Meeker (*Rubus Idaeus*, L.) Red Raspberries from Oregon and Washington. *J. Agric. Food Chem.* **2004**, *52*, 5155–5161. [CrossRef] [PubMed]
103. Hansen, A.-M.S.; Frandsen, H.L.; Fromberg, A. Authenticity of Raspberry Flavor in Food Products Using SPME-Chiral-GC-MS. *Food Sci. Nutr.* **2016**, *4*, 348–354. [CrossRef] [PubMed]
104. Ghaste, M.S. Comprehensive Mapping of Volatile Organic Compounds in Fruits. Ph.D. Thesis, University of Trento, Trento, Italy, 2015.
105. Aprea, E.; Biasioli, F.; Gasperi, F. Volatile Compounds of Raspberry Fruit: From Analytical Methods to Biological Role and Sensory Impact. *Molecules* **2015**, *20*, 2445–2474. [CrossRef] [PubMed]
106. Bautista-Rosales, P.U.; Ragazzo-Sánchez, J.A.; Ruiz-Montañez, G.; Ortiz-Basurto, R.I.; Luna-Solano, G.; Calderón-Santoyo, M. *Saccharomyces Cerevisiae* Mixed Culture of Blackberry (*Rubus Ulmifolius*, L.) Juice: Synergism in the Aroma Compounds Production. *Sci. World J.* **2014**, *2014*, e163174. [CrossRef] [PubMed]
107. Jurikova, T.; Skrovankova, S.; Mlcek, J.; Balla, S.; Snopek, L. Bioactive Compounds, Antioxidant Activity, and Biological Effects of European Cranberry (*Vaccinium Oxycoccos*). *Molecules* **2019**, *24*, 24. [CrossRef]
108. Research, C. for D.E. and Bioavailability and Bioequivalence Studies Submitted in NDAs or INDs—General Considerations. Available online: https://www.fda.gov/regulatory-information/search-fda-guidance-documents/bioavailability-and-bioequivalence-studies-submitted-ndas-or-inds-general-considerations (accessed on 10 August 2022).
109. Porrini, M.; Riso, P. Factors Influencing the Bioavailability of Antioxidants in Foods: A Critical Appraisal. *Nutr. Metab. Cardiovasc. Dis.* **2008**, *18*, 647–650. [CrossRef]
110. Winstanley, P.; Orme, M. The Effects of Food on Drug Bioavailability. *Br. J. Clin. Pharmacol.* **1989**, *28*, 621–628. [CrossRef]
111. Caldwell, J.; Gardner, I.; Swales, N. An Introduction to Drug Disposition: The Basic Principles of Absorption, Distribution, Metabolism, and Excretion. *Toxicol. Pathol.* **1995**, *23*, 102–114. [CrossRef] [PubMed]
112. Dima, C.; Assadpour, E.; Dima, S.; Jafari, S.M. Bioavailability of Nutraceuticals: Role of the Food Matrix, Processing Conditions, the Gastrointestinal Tract, and Nanodelivery Systems. *Compr. Rev. Food Sci. Food Saf.* **2020**, *19*, 954–994. [CrossRef] [PubMed]
113. Faridi Esfanjani, A.; Assadpour, E.; Jafari, S.M. Improving the Bioavailability of Phenolic Compounds by Loading Them within Lipid-Based Nanocarriers. *Trends Food Sci. Technol.* **2018**, *76*, 56–66. [CrossRef]
114. Rezaei, A.; Fathi, M.; Jafari, S.M. Nanoencapsulation of Hydrophobic and Low-Soluble Food Bioactive Compounds within Different Nanocarriers. *Food Hydrocoll.* **2019**, *88*, 146–162. [CrossRef]
115. Kohlert, C.; van Rensen, I.; März, R.; Schindler, G.; Graefe, E.U.; Veit, M. Bioavailability and Pharmacokinetics of Natural Volatile Terpenes in Animals and Humans. *Planta Med.* **2000**, *66*, 495–505. [CrossRef]
116. Verma, P.; Thakur, A.S.; Deshmukh, K.; Jha, A.K.; Verma, S. Routes of drug administration. *Int. J. Pharm. Stud. Res.* **2010**, *1*, 54–59.
117. Mun, H.; Townley, H.E. Nanoencapsulation of Plant Volatile Organic Compounds to Improve Their Biological Activities. *Planta Med.* **2021**, *87*, 236–251. [CrossRef]
118. Bouayed, J.; Bohn, T. *Nutrition, Well-Being and Health*; BoD—Books on Demand: Norderstedt, Germany, 2012; ISBN 978-953-51-0125-3.
119. Igimi, H.; Nishimura, M.; Kodama, R.; Ide, H. Studies on the Metabolism of D-Limonene (p-Mentha-1,8-Diene): I. The Absorption, Distribution and Excretion of d-Limonene in Rats. *Xenobiotica* **1974**, *4*, 77–84. [CrossRef]
120. Miyazawa, M.; Shindo, M.; Shimada, T. Metabolism of (+)- and (−)-Limonenes to Respective Carveols and Perillyl Alcohols by CYP2C9 and CYP2C19 in Human Liver Microsomes. *Drug Metab. Dispos.* **2002**, *30*, 602–607. [CrossRef]
121. Eriksson, K.; Levin, J.-O. Gas Chromatographic-Mass Spectrometric Identification of Metabolites from α-Pinene in Human Urine after Occupational Exposure to Sawing Fumes. *J. Chromatogr. B Biomed. Sci. Appl.* **1996**, *677*, 85–98. [CrossRef]
122. Papada, E.; Gioxari, A.; Amerikanou, C.; Galanis, N.; Kaliora, A.C. An Absorption and Plasma Kinetics Study of Monoterpenes Present in Mastiha Oil in Humans. *Foods* **2020**, *9*, 1019. [CrossRef] [PubMed]
123. Chadha, A.; Madyastha, K.M. Metabolism of Geraniol and Linalool in the Rat and Effects on Liver and Lung Microsomal Enzymes. *Xenobiotica* **1984**, *14*, 365–374. [CrossRef] [PubMed]
124. Madyastha, K.M.; Srivatsan, V. Biotransformations of α-Terpineol in the Rat: Its Effects on the Liver Microsomal Cytochrome P-450 System. *Bull. Environ. Contam. Toxicol.* **1988**, *41*, 17–25. [CrossRef] [PubMed]
125. El-Shemy, H. *Potential of Essential Oils*; BoD—Books on Demand: Norderstedt, Germany, 2018; ISBN 978-1-78923-779-5.
126. Erasto, P.; Viljoen, A.M. Limonene-a review: Biosynthetic, ecological and pharmacological relevance. *Nat. Prod. Commun.* **2008**, *3*, 1934578X0800300728. [CrossRef]
127. de Alvarenga, J.F.R.; Genaro, B.; Costa, B.L.; Purgatto, E.; Manach, C.; Fiamoncini, J. Monoterpenes: Current Knowledge on Food Source, Metabolism, and Health Effects. *Crit. Rev. Food Sci. Nutr.* **2021**, 1–38. [CrossRef] [PubMed]
128. Libby, P. Inflammation and Cardiovascular Disease Mechanisms. *Am. J. Clin. Nutr.* **2006**, *83*, 456S–460S. [CrossRef]
129. Ahmed, A.U. An Overview of Inflammation: Mechanism and Consequences. *Front. Biol.* **2011**, *6*, 274. [CrossRef]
130. Chi, H.; Barry, S.P.; Roth, R.J.; Wu, J.J.; Jones, E.A.; Bennett, A.M.; Flavell, R.A. Dynamic Regulation of Pro- and Anti-Inflammatory Cytokines by MAPK Phosphatase 1 (MKP-1) in Innate Immune Responses. *Proc. Natl. Acad. Sci. USA* **2006**, *103*, 2274–2279. [CrossRef]
131. Xie, C.; Kang, J.; Ferguson, M.E.; Nagarajan, S.; Badger, T.M.; Wu, X. Blueberries Reduce Pro-Inflammatory Cytokine TNF-α and IL-6 Production in Mouse Macrophages by Inhibiting NF-KB Activation and the MAPK Pathway. *Mol. Nutr. Food Res.* **2011**, *55*, 1587–1591. [CrossRef]

132. Ashley, N.T.; Weil, Z.M.; Nelson, R.J. Inflammation: Mechanisms, Costs, and Natural Variation. *Annu. Rev. Ecol. Evol. Syst.* **2012**, *43*, 385–406. [CrossRef]

133. Kyriakis, J.M.; Avruch, J. Mammalian MAPK Signal Transduction Pathways Activated by Stress and Inflammation: A 10-Year Update. *Physiol. Rev.* **2012**, *92*, 689–737. [CrossRef] [PubMed]

134. Roger, T.; David, J.; Glauser, M.P.; Calandra, T. MIF Regulates Innate Immune Responses through Modulation of Toll-like Receptor 4. *Nature* **2001**, *414*, 920–924. [CrossRef] [PubMed]

135. Lee, W.-S.; Shin, J.-S.; Jang, D.S.; Lee, K.-T. Cnidilide, an Alkylphthalide Isolated from the Roots of Cnidium Officinale, Suppresses LPS-Induced NO, PGE2, IL-1β, IL-6 and TNF-α Production by AP-1 and NF-KB Inactivation in RAW 264.7 Macrophages. *Int. Immunopharmacol.* **2016**, *40*, 146–155. [CrossRef] [PubMed]

136. Hinz, M.; Scheidereit, C. The IκB Kinase Complex in NF-KB Regulation and Beyond. *EMBO Rep.* **2014**, *15*, 46–61. [CrossRef] [PubMed]

137. Ahmed, S.M.U.; Luo, L.; Namani, A.; Wang, X.J.; Tang, X. Nrf2 Signaling Pathway: Pivotal Roles in Inflammation. *Biochim. Et Biophys. Acta (BBA)-Mol. Basis Dis.* **2017**, *1863*, 585–597. [CrossRef]

138. Bragado, P.; Armesilla, A.; Silva, A.; Porras, A. Apoptosis by Cisplatin Requires P53 Mediated P38α MAPK Activation through ROS Generation. *Apoptosis* **2007**, *12*, 1733–1742. [CrossRef]

139. Pellegrini, M.; Baldari, C.T. Apoptosis and Oxidative Stress-Related Diseases: The P66Shc Connection. *Curr. Mol. Med.* **2009**, *9*, 392–398. [CrossRef]

140. Cho, K.S.; Lim, Y.; Lee, K.; Lee, J.; Lee, J.H.; Lee, I.-S. Terpenes from Forests and Human Health. *Toxicol. Res.* **2017**, *33*, 97–106. [CrossRef]

141. Kim, T.; Song, B.; Cho, K.S.; Lee, I.-S. Therapeutic Potential of Volatile Terpenes and Terpenoids from Forests for Inflammatory Diseases. *Int. J. Mol. Sci.* **2020**, *21*, 2187. [CrossRef]

142. Amorim, J.L.; Simas, D.L.R.; Pinheiro, M.M.G.; Moreno, D.S.A.; Alviano, C.S.; da Silva, A.J.R.; Fernandes, P.D. Anti-Inflammatory Properties and Chemical Characterization of the Essential Oils of Four Citrus Species. *PLoS ONE* **2016**, *11*, e0153643. [CrossRef]

143. Chi, G.; Wei, M.; Xie, X.; Soromou, L.W.; Liu, F.; Zhao, S. Suppression of MAPK and NF-KB Pathways by Limonene Contributes to Attenuation of Lipopolysaccharide-Induced Inflammatory Responses in Acute Lung Injury. *Inflammation* **2013**, *36*, 501–511. [CrossRef] [PubMed]

144. Rufino, A.T.; Ribeiro, M.; Sousa, C.; Judas, F.; Salgueiro, L.; Cavaleiro, C.; Mendes, A.F. Evaluation of the Anti-Inflammatory, Anti-Catabolic and pro-Anabolic Effects of E-Caryophyllene, Myrcene and Limonene in a Cell Model of Osteoarthritis. *Eur. J. Pharmacol.* **2015**, *750*, 141–150. [CrossRef] [PubMed]

145. Yu, L.; Yan, J.; Sun, Z. D-Limonene Exhibits Anti-Inflammatory and Antioxidant Properties in an Ulcerative Colitis Rat Model via Regulation of INOS, COX-2, PGE2 and ERK Signaling Pathways. *Mol. Med. Rep.* **2017**, *15*, 2339–2346. [CrossRef] [PubMed]

146. de Oliveira Ramalho, T.R.; de Oliveira, M.T.P.; de Lima, A.L.A.; Bezerra-Santos, C.R.; Piuvezam, M.R. Gamma-Terpinene Modulates Acute Inflammatory Response in Mice. *Planta Med.* **2015**, *81*, 1248–1254. [CrossRef]

147. Jeong, J.B.; Jeong, H.J. Rheosmin, a Naturally Occurring Phenolic Compound Inhibits LPS-Induced INOS and COX-2 Expression in RAW264.7 Cells by Blocking NF-KB Activation Pathway. *Food Chem. Toxicol.* **2010**, *48*, 2148–2153. [CrossRef]

148. Zhang, Z.; Shen, P.; Lu, X.; Li, Y.; Liu, J.; Liu, B.; Fu, Y.; Cao, Y.; Zhang, N. In Vivo and In Vitro Study on the Efficacy of Terpinen-4-Ol in Dextran Sulfate Sodium-Induced Mice Experimental Colitis. *Front. Immunol.* **2017**, *8*, 558. [CrossRef]

149. Ning, J.; Xu, L.; Zhao, Q.; Zhang, Y.; Shen, C. The Protective Effects of Terpinen-4-Ol on LPS-Induced Acute Lung Injury via Activating PPAR-γ. *Inflammation* **2018**, *41*, 2012–2017. [CrossRef]

150. Sabogal-Guáqueta, A.M.; Osorio, E.; Cardona-Gómez, G.P. Linalool Reverses Neuropathological and Behavioral Impairments in Old Triple Transgenic Alzheimer's Mice. *Neuropharmacology* **2016**, *102*, 111–120. [CrossRef]

151. Lee, S.-C.; Wang, S.-Y.; Li, C.-C.; Liu, C.-T. Anti-Inflammatory Effect of Cinnamaldehyde and Linalool from the Leaf Essential Oil of Cinnamomum Osmophloeum Kanehira in Endotoxin-Induced Mice. *J. Food Drug Anal.* **2018**, *26*, 211–220. [CrossRef]

152. Kim, M.-G.; Kim, S.-M.; Min, J.-H.; Kwon, O.-K.; Park, M.-H.; Park, J.-W.; Ahn, H.I.; Hwang, J.-Y.; Oh, S.-R.; Lee, J.-W.; et al. Anti-Inflammatory Effects of Linalool on Ovalbumin-Induced Pulmonary Inflammation. *Int. Immunopharmacol.* **2019**, *74*, 105706. [CrossRef]

153. Ma, J.; Xu, H.; Wu, J.; Qu, C.; Sun, F.; Xu, S. Linalool Inhibits Cigarette Smoke-Induced Lung Inflammation by Inhibiting NF-KB Activation. *Int. Immunopharmacol.* **2015**, *29*, 708–713. [CrossRef] [PubMed]

154. Huo, M.; Cui, X.; Xue, J.; Chi, G.; Gao, R.; Deng, X.; Guan, S.; Wei, J.; Soromou, L.W.; Feng, H.; et al. Anti-Inflammatory Effects of Linalool in RAW 264.7 Macrophages and Lipopolysaccharide-Induced Lung Injury Model. *J. Surg. Res.* **2013**, *180*, e47–e54. [CrossRef] [PubMed]

155. Wu, Q.; Yu, L.; Qiu, J.; Shen, B.; Wang, D.; Soromou, L.W.; Feng, H. Linalool Attenuates Lung Inflammation Induced by Pasteurella Multocida via Activating Nrf-2 Signaling Pathway. *Int. Immunopharmacol.* **2014**, *21*, 456–463. [CrossRef]

156. Li, Y.; Lv, O.; Zhou, F.; Li, Q.; Wu, Z.; Zheng, Y. Linalool Inhibits LPS-Induced Inflammation in BV2 Microglia Cells by Activating Nrf2. *Neurochem Res* **2015**, *40*, 1520–1525. [CrossRef] [PubMed]

157. Kim, D.-S.; Lee, H.-J.; Jeon, Y.-D.; Han, Y.-H.; Kee, J.-Y.; Kim, H.-J.; Shin, H.-J.; Kang, J.; Lee, B.S.; Kim, S.-H.; et al. Alpha-Pinene Exhibits Anti-Inflammatory Activity Through the Suppression of MAPKs and the NF-KB Pathway in Mouse Peritoneal Macrophages. *Am. J. Chin. Med.* **2015**, *43*, 731–742. [CrossRef]

158. Porres-Martínez, M.; González-Burgos, E.; Carretero, M.E.; Gómez-Serranillos, M.P. Major Selected Monoterpenes α-Pinene and 1,8-Cineole Found in *Salvia Lavandulifolia* (Spanish Sage) Essential Oil as Regulators of Cellular Redox Balance. *Pharm. Biol.* **2015**, *53*, 921–929. [CrossRef]

159. Lemes, R.S.; Alves, C.C.F.; Estevam, E.B.B.; Santiago, M.B.; Martins, C.H.G.; Santos, T.C.L.D.; Crotti, A.E.M.; Miranda, M.L.D. Chemical Composition and Antibacterial Activity of Essential Oils from Citrus Aurantifolia Leaves and Fruit Peel against Oral Pathogenic Bacteria. *An. Acad. Bras. Ciênc.* **2018**, *90*, 1285–1292. [CrossRef]

160. Rehman, M.U.; Tahir, M.; Khan, A.Q.; Khan, R.; Oday-O-Hamiza; Lateef, A.; Hassan, S.K.; Rashid, S.; Ali, N.; Zeeshan, M.; et al. D-Limonene Suppresses Doxorubicin-Induced Oxidative Stress and Inflammation via Repression of COX-2, INOS, and NFκB in Kidneys of Wistar Rats. *Exp. Biol. Med.* **2014**, *239*, 465–476. [CrossRef]

161. Tak, P.P.; Firestein, G.S. NF-KB: A Key Role in Inflammatory Diseases. *J. Clin. Investig.* **2001**, *107*, 7–11. [CrossRef]

162. Kannan, K.; Jain, S.K. Oxidative Stress and Apoptosis. *Pathophysiology* **2000**, *7*, 153–163. [CrossRef]

163. Cutillas, A.-B.; Carrasco, A.; Martinez-Gutierrez, R.; Tomas, V.; Tudela, J. *Thymus Mastichina*, L. Essential Oils from Murcia (Spain): Composition and Antioxidant, Antienzymatic and Antimicrobial Bioactivities. *PLoS ONE* **2018**, *13*, e0190790. [CrossRef] [PubMed]

164. Kim, S.-H.; Bae, H.C.; Park, E.-J.; Lee, C.R.; Kim, B.-J.; Lee, S.; Park, H.H.; Kim, S.-J.; So, I.; Kim, T.W.; et al. Geraniol Inhibits Prostate Cancer Growth by Targeting Cell Cycle and Apoptosis Pathways. *Biochem. Biophys. Res. Commun.* **2011**, *407*, 129–134. [CrossRef] [PubMed]

165. Lee, S.; Park, Y.R.; Kim, S.-H.; Park, E.-J.; Kang, M.J.; So, I.; Chun, J.N.; Jeon, J.-H. Geraniol Suppresses Prostate Cancer Growth through Down-Regulation of E2F8. *Cancer Med.* **2016**, *5*, 2899–2908. [CrossRef] [PubMed]

166. Yang, H.; Liu, G.; Zhao, H.; Dong, X.; Yang, Z. Inhibiting the JNK/ERK Signaling Pathway with Geraniol for Attenuating the Proliferation of Human Gastric Adenocarcinoma AGS Cells. *J. Biochem. Mol. Toxicol.* **2021**, *35*, e22818. [CrossRef]

167. Zhao, Y.; Chen, R.; Wang, Y.; Qing, C.; Wang, W.; Yang, Y. In Vitro and In Vivo Efficacy Studies of Lavender Angustifolia Essential Oil and Its Active Constituents on the Proliferation of Human Prostate Cancer. *Integr. Cancer Ther.* **2017**, *16*, 215–226. [CrossRef]

168. Mirza, M.B.; Elkady, A.I.; Al-Attar, A.M.; Syed, F.Q.; Mohammed, F.A.; Hakeem, K.R. Induction of Apoptosis and Cell Cycle Arrest by Ethyl Acetate Fraction of *Phoenix Dactylifera*, L. (Ajwa Dates) in Prostate Cancer Cells. *J. Ethnopharmacol.* **2018**, *218*, 35–44. [CrossRef]

169. Zhao, Y.; Cheng, X.; Wang, G.; Liao, Y.; Qing, C. Linalool Inhibits 22Rv1 Prostate Cancer Cell Proliferation and Induces Apoptosis. *Oncol. Lett.* **2020**, *20*, 289. [CrossRef]

170. Ou-Yang, F.; Tsai, I.-H.; Tang, J.-Y.; Yen, C.-Y.; Cheng, Y.-B.; Farooqi, A.A.; Chen, S.-R.; Yu, S.-Y.; Kao, J.-K.; Chang, H.-W. Antiproliferation for Breast Cancer Cells by Ethyl Acetate Extract of *Nepenthes Thorellii* × (*Ventricosa* × *Maxima*). *Int. J. Mol. Sci.* **2019**, *20*, 3238. [CrossRef]

171. Abou Baker, D.H. Achillea Millefolium, L. Ethyl Acetate Fraction Induces Apoptosis and Cell Cycle Arrest in Human Cervical Cancer (HeLa) Cells. *Ann. Agric. Sci.* **2020**, *65*, 42–48. [CrossRef]

172. Noor, R.; Astuti, I. Cytotoxicity of α-terpineol in HeLa cell line and its effects to apoptosis and cell cycle. *J. Med. Sci. (Berk. Ilmu Kedokt.)* **2014**, *46*, 1–9.

173. Chang, M.-Y.; Shieh, D.-E.; Chen, C.-C.; Yeh, C.-S.; Dong, H.-P. Linalool Induces Cell Cycle Arrest and Apoptosis in Leukemia Cells and Cervical Cancer Cells through CDKIs. *Int. J. Mol. Sci.* **2015**, *16*, 28169–28179. [CrossRef] [PubMed]

174. Pan, W.; Zhang, G. Linalool monoterpene exerts potent antitumor effects in OECM 1 human oral cancer cells by inducing sub-G1 cell cycle arrest, loss of mitochondrial membrane potential and inhibition of PI3K/AKT biochemical pathway. *J BUON* **2019**, *24*, 323–328. [PubMed]

175. Pushpalatha, R.; Sindhu, G.; Sharmila, R.; Tamizharasi, G.; Vennila, L.; Vijayalakshmi, A. Linalool Induces Reactive Oxygen Species Mediated Apoptosis in Human Oral Squamous Carcinoma Cells. *Indian J. Pharm. Sci.* **2021**, *83*, 906–917. [CrossRef]

176. Rodenak-Kladniew, B.; Castro, M.A.; Crespo, R.; Galle, M.; de Bravo, M.G. Anti-Cancer Mechanisms of Linalool and 1,8-Cineole in Non-Small Cell Lung Cancer A549 Cells. *Heliyon* **2020**, *6*, e05639. [CrossRef] [PubMed]

177. Yu, X.; Lin, H.; Wang, Y.; Lv, W.; Zhang, S.; Qian, Y.; Deng, X.; Feng, N.; Yu, H.; Qian, B. D-Limonene Exhibits Antitumor Activity by Inducing Autophagy and Apoptosis in Lung Cancer. *OTT* **2018**, *11*, 1833–1847. [CrossRef] [PubMed]

178. Chidambara Murthy, K.N.; Jayaprakasha, G.K.; Patil, B.S. D-Limonene Rich Volatile Oil from Blood Oranges Inhibits Angiogenesis, Metastasis and Cell Death in Human Colon Cancer Cells. *Life Sci.* **2012**, *91*, 429–439. [CrossRef]

179. Ye, Z.; Liang, Z.; Mi, Q.; Guo, Y. Limonene terpenoid obstructs human bladder cancer cell (T24 cell line) growth by inducing cellular apoptosis, caspase activation, G2/M phase cell cycle arrest and stops cancer metastasis. *J BUON/Off. J. Balk. Union. Oncol.* **2020**, *25*, 280–285.

180. Viana, A.F.S.C.; Lopes, M.T.P.; Oliveira, F.T.B.; Nunes, P.I.G.; Santos, V.G.; Braga, A.D.; Silva, A.C.A.; Sousa, D.P.; Viana, D.A.; Rao, V.S.; et al. (−)-Myrtenol Accelerates Healing of Acetic Acid-Induced Gastric Ulcers in Rats and in Human Gastric Adenocarcinoma Cells. *Eur. J. Pharmacol.* **2019**, *854*, 139–148. [CrossRef]

181. Wu, Z.; Li, Z.; Liang, Y. Myrcene Exerts Anti-Tumor Effects on Oral Cancer Cells in Vitro via Induction of Apoptosis. *Trop. J. Pharm. Res.* **2022**, *21*, 933–938. [CrossRef]

182. Wang, X.; Zhao, S.; Su, M.; Sun, L.; Zhang, S.; Wang, D.; Liu, Z.; Yuan, Y.; Liu, Y.; Li, Y. Geraniol Improves Endothelial Function by Inhibiting NOX-2 Derived Oxidative Stress in High Fat Diet Fed Mice. *Biochem. Biophys. Res. Commun.* **2016**, *474*, 182–187. [CrossRef]

183. de Sousa, G.M.; Cazarin, C.B.B.; Junior, M.R.M.; de Lamas, C.A.; Quitete, V.H.A.C.; Pastore, G.M.; Bicas, J.L. The Effect of α-Terpineol Enantiomers on Biomarkers of Rats Fed a High-Fat Diet. *Heliyon* **2020**, *6*, e03752. [CrossRef] [PubMed]

184. Li, D.; Wu, H.; Dou, H.; Guo, L.; Huang, W. Microcapsule of Sweet Orange Essential Oil Changes Gut Microbiota in Diet-Induced Obese Rats. *Biochem. Biophys. Res. Commun.* **2018**, *505*, 991–995. [CrossRef] [PubMed]
185. Lone, J.; Yun, J.W. Monoterpene Limonene Induces Brown Fat-like Phenotype in 3T3-L1 White Adipocytes. *Life Sci.* **2016**, *153*, 198–206. [CrossRef] [PubMed]
186. Ayala-Ruiz, L.A.; Ortega-Pérez, L.G.; Piñón-Simental, J.S.; Magaña-Rodriguez, O.R.; Meléndez-Herrera, E.; Rios-Chavez, P. Role of the Major Terpenes of Callistemon Citrinus against the Oxidative Stress during a Hypercaloric Diet in Rats. *Biomed. Pharmacother.* **2022**, *153*, 113505. [CrossRef]
187. Bacanlı, M.; Anlar, H.G.; Aydın, S.; Çal, T.; Arı, N.; Ündeğer Bucurgat, Ü.; Başaran, A.A.; Başaran, N. D-Limonene Ameliorates Diabetes and Its Complications in Streptozotocin-Induced Diabetic Rats. *Food Chem. Toxicol.* **2017**, *110*, 434–442. [CrossRef]
188. Murali, R.; Saravanan, R. Antidiabetic Effect of D-Limonene, a Monoterpene in Streptozotocin-Induced Diabetic Rats. *Biomed. Prev. Nutr.* **2012**, *2*, 269–275. [CrossRef]
189. More, T.A.; Kulkarni, B.R.; Nalawade, M.L.; Arvindekar, A.U. Antidiabetic activity of linalool and limonene in streptozotocin-induced diabetic rat: A combinatorial therapy approach. *Int. J. Pharm. Pharm. Sci.* **2014**, *6*, 159–163.
190. Murali, R.; Karthikeyan, A.; Saravanan, R. Protective Effects of D-Limonene on Lipid Peroxidation and Antioxidant Enzymes in Streptozotocin-Induced Diabetic Rats. *Basic Clin. Pharmacol. Toxicol.* **2013**, *112*, 175–181. [CrossRef]
191. El-Bassossy, H.M.; Ghaleb, H.; Elberry, A.A.; Balamash, K.S.; Ghareib, S.A.; Azhar, A.; Banjar, Z. Geraniol Alleviates Diabetic Cardiac Complications: Effect on Cardiac Ischemia and Oxidative Stress. *Biomed. Pharmacother.* **2017**, *88*, 1025–1030. [CrossRef]
192. Kamble, S.P.; Ghadyale, V.A.; Patil, R.S.; Haldavnekar, V.S.; Arvindekar, A.U. Inhibition of GLUT2 Transporter by Geraniol from Cymbopogon Martinii: A Novel Treatment for Diabetes Mellitus in Streptozotocin-Induced Diabetic Rats. *J. Pharm. Pharmacol.* **2020**, *72*, 294–304. [CrossRef]
193. Babukumar, S.; Vinothkumar, V.; Sankaranarayanan, C.; Srinivasan, S. Geraniol, a Natural Monoterpene, Ameliorates Hyperglycemia by Attenuating the Key Enzymes of Carbohydrate Metabolism in Streptozotocin-Induced Diabetic Rats. *Pharm. Biol.* **2017**, *55*, 1442–1449. [CrossRef]
194. El-Said, Y.A.M.; Sallam, N.A.A.; Ain-Shoka, A.A.-M.; Abdel-Latif, H.A.-T. Geraniol Ameliorates Diabetic Nephropathy via Interference with MiRNA-21/PTEN/Akt/MTORC1 Pathway in Rats. *Naunyn-Schmiedeberg's Arch Pharm.* **2020**, *393*, 2325–2337. [CrossRef] [PubMed]
195. El-Bassossy, H.M.; Elberry, A.A.; Ghareib, S.A. Geraniol Improves the Impaired Vascular Reactivity in Diabetes and Metabolic Syndrome through Calcium Channel Blocking Effect. *J. Diabetes Complicat.* **2016**, *30*, 1008–1016. [CrossRef] [PubMed]
196. Mahdavifard, S.; Nakhjavani, M. Effect of linalool on the activity of glyoxalase-I and diverse glycation products in rats with type 2 diabetes. *J. Maz. Univ. Med. Sci.* **2020**, *30*, 24–33.
197. Deepa, B.; Venkatraman Anuradha, C. Effects of Linalool on Inflammation, Matrix Accumulation and Podocyte Loss in Kidney of Streptozotocin-Induced Diabetic Rats. *Toxicol. Mech. Methods* **2013**, *23*, 223–234. [CrossRef] [PubMed]
198. Xuemei, L.; Qiu, S.; Chen, G.; Liu, M. Myrtenol Alleviates Oxidative Stress and Inflammation in Diabetic Pregnant Rats via TLR4/MyD88/NF-KB Signaling Pathway. *J. Biochem. Mol. Toxicol.* **2021**, *35*, e22904. [CrossRef] [PubMed]

Article

# Compound Identification from *Bromelia karatas* Fruit Juice Using Gas Chromatography–Mass Spectrometry and Evaluation of the Bactericidal Activity of the Extract

Benjamín A. Ayil-Gutiérrez [1], Karla Cecilia Amaya-Guardia [2], Arturo A. Alvarado-Segura [3], Glendy Polanco-Hernández [2], Miguel Angel Uc-Chuc [2], Karla Y. Acosta-Viana [2], Eugenia Guzmán-Marín [2], Blancka Yesenia Samaniego-Gámez [4], Wilberth Alfredo Poot-Poot [5], Gabriel Lizama-Uc [6] and Hernán de Jesús Villanueva-Alonzo [7,*]

[1] CONACYT-Biotecnología Vegetal, Centro de Biotecnología Genómica, Instituto Politécnico Nacional, Blvd. del Mtro, s/n, Esq. Elías Piña, Reynosa 88710, Mexico; bayil@ipn.mx
[2] Laboratorio de Biología Celular, Centro de Investigaciones Regionales "Dr. Hideyo Noguchi", Universidad Autónoma de Yucatán, Av. Itzáes, núm. 490 x calle 59, col. Centro, Mérida 97000, Mexico; karla.amaya@correo.uady.mx (K.C.A.-G.); glendy.polanco@correo.uady.mx (G.P.-H.); ma.uc@outlook.com (M.A.U.-C.); aviana@correo.uady.mx (K.Y.A.-V.); gmarin@correo.uady.mx (E.G.-M.)
[3] Tecnológico Nacional México, Instituto Tecnológico Superior del Sur del Edo. de Yucatán. Carr. Muna-Felipe Carrillo Puerto, tramo Oxkutzcab-Akil, Km. 41+400, Oxkutzcab 97880, Mexico; aalvarado@suryucatan.tecnm.mx
[4] Facultad de Ingeniería y Ciencias, Instituto de Ciencias Agrícolas, Universidad Autónoma de Baja California, Carretera a Ejido Delta, Mexicali 21705, Mexico; samaniego.blancka@uabc.edu.mx
[5] Laboratorio de Biotecnología Vegetal, Facultad de Ingeniería y Ciencias, Universidad Autónoma de Tamaulipas, Ciudad Victoria 87149, Mexico; waflaco@yahoo.com.mx
[6] Tecnológico Nacional de México, Instituto Tecnológico de Mérida, Unidad de Posgrado e Investigación Av. Tecnológico, Km. 4.5 S/N, Mérida 97000, Mexico; gabriel.lu@merida.tecnm.mx
[7] CONACYT-Laboratorio de Biología Celular, Centro de Investigaciones Regionales "Dr. Hideyo Noguchi", Universidad Autónoma de Yucatán, Av. Itzáes, núm. 490 x calle 59, col. Centro, Mérida 97000, Mexico
* Correspondence: hernan.villanueva@correo.uady.mx; Tel.: +52-999-924-5755

**Citation:** Ayil-Gutiérrez, B.A.; Amaya-Guardia, K.C.; Alvarado-Segura, A.A.; Polanco-Hernández, G.; Uc-Chuc, M.A.; Acosta-Viana, K.Y.; Guzmán-Marín, E.; Samaniego-Gámez, B.Y.; Poot-Poot, W.A.; Lizama-Uc, G.; et al. Compound Identification from *Bromelia karatas* Fruit Juice Using Gas Chromatography–Mass Spectrometry and Evaluation of the Bactericidal Activity of the Extract. *Appl. Sci.* **2022**, *12*, 7275. https://doi.org/10.3390/app12147275

Academic Editors: Luca Mazzoni, Maria Teresa Ariza Fernández and Franco Capocasa

Received: 21 May 2022
Accepted: 17 July 2022
Published: 20 July 2022

**Publisher's Note:** MDPI stays neutral with regard to jurisdictional claims in published maps and institutional affiliations.

**Copyright:** © 2022 by the authors. Licensee MDPI, Basel, Switzerland. This article is an open access article distributed under the terms and conditions of the Creative Commons Attribution (CC BY) license (https://creativecommons.org/licenses/by/4.0/).

**Abstract:** Fruits of species of the genus *Bromelia* contain compounds with health benefits and potential biotechnological applications. For example, *Bromelia karatas* fruits contain antioxidants and proteins with bactericidal activity, but studies regarding the activity of these metabolites and potential benefits are required. We evaluated the bactericidal activity of the methanolic extract (treated and not treated with activated charcoal) and its fractions (hexane, ethyl acetate, and methanol) from ripe *B. karatas* fruit (8 °Brix) against *Escherichia coli*, *Enterococcus faecalis*, *Salmonella enteritidis*, and *Shigella flexneri*. The methanolic extract (ME) minimum inhibitory concentration (MIC) and minimum bactericidal concentration (MBC) were determined at eight concentrations. The methanolic extract MIC was 5 mg/mL for *E. faecalis* and 10 mg/mL for the other bacteria; the MBC was 20 mg/mL for *E. coli* and *E. faecalis*, and 40 mg/mL for *S. enteritidis* and *S. flexneri*. Through gas chromatography–mass spectrometry, 131 compounds were identified, some of which had previously been reported to have biological activities, such as bactericidal, fungicide, anticancer, anti-inflammatory, enzyme inhibiting, and anti-allergic properties. The most abundant compounds found in the ME of *B. karatas* fruits were maleic anhydride, 5-hydroxymethylfurfural, and itaconic anhydride. This study shows that *B. karatas* fruits contain metabolites that are potentially beneficial for health.

**Keywords:** bactericidal activity; *Bromelia karatas*; isocitrate lyase; itaconic anhydride; maleic anhydride

## 1. Introduction

Functional foods, which are consumed regularly as part of humans' diets, can be defined as foods with bioactive components that, in addition to their impact on basic nutrition, have been scientifically proven to reduce the risk of disease [1,2]. The consumption of fruits as functional foods and the evaluation of the different types of biological activities of their secondary metabolites have been increasing due to consumers' awareness of the health benefits that they provide [3,4]. For example, there is evidence that for healthy women with recurrent urinary tract infections, the consumption of an optimal dose of cranberry extract can contribute to their prevention [5].

As Mexico is a megadiverse country with different vegetation types, it is home to a wide variety of wild fruits that are often used by one or more ethnic groups, such as the fruits of different species of bromeliads [6]. The genus *Bromelia* includes species that grow wild in Mexico and their fruits are known for their antifungal [7], bactericidal [6,8], and anthelmintic [9] activity. *Bromelia karatas* fruits contain antioxidants, proteases, and phenolic compounds, such as flavonoids, phenylpropanoids, terpenes, and coumarins [10,11]. Throughout tropical America, *B. karatas* fruits are consumed as foods and beverages, and their stems can be used as living fences. Traditional medicinal applications of this species include the treatment of helminth infections and certain types of ulcers [12,13]. It has been shown that prepurified proteases from *B. karatas* and *B. pinguin* fruits exhibited bactericidal activity against *Escherichia coli* and *Staphylococcus aureus*. However, it was shown that this activity is significantly reduced when the proteolytic extract is heated to 80 °C/15 min [14]. On the other hand, soluble protein extracts of *B. karatas* have been found to have a dose-dependent inhibitory effect on the growth of *Salmonella typhimurium* and *Listeria monocytogenes*, and proteolytic activity is suggested to play a role in the inhibition of *S. typhimurium* [15]. However, other studies regarding Bromeliads fruits have shown that methanolic extracts also have bactericidal activity [6].

This plant is a promising alternative and should be cultivated in tropical America and recognized as a functional food, thereby motivating the increase in its consumption. However, the specific compounds they contain and their biological activities that may benefit consumers' health are not yet fully understood.

We evaluated bactericidal activity from a methanolic extract of *B. karatas* fruits and its fractions versus *Escherichia coli*, *Enterococcus faecalis*, *Salmonella enteritidis*, and *Shigella flexneri*. Finally, we identified compounds with previously reported biological activity that are potentially beneficial to the health of *B. karatas* fruit consumers.

## 2. Materials and Methods

### 2.1. Fruit Collection and Juice Extraction

Ripe, wild *B. karatas* fruits were collected (Brix = 8 °Bx) during October near the town of Sitpach, in Merida, Yucatan, Mexico. The juice was extracted manually in the laboratory by cutting and squeezing the fruit. The pH of the juice was measured with an Extech ExStik®. The protein concentration was calculated following the Bradford method using the Quick StarTM Bradford Protein Assay commercial kit (Bio-Rad, Hercules, CA, USA), according to the manufacturer's instructions.

### 2.2. Antimicrobial Activity in Agar Well Diffusion

In a Petri dish with Mueller–Hinton II agar culture medium, 100 µL of a bacterial solution with turbidity equivalent to that of a 0.5 McFarland standard ($1.5 \times 10^8$ CFU/mL) was deposited, and with a sterile swab, the sample was distributed. Subsequently, 50 µL of the methanolic extract (50 mg/mL) was deposited in the wells (5 mm diameter). It was incubated for 24 h at 37 °C.

### 2.3. Organic Extract

The organic extract was produced from 240 mL of juice. The juice was dried by placing it on a tray and heating it in an MS hybridization shaker oven (MO-AOR (Orbital)) at 50 °C for 24 h. Subsequently, the dried juice was pulverized, deposited in a 100 mL methanol flask, and left at room temperature. Methanol was replaced via filtration every 24 h for three days. Methanol from each of the three changes was collected in a flask and evaporated to obtain the methanolic extract (ME). The methanolic extract was partitioned successively with hexane (HF), ethyl acetate (EAF), and methanol (MF) (Figure 1). The solvents were removed via evaporation and the fractions were dissolved in dimethyl sulfoxide (DMSO) at 0.5%.

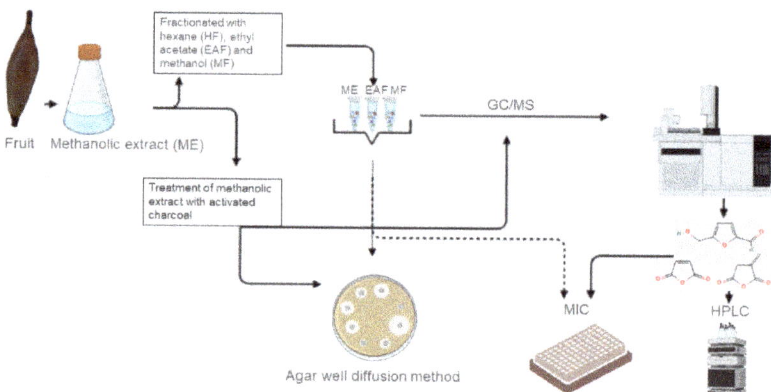


**Figure 1.** Identification strategy for *Bromelia karatas* fruit compounds and their bactericidal activity evaluation.

### 2.4. Minimum Inhibitory Concentration (MIC) and Minimum Bactericidal Concentration (MBC)

The microorganisms evaluated were *Escherichia coli* ATCC 35401, *Enterococcus faecalis*, *Salmonella enteritidis* ATCC 13076, and *Shigella flexneri* ATCC 12022. Both the MIC and MBC were determined using a 96-well plate. The four evaluated bacterial strains were cultured in Petri dishes on Mueller–Hinton medium at 37 °C for 16 h. One to two colonies were added to a test tube containing saline solution (0.85% NaCl) until turbidity equivalent to that of a 0.5 McFarland standard was reached ($1.5 \times 10^8$ CFU/mL). This solution was diluted to $1.5 \times 10^6$ CFU/mL, and 50 µL was added to each well, followed by the addition of 50 µL of the solution (ME, EAF, or MF) to be tested at one of eight concentrations (0.625, 1.25, 2.5, 5, 10, 20, 40, or 80 mg/mL). In the case of synthetic metabolites (itaconic anhydride (ITAN), maleic anhydride (MA), and 5-hydroxymethylfurfural (5-HMF)), the concentrations were: 0.39, 0.78, 1.56, 3.125, 6.25, 12.5, and 25 mg/mL. The positive control was ampicillin, the negative control was bacteria alone, and the toxicity of 0.5% DMSO was evaluated. The 96-well plate was incubated at 37 °C for 18 h. The lowest concentration that did not present bacterial growth was taken as the MIC value. The MBC was measured by taking a sample from each well without turbidity, inoculating it in a Petri dish containing Mueller–Hinton medium, and incubating it at 37 °C for 18 h. The dish without observable bacterial growth was taken as the MBC.

### 2.5. Bactericidal Activity of the ME after Heating and Compound Removal with Activated Charcoal

Bactericidal thermostability was evaluated by heating 100 mg/mL of ME in an autoclave at 121 °C for 15 min, and then, bactericidal activity was measured through the agar well diffusion method. A second assay was performed to evaluate the bactericidal activity of the ME after treatment with activated charcoal (MEC). ME (100 mg) was dissolved in 1 mL of distilled water with 10% activated charcoal in a test tube. The mixture was vortexed for 5 min and centrifuged for 5 min at 12,000 g; bactericidal activity agar well diffusion was assessed using the supernatant. For further analysis, the MEC was dried by placing it on a tray and heating it in an MS hybridization shaker oven (MO-AOR (Orbital)) at 50 °C for 24 h.

*2.6. Gas Chromatography–Mass Spectrometry of Organic Extract*

The organic extract (30 mg extract/mL $CH_3OH$) was filtered through 0.2 µm nylon membranes (Thermo Scientific Cat. No. 726-2520). Then, 2 mL of the filtered solutions were injected into a gas chromatographer–mass spectrometer (GC-MS, Agilent 7890A) equipped with a hydrogen flame ionization detector for compound identification. Compound separation was conducted with an HP5MS column (30 m × 0.250 mm, 0.25 µm, Cat. No. 190915-435). The injector temperature was set at 250 °C and the initial oven temperature was 70 °C for 3 min, which increased at 5 °C/min to 250 °C.

*2.7. Metabolites Quantification*

High-performance liquid chromatography (HPLC) was used to determine the concentration of itaconic anhydride (ITAN), maleic anhydride (MA), and 5-hydroxymethylfurfural (5-HMF). Synthetic versions (Sigma-Aldrich, Saint Louis, MO, USA) of these three compounds were used as standards to make the calibration curve. The chromatographic system HPLC 1100 consisted of a quaternary system of pumps (Agilent Technologies G1310A, Waldbronn, Germany) connected to an automated sample injector (Agilent Technologies G1313A, Waldbronn, Germany). The compounds were detected using a wavelength of 210 nm (Agilent Technologies G1314A, Hachioji, Tokyo, Japan), and a reverse phase C18 column (Polaris 5 µm, 4.6 mm inner diameter × 250 Santa Clara, CA, USA). For the detection of ITA, MA, and 5-HMF, the mobile phase was $H_2O$/acetic acid at 0.05%. The flow rate for all of the samples was 0.625 mL/min and the column temperature was kept constant at 25 °C. The water used was previously degassed and filtered (ultrapure water; Milli-Q® ZMQS6V001).

*2.8. Molecular Docking of the Mycobacterium tuberculosis Isocitrate Lyase (MtICL) with 5-HMF, ITAN, and MA*

The crystal structure of the MtICL (PDB code = 6XPP) enzyme was used for docking analysis. The three-dimensional (3D) structure was downloaded from the Protein Data Bank database (https://www.rcsb.org/, accessed on 9 February 2022). The 3D structure was characterized and represented using X-ray crystallography [16].

A molecular docking experiment was conducted to evaluate the docking properties of four compounds (5-HMF, ITAN, and MA) with the MtICL enzyme. Docking was carried out with the HDOCK server (http://hdock.phys.hust.edu.cn/, accessed on 11 February 2022), a multi-component integrated package [17,18]. The molecular ligand formulas were downloaded from PubChem (https://pubchem.ncbi.nlm.nih.gov/, accessed on 11 February 2022), and a blind molecular docking grid box was used to define the docking region. The parameters were the default values in HDOCK, and the structures had a root mean square deviation (RMSD) of up to 3 Å. All of the results were analyzed and visualized using the UCSF Chimera 1.14 Molecular Graphics Systems [19].

## 3. Results

The ripe fruits collected had 8 °Bx; the average fruit weighed 19.2 g with 44% of juice, its pH was 3.25, and the protein content was 1870 mg/mL. The juice contained 5.8% ME, which exhibited bactericidal activity. This bactericidal activity remained unchanged after the ME was heat sterilized (at 121 °C, 15 min). The ME treated with 10% activated carbon generated an inhibition halo of 1.76 $cm^2$ in a bacterial culture, while the ME that did not undergo treatment generated an inhibition halo of 3.14 $cm^2$; this means that there was a 45% smaller inhibition halo when the ME was treated with activated charcoal.

The ME was fractionated using hexane (low polarity), ethyl acetate (medium polarity), and methanol (high polarity). It resulted in a 3% hexane fraction (FH), a 3% ethyl acetate fraction (EAF), a 68% methanol fraction (MF), and a 26% residue fraction (RF). The EAF and MF fractions exhibited bactericidal activity (Table 1).

**Table 1.** Minimum inhibitory concentration (MIC) and minimum bactericidal concentration (MBC) of methanolic extract (ME) from *B. karatas* fruit, its fractions, and synthetic compounds (ITAN, MA, and 5-HMF).

| | Bacteria | | | |
| --- | --- | --- | --- | --- |
| | *E. coli* | *E. faecalis* | *S. enteritidis* | *S. flexneri* |
| MIC (mg/mL) | | | | |
| ME | 10 | 5 | 10 | 10 |
| EAF | 10 | 10 | 10 | 10 |
| MF | 10 | 10 | 10 | 10 |
| ITAN | 0.78 | 0.78 | 0.78 | 0.78 |
| MA | 0.78 | 0.78 | 0.78 | 0.78 |
| 5-HMF | 6.25 | 6.25 | 6.25 | 3.12 |
| MBC (mg/mL) | | | | |
| ME | 20 | 20 | 40 | 40 |
| EAF | 20 | 20 | 20 | 20 |
| MF | 20 | 20 | 20 | 20 |
| ITAN | 1.56 | 1.56 | 1.56 | 1.56 |
| MA | 1.56 | 1.56 | 1.56 | 1.56 |
| 5-HMF | 12.5 | 12.5 | 12.5 | 6.25 |

To identify the metabolites that contributed to the antimicrobial activity and those with a possible biological activity that could affect fruit consumers, the methanolic extract (ME), methanolic extract treated with activated charcoal (MEC), and fractions with bactericidal activity (EAF and MF) were analyzed using the gas chromatography–mass (GC-MS) spectrometry (GS-MS) method. One hundred thirty-one compounds were identified, of which forty-nine were identified in the ME; in MEC, thirty-seven compounds were removed by activated charcoal and twenty-one new ones were detected (Table 2). On the other hand, in the ethyl acetate and methanolic fractions, compounds were detected which were not detected in the ME: 33 in EAF and 23 MF. Additionally, common and more abundant compounds in the ME, MEC, EAF, and MF were maleic anhydride (MA), 2,5 furandione, dihydro-3-methylene (itaconic anhydride; ITAN), and 5-hydroxymethylfurfural (5-HMF) (Supplementary Material Tables S1–S4).

**Table 2.** Compounds detected in *Bromelia karatas* fruit juice extract.

| No. | Compound Name | Area % | No. | Compound Name | Area % |
| --- | --- | --- | --- | --- | --- |
| 1 | Maleic anhydride | 18.98 | 26 | Thymine | 2.33 |
| *2 | 2-Methylpentyl formate | 0.22 | 27 | 4H-Pyran-4-one,2,3-dihydro-3,5-dihydroxy-6-methyl | 2.05 |
| *3 | 1 H-Imidazole, 4,5 dihydro-2-methyl | 0.28 | *28 | 2-Cyclopenten-1-one,5-hydroxy-2,3-dimethyl | 0.29 |
| *4 | Trans-2-Pentenoic acid | 0.30 | *29 | Bicyclo [3.1.0] hexan -2-ol | 0.20 |
| *5 | 1 H-Tetrazole 1-methyl | 0.20 | *30 | Isobutyl nonyl carbonate | 0.33 |
| *6 | N-(n-Butoxymethyl) acrylamide | 0.26 | *31 | 2-Vinyl-9-[3-deoxy-beta-d-ribofuranosyl] hypoxanthine | 0.35 |
| *7 | 2-Heptanol, 5-ethyl | 0.26 | 32 | 5-Hydroxymethylfurfural | 18.09 |
| 8 | Itaconic anhydride | 22.47 | *33 | Thymol | 0.47 |
| 9 | 2,4-Dihydroxy-2,5-dimethyl-3(2H)-furan-3-one | 1.42 | *34 | 4-Hydroxy-3-methylacetophenone | 0.46 |
| *10 | 2H-Pyran, 3,4-dihydro | 0.67 | *35 | 2-Methoxy-4-vinylphenol | 0.81 |
| *11 | cis-1,4-Dimethylcyclohexane | 0.56 | *36 | 3-Methoxyacetophenone | 0.73 |
| 12 | 2 (3H)-Furanone | 0.71 | *37 | 3,4-Diethylphenol | 0.33 |

**Table 2.** *Cont.*

| No. | Compound Name | Area % | No. | Compound Name | Area % |
|---|---|---|---|---|---|
| 13 | 2,5-Furandione, 3,4-dimethyl | 0.22 | *38 | Ethyl propionylacetate | 1.22 |
| *14 | Pent-2-ynal | 0.29 | *39 | Methyl 3-hydroxypentanoate | 1.12 |
| *15 | 1,2-Butadiene | 0.64 | *40 | 3-Methoxy-hexane-1,6-diol | 0.77 |
| 16 | Hexan-3-yl acetate | 1.41 | *41 | Heptyl butyrate | 2.14 |
| *17 | Pentanoic acid, 4-oxo | 1.45 | *42 | Triethylene glycol monododecyl ether | 1.21 |
| *18 | 1,5-Diacetoxypentane | 1.23 | *43 | Malic Acid | 1.49 |
| *19 | 2-Propanamine, N-methyl-N-nitroso | 1.72 | *44 | 2-Ethyl-3-hydroxyhexyl 2-methylpropanoate | 1.37 |
| *20 | Methyl furan-3-carboxylate | 1.08 | *45 | 2,3,5,6-Tetrafluoroanisole | 0.60 |
| 21 | Furyl hydroxymethyl ketone | 1.26 | *46 | Ethanone, 1-(2,5-dimethoxyphenyl) | 0.60 |
| 22 | Methyl 2-furoate | 1.69 | *47 | 1,2,4-Cyclopentanetrione, 3-(2-pentenyl) | 0.40 |
| 23 | Cyclopentene | 1.11 | *48 | m-Ethyl aminophenol | 0.25 |
| *24 | 2-Cyclopentene-1-one, 2-methyl | 0.77 | *49 | 2,4,6 (3H)-Pteridinetrione, 1,5-dihydro | 0.39 |
| *25 | 3H-Pyrazol-3-one, 2,4 –dihydro-2,4,5-trimethyl | 2.55 | | | |

Note: Items with an asterisk were removed with activated charcoal treatment and aromatic compounds are underlined.

To evaluate whether common and more abundant compounds in the fractions and the ME contributed to the bactericidal activity, synthetic versions (Sigma-Aldrich) of them were tested, and they exhibited activity against all of the bacteria tested (Table 1). The concentrations of these three compounds in the extract and the fractions with bactericidal activity can be seen in Figure 2.

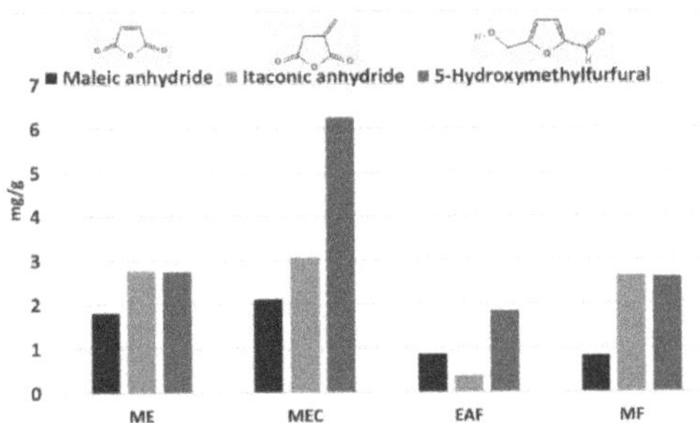


**Figure 2.** HPLC quantification of common compounds in the methanol extract (ME), the methanolic extract treated with activated charcoal (MEC), the ethyl acetate fraction (EAF), and the methanol fraction (MF).

## 4. Discussion

We determined the bactericidal activity of the *B. karatas* fruits' methanolic extract (ME) against *E. coli, S. enteritidis, S. flexneri,* and *E. faecalis* (causal agents of human infections) [5,20]; it was shown that these results are consistent with those reported in *Bromelia pinguin* [6]. We found that the bactericidal activity resisted the sterilization temperature, meaning its potential benefits are still maintained when the fruits are consumed cooked since cooking is necessary to avoid oral injuries caused by their proteolytic activity [11]. In another study, it was determined that the protein extract of the fruits had bactericidal activity, but this was not resistant to heat [14,15], which differs from the heat-resistant bactericidal activity of the methanolic extract found in the present study.

The decrease of 45% in bactericidal activity after the treatment with activated charcoal can be explained by the ability of activated charcoal to remove compounds by adsorption, and the removal effectiveness is influenced by the molecular size, polarity, and branching of the molecule. For example, branched aromatic compounds are more effectively adsorbed on activated charcoal than linear and small molecules such as ethanol and methanol [21,22].

In the GC-MS analysis, 37 of the 49 compounds detected in the ME before charcoal treatment were removed, including 9 aromatic compounds. It was also possible to detect 26 compounds that were not detected in the ME; in this sense, activated charcoal is also likely to remove substances that were not detected in ME. The recovery of these removed compounds can be considered a method to isolate part of the antibacterial substances from *B. Karatas* fruits. However, the bactericidal activity of the substances removed still needs to be identified and evaluated.

On the other hand, the detection of new compounds in the MEC could have been because its concentration increased or the substances that prevented their detection, before treatment, were removed. In the same way, compounds were detected in the ethyl acetate and methanolic fractions that were not detected in the methanolic extract; this could be due to the compound enrichment and the removal of substances that interfere in the detection process [23]. Activated charcoal treatment and the partitioning of the methanolic extract into fractions of different polarities allowed for the broader detection of compounds.

The bactericidal activity level of the *B. karatas* fruits' ME was too low (the MBC values were 20 mg/mL for *E. coli* and *E. faecalis* and 40 mg/mL for *S. enteritidis* and *S. flexneri*) for them to be considered sources of pure antibiotic activity [24]. However, being an edible fruit, the consumption of 10 fruits, with approximately 85 g of fruit juice (equivalent to 5 g of ME), may be enough to prevent mild bacterial infections, similar to the effect of consumption of cranberry fruit extracts [5,6,25]. However, this still needs to be evaluated.

In the GC-MS analysis of the extract and of the fractions with bactericidal activity (ME and MEC, EAF, and MF), we found that the common compounds were: maleic anhydride (MA), 2,5 furandione, dihydro-3-methylene (itaconic anhydride, ITAN), and 5-hydroxymethylfurfural (5-HMF). An increase in the concentration of 5-HMF in the MEC could be observed (Figure 2), possibly caused by the removal of other compounds; furthermore, 5-HMF could be produced during the drying process (50 °C/ 24 h) [26].

Maleic anhydride (MA) has been reported to be present in the copolymer manufacturing of biopolymers with activity against pathogens such as *S. enteritidis*, *S. faecalis*, and *E. coli* [27]. However, the results of the present study indicated that it exhibits bactericidal activity as a monomer, and 5-HMF is known to be an intermediary in the Maillard reaction, formed by the degradation of sugars at high temperatures. It has been found in various food products, including nuts, fruit juices, caramel products, coffee, bakery products, malt, and vinegar [28,29]. 5-HMF has anti-quorum sensing and anti-biofilm activity against *Pseudomonas aeruginosa* [30]. It is also reported to inhibit growth in *E. coli* LY01 [31], which is consistent with our results (Table 1). MA, ITAN, and 5-HMF showed activity against the bacteria evaluated (Table 1). However, their activity levels were relatively low [24] and their concentrations were not sufficient (Figure 2) to explain the ME activity against the bacteria analyzed. Therefore, further studies are required to identify the other bactericidal compounds of the ME.

We did not find evidence of the bactericidal activity of ITAN in the literature reports; however, it is reported to be an inhibitor of *M. tuberculosis* isocitrate lyase (MtICL) [32]. These authors showed that 0.03 mg/mL of ITAN can inhibit 90% of MtICL. This enzyme can promote the persistence and virulence of bacteria in macrophages, develop resistance to antibiotics, and promote the growth and survival of *M. tuberculosis* during its latent infection [33–35]. Considering their crucial roles, MtICLs are current inhibition targets for the development of new antibiotics to treat tuberculosis [35]. Considering that the structures of ITAN, MA, and 5-HMF have a certain degree of similarity, using molecular docking analysis, we evaluated whether MA and 5-HMF interact molecularly at the MtICL catalytic site as does ITAN. In Figure 3, it can be observed that MA, ITAN, and 5-HMF have similar binding patterns to the catalytic site. These results could suggest the possibility that MA and 5-HMF have roles in the inhibition of MtICL function. To fully understand this aspect, it is necessary to carry out enzyme activity assays; however, this was not the objective of this study.

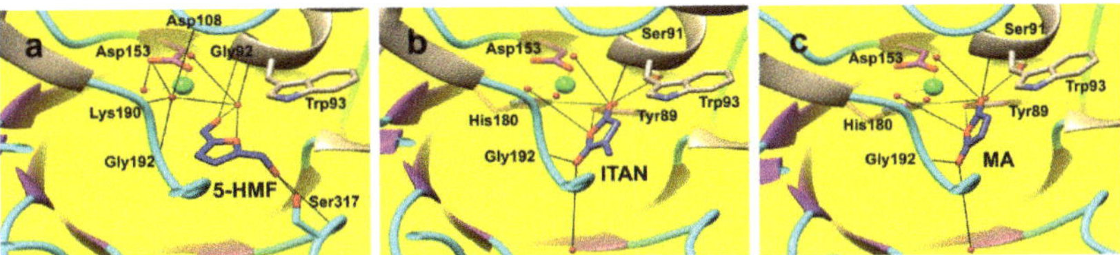


**Figure 3.** Molecular docking of MtICL with (**a**) 5-HMF; (**b**) ITAN, and (**c**) MA. A-C, all three compounds bind to the catalytic site of MtICL. Residues Trp 93, Asp153, and Gly192 form hydrogen bonds with the three compounds analyzed. Panel (**a**), 5-HMF binding pose. Panel (**b**), ITAN binding pose. Panel (**c**), MA binding pose. The black lines represent the hydrogen bonds. The total energy in Kcal/mol of the binding affinity of each compound to the catalytic site of MtICL is: for 5-HMF, −90.69; ITAN, −84.33; and MA, −78.20. The small red spheres correspond to water molecules and the large green spheres are magnesium ions.

Metabolites from *B. karatas* fruits have other biological activities in addition to their bactericidal activity (Table 3). This strengthens the proposal that the fruit of *B. karatas* can be a functional food.

**Table 3.** Compounds with biological activity in *Bromelia karatas* juice.

| Compound | Biological Activity | Reference |
|---|---|---|
| 2-Heptanol, 5-ethyl | Possible alpha-amylase inhibitor. | [36] |
| 2 (3H)-Furanone | It induces DNA damage and is possibly an anticancer. | [37] |
| Thymol | Bactericidal, fungicidal, anticarcinoma. | [38,39] |
| 5-HMF | Inhibition of alcoholic hepatic oxidative injury, some toxicological effects, anti-inflammatory, and anti-allergic effects. | [40–43] |
| 2-Methoxy-4-vinylphenol | Anti-inflammatory and possible anticancer. | [44,45] |
| 1,2,3-Benzenetriol | anti-allergic | [46] |

## 5. Conclusions

In this study, in the methanolic extract of *B. karatas* fruits, we identified the presence of several metabolites with a wide range of bioactivities, including bactericidal, fungicidal, anticancer, anti-inflammatory, enzyme inhibiting, and anti-allergic properties. Additionally, the ME was found to have activity against *E. coli*, *E. faecalis*, *S. enteritidis*, and *S. flexneri*. It was shown that the most abundant compounds of the ME are maleic anhydride, 5-hydroxymethylfurfural, and itaconic anhydride, which were active against the bacteria tested. This study shows the potential effects that *B. karatas* fruits could have on consumers' health. However, more studies are still required.

**Supplementary Materials:** The following supporting information can be downloaded at: https://www.mdpi.com/article/10.3390/app12147275/s1, Table S1: Identified compounds in methanol extract; Table S2: Compounds identified in the methanolic extract treated with activated charcoal; Table S3: Compounds identified in the methanol fraction; Table S4: Compounds identified in the ethyl acetate fraction.

**Author Contributions:** Conceptualization, B.A.A.-G. and H.d.J.V.-A.; methodology, B.A.A.-G., H.d.J.V.-A. and K.C.A.-G.; software, M.A.U.-C.; formal analysis, B.A.A.-G. and H.d.J.V.-A.; writing—original draft preparation, A.A.A.-S., B.A.A.-G. and H.d.J.V.-A.; writing—review and editing, K.Y.A.-V., G.P.-H., E.G.-M., B.Y.S.-G., W.A.P.-P., G.L.-U.; supervision, H.d.J.V.-A. All authors have read and agreed to the published version of the manuscript.

**Funding:** This research received no external funding.

**Institutional Review Board Statement:** Not applicable.

**Informed Consent Statement:** Not applicable.

**Data Availability Statement:** Not applicable.

**Conflicts of Interest:** The authors declare no conflict of interest.

# References


1. Siró, I.; Kápolna, E.; Kápolna, B.; Lugasi, A. Functional food. Product development marketing and consumer acceptance a review. *Appetite* **2008**, *51*, 456–467. [CrossRef]
2. Bigliardi, B.; Galati, F. Innovation trends in the food industry: The case of functional foods. *Trends Food Sci. Technol.* **2013**, *31*, 118–129. [CrossRef]
3. Willett, W.C. Diet and health: What should we eat? *Science* **1994**, *264*, 532–537. [CrossRef]
4. Barros, L.; Carvalho, A.M.; Ferreira, I.C.F.R. Exotic fruits as a source of important phytochemicals: Improving the traditional use of *Rosa canina* fruits in Portugal. *Food Res. Int.* **2011**, *44*, 2233–2236. [CrossRef]
5. Asma, B.; Vicky, L.; Stephanie, D.; Yves, D.; Amy, H.; Sylvie, D. Standardised high dose versus low dose cranberry Proanthocyanidin extracts for the prevention of recurrent urinary tract infection in healthy women [PACCANN]: A double blind randomised controlled trial protocol. *BMC Urol.* **2018**, *18*, 29. [CrossRef]
6. Pío-León, J.F.; López-Angulo, G.; Paredes-López, O.; Uribe-Beltran, M.J.; Díaz-Camacho, S.P.; Delgado-Vargas, F. Physicochemical, nutritional and antibacterial characteristics of the fruit of *Bromelia pinguin*. *Plant Foods Hum. Nutr.* **2009**, *64*, 181–187. [CrossRef]
7. Camacho-Hernández, I.L.; Chávez-Velázquez, J.A.; Uribe-Beltrán, M.J.; Ríos-Morgan, A.; Delgado-Vargas, F. Antifungal activity of fruit pulp extract from *Bromelia pinguin*. *Phytotherapies* **2009**, *73*, 411–413. [CrossRef]
8. Coelho, R.G.; Honda, N.K.; Vieira, M.; Brum, R.L.; Pavan, F.R.; Leite, C.Q.; Cardoso, C.A. Chemical composition and antioxidant and antimycobacterial activities of *Bromelia balansae* (Bromeliaceae). *J. Med. Food* **2010**, *13*, 1277–1280. [CrossRef]
9. Abreu-Payrol, J.; González-Mosquera, D.M.; Meneses, A.; Cruz-de-la-Cruz, M.E.; Banze, F.; Miranda-Martínez, M.; Ros-López, O. Determinación de parámetros farmacognósticos y bromatológicos y evaluación de la actividad antiparasitaria de una preparación obtenida del fruto de *Bromelia pinguin* L. que crece en Cuba. *Acta Farm. Bonaer.* **2005**, *24*, 377–382.
10. Moyano, D.D.; Osorio, M.R.; Murillo, E.P.; Murillo, W.A.; Solanilla, J.D.; Méndez, J.A.; Aristizabal, J.S. Evaluación de parámetros bromatológicos, fitoquímicos y funcionalidad antioxidante de frutos de *Bromelia karatas* (Bomeliaceae). *Vitae* **2012**, *19*, S439–S441.
11. Villanueva-Alonzo, H.J.; Polanco-Hernández, G.M.; Lizama-Uc, G.; Acosta-Viana, K.Y.; Alvarado-Segura, A.A. Proteolytic activity of wild fruits of *Bromelia karatas* L. of Yucatán, México. *Rev. Chap. Ser. Cienc. For. Ambient.* **2019**, *25*, 157–168. [CrossRef]
12. Hornung-Leoni, C.T. Avances sobre usos etnobotánicos de las Bromeliaceae en Latinoamérica. *B Latinoam Caribe PL* **2011**, *10*, 297–314.
13. Mondragón-Chaparro, D.M.; Ramírez-Morillo, I.; Flores-Cruz, M.M.; García-Franco, J.G. La familia Bromeliaceae en México. *Bot. Sci.* **2011**, *96*, 553–554. [CrossRef]
14. Montalvo-González, E.; Anaya-Esparza, L.M.; Martínez-Olivo, A.O.; Abreu-Payrol, J.; Sánchez-Burgos, J.A.; García-Magaña, M.L. Physiological and physicochemical behavior of guamara (*Bromelia pinguin*) and cocuixtle (*Bromelia karatas*) fruits, as well as the antibacterial effect of their pre-purified proteases. *Emir. J. Food Agric.* **2021**, *33*, 277–286. [CrossRef]
15. Ávalos-Flores, E.; López-Castillo, L.M.; Wielsch, N.; Hupfer, Y.; Winkler, R.; Magaña-Ortiz, D. Protein extract of *Bromelia karatas* L. rich in cysteine proteases (ananain- and bromelain-like) has antibacterial activity against foodborne pathogens *Listeria monocytogenes* and *Salmonella* Typhimurium. *Folia Microbiol.* **2022**, *67*, 1–13. [CrossRef]
16. Kwai, B.X.C.; Collins, A.J.; Middleditch, M.J.; Sperry, J.; Bashiri, G.; Leung, I.K.H. Itaconate is a covalent inhibitor of the *Mycobacterium tuberculosis* isocitrate lyase. *RSC Med. Chem.* **2020**, *12*, 57–61. [CrossRef]
17. Yan, Y.; Zhang, D.; Zhou-Li, B.; Sheng-You, H. HDOCK: A web server for protein-protein and protein-DNA/RNA docking based on a hybrid strategy. *Nucleic Acids Res.* **2017**, *3*, W365–W373. [CrossRef]
18. Yan, Y.; Tao, H.; He, J.; Sheng-You, H. The HDOCK server for integrated protein-protein docking. *Nat. Protoc.* **2020**, *15*, 1829–1852. [CrossRef]
19. Pettersen, E.F.; Goddard, T.D.; Huang, C.C.; Couch, G.S.; Greenblatt, D.M.; Meng, E.C.; Ferrin, T.E. UCSF Chimera—A visualization system for exploratory research and analysis. *J. Comput. Chem.* **2004**, *25*, 1605–1612. [CrossRef]
20. Ziadeh, C.; Clark, L.L.; Williams, V.F.; Stahlman, S. Update: Incidence of acute gastrointestinal infections and diarrhea, active component, U.S. Armed Forces, 2010–2019. *MSMR* **2020**, *27*, 9–14.
21. Neuvonen, P.J.; Olkkola, K.T. Oral activated charcoal in the treatment of intoxications. Role of single and repeated doses. *Med. Toxicol. Advers. Drug Exp.* **1998**, *3*, 33–58. [CrossRef]
22. Cecen, F.; Aktas, O. *Activated Carbon for Water and Wastewater Treatment: Integration of Adsorption and Biological Treatment*, 1st ed.; Wiley-VCH Verlag GmbH & Co. KGaA: Weinheim, Germany, 2011; pp. 303–304.
23. El-Ghorab, A.; Shaaban, H.A.; El-Massry, K.F.; Shibamoto, T. Chemical composition of volatile extract and biological activities of volatile and less-volatile extracts of juniper berry (*Juniperus drupacea* L.) fruit. *J. Agric. Food Chem.* **2008**, *56*, 5021–5025. [CrossRef]
24. Ríos, J.L.; Recio, M.C. Medicinal plants and antimicrobial activity. *J. Ethnopharmacol.* **2005**, *100*, 80–84. [CrossRef]
25. McMurdo, M.E.; Argo, I.; Phillips, G.; Daly, F.; Davey, P. Cranberry or trimethoprim for the prevention of recurrent urinary tract infections? A randomized controlled trial in older women. *J. Antimicrob. Chemother.* **2009**, *63*, 389–395. [CrossRef]
26. Bao, T.; Hao, X.; Shishir, M.; Karim, N.; Chen, W. Cold plasma: An emerging pretreatment technology for the drying of jujube slices. *Food Chem.* **2021**, *337*, 127783. [CrossRef]

27. Nagaraja, A.; Jalageri, M.D.; Puttaiahgowda, Y.M.; Raghava Reddy, K.; Raghu, A.V. A review on various maleic anhydride antimicrobial polymers. *J. Microbiol. Methods* **2019**, *163*, 105650. [CrossRef]
28. Rada-Mendoza, M.; Sanz, M.; Olano, A.; Villamiel, M. Formation of hydroxymethylfurfural and furosine during the storage of jams and fruit-based infant foods. *Food Chem.* **2004**, *85*, 605–609. [CrossRef]
29. Teixidó, E.; Santos, F.J.; Puignou, L.; Galceran, M.T.J. Analysis of 5-hydroxymethylfurfural in foods by gas chromatography–mass spectrometry. *J. Chromatogr. A* **2006**, *1135*, 85–90. [CrossRef]
30. Rajkumari, J.; Borkotoky, S.; Reddy, D.; Mohanty, S.K.; Kumavath, R.; Murali, A.; Suchiang, K.; Busi, S. Anti-quorum sensing and anti-biofilm activity of 5-hydroxymethylfurfural against *Pseudomonas aeruginosa* PAO1: Insights from in vitro, in vivo and in silico studies. *Microbiol. Res.* **2019**, *226*, 19–26. [CrossRef]
31. Zaldivar, J.; Martinez, A.; Ingram, L.O. Effect of selected aldehydes on the growth and fermentation of ethanologenic *Escherichia coli*. *Biotechnol. Bioeng.* **1999**, *65*, 24–33. [CrossRef]
32. Bai, B.; Xie, J.; Yan, J.F.; Wang, H.H.; Hu, C.H. A High Throughput screening approach to identify Isocitrate Lyase inhibitors from traditional Chinese medicine sources. *Drug Dev. Res.* **2006**, *67*, 818–823. [CrossRef]
33. Wayne, L.G.; Lin, K.Y. Glyoxylate metabolism and adaptation of *Mycobacterium tuberculosis* to survival under anaerobic conditions. *Infect Immun.* **1982**, *37*, 1042–1049. [CrossRef]
34. Chew, S.Y.; Yang-Chee, W.J.; Lung-Than, L.T. The glyoxylate cycle and alternative carbon metabolism as metabolic adaptation strategies of *Candida glabrata*: Perspectives from *Candida albicans* and *Saccharomyces cerevisiae*. *J. Biomed. Sci.* **2019**, *26*, 52. [CrossRef]
35. Sharma, R.; Das, O.; Damle, S.G.; Sharma, A.K. Isocitrate lyase: A potential target for anti-tubercular drugs. *Recent Pat. Inflamm. Allergy Drug Discov.* **2013**, *7*, 114–123. [CrossRef]
36. Sudha, P.; Zinjarde, S.S.; Bhargava, S.Y.; Kumar, A.R. Potent α-amylase inhibitory activity of Indian Ayurvedic medicinal plants. *BMC Complementary Altern. Med.* **2011**, *11*, 5. [CrossRef]
37. Calderón-Montaño, J.M.; Burgos-Morón, E.; Orta, M.L.; Pastor, N.; Austin, C.A.; Mateos, S.; López-Lázaro, M. Alpha, beta-unsaturated lactones 2-furanone and 2-pyrone induce cellular DNA damage, formation of topoisomerase I- and II-DNA complexes and cancer cell death. *Toxicol. Lett.* **2013**, *222*, 64–71. [CrossRef]
38. Marchese, A.; Orhan, I.E.; Daglia, M.; Barbieri, R.; Di Lorenzo, A.; Nabavi, S.F.; Gortzi, O.; Izadi, M.; Nabavi, S.M. Antibacterial and antifungal activities of thymol: A brief review of the literature. *Food Chem.* **2016**, *210*, 402–414. [CrossRef]
39. De La Chapa, J.J.; Singha, P.K.; Lee, D.R.; Gonzales, C.B. Thymol inhibits oral squamous cell carcinoma growth via mitochondria-mediated apoptosis. *J. Oral Pathol. Med.* **2018**, *47*, 674–682. [CrossRef]
40. Kong, F.; Lee, B.H.; Wei, K. 5-Hydroxymethylfurfural Mitigates Lipopolysaccharide-Stimulated Inflammation via Suppression of MAPK, NF-κB and mTOR Activation in RAW 264.7 Cells. *Molecules* **2019**, *24*, 275. [CrossRef]
41. Durling, L.J.; Busk, L.; Hellman, B.E. Evaluation of the DNA damaging effect of the heat-induced food toxicant 5-hydroxymethylfurfural (HMF) in various cell lines with different activities of sulfotransferases. *Food Chem. Toxicol.* **2009**, *47*, 880–884. [CrossRef]
42. Svendsen, C.; Husøy, T.; Glatt, H.; Paulsen, J.E.; Alexander, J. 5-Hydroxymethylfurfural and 5-Sulfooxymethylfurfural Increase Adenoma and Flat ACF Number in the Intestine of Min/+ Mice. *Anticancer Res.* **2009**, *29*, 1921–1926.
43. Li, W.; Qu, X.N.; Han, Y.; Zheng, S.W.; Wang, J.; Wang, Y.P. Ameliorative effects of 5-Hydroxymethyl-2-furfural (5-HMF) from *Schisandra chinensis* on alcoholic liver oxidative injury in mice. *Int. J. Mol. Sci.* **2015**, *16*, 2446–2457. [CrossRef]
44. Kim, D.H.; Han, S.I.; Go, B.; Oh, U.H.; Kim, C.S.; Jung, Y.H.; Lee, J.; Kim, J.H. 2-Methoxy-4-vinylphenol attenuates migration of human pancreatic cancer cells via blockade of FAK and AKT signaling. *Anticancer Res.* **2019**, *39*, 6685–6691. [CrossRef]
45. Jeong, J.B.; Hong, S.C.; Jeong, H.J.; Koo, J.S. Anti-inflammatory effect of 2-Methoxy-4-Vinylphenol via the suppression of NF-κB and MAPK activation, and acetylation of histone H3. *Arch. Pharm. Res.* **2011**, *34*, 2109–2116. [CrossRef]
46. Nakano, T.; Ikeda, M.; Wakugawa, T.; Kashiwada, Y.; Kaminuma, O.; Kitamura, N.; Yabumoto, M.; Fujino, H.; Kitamura, Y.; Fukui, H.; et al. Identification of pyrogallol from Awa-tea as an anti-allergic compound that suppresses nasal symptoms and IL-9 gene expression. *J. Med. Investig.* **2020**, *67*, 289–297. [CrossRef]

*Article*

# Variation of Nutritional Quality Depending on Harvested Plant Portion of Broccoli and Black Cabbage

Bruno Mezzetti [1], Francesca Biondi [2], Francesca Balducci [1], Franco Capocasa [1], Elena Mei [2], Massimo Vagnoni [2], Marino Visciglio [2] and Luca Mazzoni [1,*]

1 Department of Agricultural, Food and Environmental Sciences (D3A), Università Politecnica delle Marche, Via Brecce Bianche 10, 60131 Ancona, Italy; b.mezzetti@staff.univpm.it (B.M.); francesca.balducci@staff.univpm.it (F.B.); f.capocasa@staff.univpm.it (F.C.)
2 SOC.AGR. VALLI DI MARCA S.S., Contrada Valle 2, 63068 Montalto delle Marche, Italy; fra90ap@hotmail.it (F.B.); maddymei@hotmail.it (E.M.); massimo.agv@gmail.com (M.V.); marino.visciglio@gmail.com (M.V.)
* Correspondence: l.mazzoni@staff.univpm.it

**Abstract:** Brassicaceae plants are rich with antioxidant compounds that play a key role for human health. This study wants to characterize two Italian broccoli cultivars (Roya and Santee) and black cabbage, evaluating the variation of antioxidants in different portion and at different developmental stage of the plants: for broccoli, heads and stems were sampled, while for black cabbage, leaves and seeds were analyzed. Roja cultivar was also analyzed at the first and second harvest to evaluate the variation of phytochemical compounds over time. Nutritional and sensorial qualities were investigated. Black cabbage seeds showed higher value of total antioxidants, total phenols, and total anthocyanins than leaves. Similarly, phenolics and anthocyanins content in head was higher than in stem in broccoli. In Roja cultivar, the harvest date seemed to influence the antioxidant capacity and the phytochemical compounds content, with broccoli sampled in the second harvest showing better results for all the nutritional parameters. These local vegetables represent a significant source of antioxidants and may contribute to health benefits of the consumers.

**Keywords:** broccoli; black cabbage; sensorial quality; nutritional quality; antioxidant activity; total phenols; anthocyanins; plant portion; stage development

**Citation:** Mezzetti, B.; Biondi, F.; Balducci, F.; Capocasa, F.; Mei, E.; Vagnoni, M.; Visciglio, M.; Mazzoni, L. Variation of Nutritional Quality Depending on Harvested Plant Portion of Broccoli and Black Cabbage. *Appl. Sci.* **2022**, *12*, 6668. https://doi.org/10.3390/app12136668

Academic Editor: Catarina Guerreiro Pereira

Received: 11 May 2022
Accepted: 28 June 2022
Published: 1 July 2022

**Publisher's Note:** MDPI stays neutral with regard to jurisdictional claims in published maps and institutional affiliations.

**Copyright:** © 2022 by the authors. Licensee MDPI, Basel, Switzerland. This article is an open access article distributed under the terms and conditions of the Creative Commons Attribution (CC BY) license (https://creativecommons.org/licenses/by/4.0/).

## 1. Introduction

Italy is known in the world for its high-quality horticulture; among these, *Brassica* vegetables cover an important part of the market, with a total production of 420,630 t in 2020 (https://www.fao.org/faostat/en/#data/QCL, accessed on 12 March 2022). *Brassica* Italian crops production registered an increase of about 70,000 t since 2010.

A lot of studies showed a relation between the high consumption of *Brassica* vegetables and the reduction in the risk of age-related chronic illness, such as cardiovascular and other degenerative diseases [1]; they also reduce the risk of several types of cancer [2]. Broccoli and black cabbage are among the vegetable food with the highest antioxidant potential [3]. The antioxidant and antiradical activity in *Brassica* vegetables is mainly represented by the large group of polyphenols, constituted by flavonoids (mainly flavonols and anthocyanins) and hydroxycinnamic acids [4]. These secondary metabolites have several functions in the plant, such as UV protection, pigmentation, and disease control, such as glucosinolates [5]. In *Brassica* genus, the main represented flavonols are quercetin, kaempferol, and isorhamnetin. Anthocyanins, in addition to conferring the blue and red pigmentation in broccoli sprouts and red cabbage, possess high antioxidant capacity [6]; among them, cyanidin-3-glucoside is the most represented in *Brassica* crops [7].

Several studies described the variation of the antioxidant activity and phenolics content in the plant portion: for example, in turnip plants, flower buds are the most active portion, followed by leaves and stems, while roots seem to have minor phenolic concentration [8]. Ferreres et al. [9] revealed that kale seeds have higher antioxidant potential than kale leaves, but leaves are the richest in phenolics. In *Brassica rapa*, Francisco et al. [10] reported that phenolic compounds content is higher in leaves harvested during the vegetative period than fructiferous stems (composed of flower, buds, and surrounding leaves), and flavonoids are the main represented compounds. These findings were confirmed by the same group [11], which found that total phenolic content is higher in green turnip than in turnip top. On the contrary, Schmidt et al. [12] reported that harvest period has only a minimal influence on flavonoid concentration in kale. Broccoli cultivars register an increase in phenolic compounds concentration during the inflorescence's development, with an interesting antioxidant capacity in flower, stem, and leaves [13,14].

The aim of this work was to analyze the content of phytochemical compounds in different edible plant portions of *Brassica* species at different developmental stages. In particular, in broccoli heads and stems were analyzed, while black cabbage leaves and seeds were evaluated.

To investigate the interaction between genotype and plant portion, two different cultivars of broccoli were compared. The effect of plant developmental stage on phytochemicals amount was also considered, comparing broccoli of the cultivar Roja from first and second harvest period.

## 2. Materials and Methods

### 2.1. Plant Material

Vegetable materials used for these trials were provided by Valli di Marca Agricultural Company, located in the south of Marche Region (Italy), at 136 m.a.s.l. *Brassica* plants (broccoli and black cabbages) were cultivated in open field conditions, according to the typical cultivation system adopted in this area. Broccoli samples belonged to Roja and Santee cultivars and were harvested on 8 February, while Roja-II (second harvest) was sampled on the 26 February. After harvesting, samples were packaged in 300-g packs and immediately frozen at $-20$ °C. In a second step, broccoli samples were divided into heads and stems (approximately representing 25% and 75% of total fresh weight, respectively) for analyzing the antioxidant capacity, the phenols content, and the anthocyanins content in the two portions. Black cabbage seeds and leaves were also harvested (seeds were sampled during previous April). The list of plant samples collected and analyzed is reported in Table 1.

**Table 1.** Plant materials sampled.

| Species | Samples (300 g) |
| --- | --- |
| *Brassica oleracea* **L. var.** *italica* | 5 samples of Broccoli var. Roja* divided into heads and stems<br>5 samples of Broccoli var. Santee F1* divided into heads and stems<br>3 samples of Broccoli var. Roja*-II harvest divided into heads and stems |
| *Brassica oleracea* **L. var.** *Acephala* **subvar.** *Laciniata* **L.** | 2 samples of black cabbage leaves<br>2 samples of black cabbage seeds |

### 2.2. Vegetables Sensorial Quality

The sensorial quality analysis was carried out on a vegetable juice extract obtained by the fresh material and prepared with a centrifuge for food (Bosch, Munich, Germany) (except for the seeds) at $20.0 \pm 0.5$ °C. The measurement of vegetables soluble solids content (SSC) was performed using a hand-held refractometer (Atago, Tokio, Japan). For each sample, few drops of the previously obtained juice were put on the refractometer prism, and the SSC was recorded as °Brix. The refractometer prism was cleaned with distilled water after each sample. Vegetables Titratable Acidity (TA) was determined from 5 mL of the previously obtained juice diluted with 45 mL of distilled water. This solution was titrated by an automatic titrator (HI 84532 Fruit Juice—Titratable acidity—Hanna Instruments,

Woonsocket, RI, USA), until the vegetable juice aqueous solution reached the neutral pH. The titratable acidity was expressed as % citric acid.

### 2.3. Vegetables Nutritional Quality

#### 2.3.1. Chemicals

Methanol (99%, ACS-ISO) was purchased from Carlo Erba Reagents (Milan, Italy). Folin–Ciocalteu reagent, sodium carbonate (anhydrous), potassium chloride, sodium acetate, chloridric acid, glacial acetic acid, ferric chloride hexahydrate, dihydrogen potassium phosphate, dipotassium hydrogen phosphate, 2,4,6-tris(2-pyridyl)-striazine (TPTZ, 99%), 6-hydroxy-2,5,7,8-tetramethylchromane-2-carboxylic acid (Trolox), ferrous sulphate heptahydrate, potassium persulphate, 3,4,5-trihydroxybenzoic acid (gallic acid), and sodium hydroxide, were purchased from Sigma-Aldrich (Sigma-Aldrich s.r.l., Milan, Italy).

#### 2.3.2. Vegetables Extraction

Fresh materials stored at $-20\ ^{\circ}\mathrm{C}$ were sliced for obtaining 10 g of representative samples and placed into test tubes. 100 mL of methanol extracting solution, constituted by 20:80 water:methanol and 1% of acetic acid, were added to the samples. Samples were then homogenized using an Ultraturrax T25 homogenizer (Janke and Kunkel, IKA Labortechnik, Staufen, Germany). The homogenized suspensions were placed in a fridge at $4\ ^{\circ}\mathrm{C}$ in the dark. After 48 h, the suspensions were centrifuged at 2500 rpm for 15 min (Thermo Fisher Scientific Heraeus Megafuge 16R Centrifuge, Waltham, MA, USA) and the supernatants were collected and stored in six amber vials (for each sample), of 4 mL each, at $-20\ ^{\circ}\mathrm{C}$ [15,16].

#### 2.3.3. Total Antioxidant Capacity (TAC)

Vegetables TAC was evaluated using a FRAP (Ferric Reducing Antioxidant Power) assay, a method based on the rapid reduction in ferric-tripyridyltriazine (FeIII-TPTZ) in the blue-colored ferrous-tripyridyltriazine (FeII-TPTZ) by antioxidants present in the samples. The reduction in ferric–TPTZ was measured by the method of Benzie and Strain [17], modified by Deighton et al. [18] and optimized for *Brassica* vegetables. The FRAP reagent was freshly prepared by mixing 10:1:1 ($v/v/v$) of sodium acetate (300 mM acidified with acetic acid until pH 3.6), ferric chloride (20 mM), and TPTZ (10 mM in 40 mM HCl). Briefly, the vegetable methanol extract was diluted 1:5 and vortexed. This solution was further diluted 1:10 adding the FRAP solution previously prepared, vortexed, and incubated in darkness for 4 min; after this process, the absorbance was measured at 593 nm by spectrophotometer (UV-1800 Shimadzu, Kyoto, Japan). The results were expressed as mM Trolox Equivalent per kg of fresh weight (mM TE/kg fw). The calibration was calculated by linear regression from the dose–response curve of the Trolox standards.

#### 2.3.4. Total Phenol Content (TPH)

The vegetable total phenol content was evaluated using the Folin–Ciocalteu reagent method [19], with gallic acid as the standard for the calibration curve. Briefly, glass test-tubes were filled with 3.5 mL water, and 150 µL of water-diluted vegetable methanolic extract (1:3) was added. The absorbance of the samples was measured at 760 nm by spectrophotometer (UV-1800 Shimadzu, Kyoto, Japan) after 60 min. The data were calculated and expressed as mg gallic acid per kg of fresh weight (mg GA/kg fw).

#### 2.3.5. Total Anthocyanin Content (ACY)

The vegetable total anthocyanin content was measured using the pH differential shift method [20]. This assay is based on the characteristic change in intensity of the hue of the anthocyanins, according to the pH shift method. Briefly, the vegetable methanolic extracts were diluted (1:1) with potassium-chloride (pH 1.00) and with sodium acetate (pH 4.50). Then, the corresponding maximum absorbance of both solutions was measured at 520 nm and 700 nm of wavelength by spectrophotometer (UV-1800 Shimadzu, Kyoto, Japan).

The data are expressed as mg cyanidin 3-glucoside (the most represented anthocyanin in broccoli) per kg of fresh weight (mg Cya-Glu/kg fw).

*2.4. Statistical Analysis*

Vegetable sensorial and nutritional parameters were analyzed in triplicate for each sample. The data for the titratable acidity, soluble solids content, FRAP, TPH, and ACY were all analyzed through the STATISTICA 7 software (Stat Soft, Tulsa, OK, USA), using one-way analysis of variance (ANOVA), with each genotype or plant parts as an independent variable. Significant differences within genotypes or plant parts were calculated according to Student's Newman–Keuls tests, and differences for $p \leq 0.05$ were considered significant.

## 3. Results and Discussion

### 3.1. Vegetables Sensorial Quality

Sensorial quality was analyzed on fresh material considering the whole samples of broccoli. The results of sensorial parameters are reported in Table 2. Broccoli var. Santee had the statistically highest value of SSC. Regarding the Titratable Acidity, the statistical analysis underlined how black cabbage samples had the highest values, while Roja-II samples showed the significantly highest pH value. Broccoli data confirm the results found by Nicoletto et al. [21]; they reported similar value for SSC content (8.2–9.3 °Brix), and higher value for what concern the TA (0.40–0.43% citric acid), and lower for pH (5.67–5.83).

**Table 2.** Average values of total soluble solids content, titratable acidity, and pH of *Brassica* samples.

| Vegetables | SSC [1] (°Brix) | TA [2] (% Citric Acid) | pH |
|---|---|---|---|
| Black Cabbage | 9.55 ± 0.31 [b] | 0.23 ± 0.02 [a] | 5.88 ± 0.06 [d] |
| Roja-II [3] | 8.97 ± 0.26 [b] | 0.15 ± 0.01 [b] | 6.77 ± 0.06 [a] |
| Roja [4] | 9.56 ± 0.21 [b] | 0.15 ± 0.01 [b] | 6.53 ± 0.07 [b] |
| Santee [5] | 10.62 ± 0.26 [a] | 0.16 ± 0.01 [b] | 6.35 ± 0.02 [c] |

Average of [1] SSC: Soluble Solids Content; [2] TA: Titratable Acidity and pH ± standard deviation; [3] Roja-II: Broccoli var. Roja II harvest; [4] Roja: Broccoli var. Roja I harvest: [5] Santee: Broccoli var. Santee. Values with the same letter were not significantly different (Test SNK $p \leq 0.05$). $n = 3$.

Black cabbage belongs to the *Brassica oleracea* var. *acephala* group that includes kales. A lot of studies considered the sensorial aspects of kale family; Armesto et al. [22,23] and Martìnez et al. [24] reported similar amount of SSC, TA, and pH in kale.

According to these results, sensorial quality is affected by genotype. Regarding the developmental stage, the comparison between Roja and Roja-II did not show any significant difference for the SSC and TA parameters. The only significant difference was registered for the pH value, with Roja-II value being higher than Roja.

### 3.2. Vegetables Nutritional Quality

3.2.1. Black Cabbage and Whole Broccoli Nutritional Value

In Table 3, results about the average values of nutritional parameters for black cabbage seeds and leaves are reported. Furthermore, average values of nutritional parameters for each cultivar of broccoli are showed, comprising Roja harvested at two different developmental stages.

Black cabbage seeds showed a higher content of both the investigated phytochemical compounds (TPH and ACY) than leaves and, consequently, a statistically higher TAC value. These results were confirmed by Ferreres et al. [9], who performed characterization trials on kale, analysing the antioxidant activity of kale seeds and leaves. They found that seeds were rich in quercetin and isorhamnetin derivatives, not found in leaves, and in phenolic acids, conferring them a higher antioxidant capacity than leaves, which were rich in flavonols. The higher antioxidant capacity found in seeds is bound to their physiological functions as germination, permeability to water, protection by pathogen and insect attacks, storage, and protection of lipids from oxidation.

**Table 3.** Average values of total antioxidant capacity, phenolics and anthocyanins content of the different vegetables analyzed.

| Vegetables | TEAC [1] (mM TE/kg fw) | TPH [2] (mg GA/kg fw) | ACY [3] (mg CYA-3-GLU/kg fw) |
|---|---|---|---|
| **Black Cabbage Leaves** | 5.68 ± 0.21 [b] | 1444.60 ± 23.82 [b] | 3.25 ± 0.24 [b] |
| **Black Cabbage Seeds** | 8.39 ± 0.53 [a] | 1767.79 ± 12.67 [a] | 16.87 ± 0.24 [a] |
| | | | $n = 8$ |
| **Santee [4]** | 4.65 ± 0.13 [B] | 1038.74 ± 26.48 [B] | 18.15 ± 2.33 [B] |
| **Roja [5]** | 4.31 ± 0.15 [B] | 1046.20 ± 28.73 [B] | 23.42 ± 2.75 [B] |
| **Roja-II [6]** | 8.86 ± 0.24 [A] | 2021.17 ± 117.64 [A] | 72.61 ± 9.05 [A] |

Average of [1] TAC: Total Antioxidant Capacity; [2] TPH: Total Phenol Content; [3] ACY: Total Anthocyanin Content ± standard deviation; [4] Santee: Broccoli var. Santee; [5] Roja: Broccoli var. Roja I harvest; [6] Roja-II: Broccoli var. Roja-II harvest. Values with the same lowercase letter were not significantly different (Test SNK $p \leq 0.05$, $n = 3$). Lowercase letters refer to black cabbage analyses. Values with the same uppercase letter were not significantly different (Test SNK $p \leq 0.05$, $n = 3$). Uppercase letters refer to broccoli analyses.

In our study, the effect of genotype on the phytochemical composition and the antioxidant capacity was not evident, given that Roja (first harvest) and Santee cultivars did not show any significant difference. This was due, probably, to the low amount of antioxidant compounds accumulated in the tissues of these cultivars. On the contrary, results on the effect of developmental stage on antioxidant capacity and phytochemical composition were very interesting. In fact, Broccoli Roja-II harvest showed a statistically higher content of TPH and ACY than Roja (first harvest), associated with doubled value of antioxidant capacity. The influence of the harvest date on the antioxidant capacity of these type of plants was reported by Soengas et al. [25]. Vallejo et al. [13] reported that the content of phytochemical compounds, in particular phenolic compounds, increased with the exposure to sunlight. In fact, the growth of broccoli is very sensitive to climate and light conditions [2].

### 3.2.2. Broccoli's Head and Stem Nutritional Quality

In Figure 1, FRAP analysis showed contrasting results: in fact, Roja and Santee possessed the highest concentration of antioxidant in stem, while in Roja-II head was the plant part with the highest antioxidant capacity (9.91 ± 0.16 mM Trolox/kg fw), thus highlighting that with maturation there is a greater accumulation of nutritional substances in this part of plants. Similar FRAP results were obtained by Kaur et al. [26] (on heads) and Pellegrini et al. [27] (on whole material).

The total average of different plant parts showed that stem possessed a higher FRAP value than head. In the literature, Fernandes et al. [8] found an opposite trend in turnip, reporting that flower buds possess the highest antioxidant capacity, followed by leaves and stems, where flavonols were the most represented compound.

In the analyzed broccoli, the range of phenol concentration varied from 891.26 to 2574.77 mg GA/kg of fw (Figure 2). Total phenol content, that is responsible for the 80% of total antioxidant capacity, resulted to be highest in the broccoli heads, with an average value of 1499.35 mg GA/kg of fw. This trend was confirmed in all the analyzed genotypes, where TPH of heads were always statistically higher than the corresponding stems. Roja-II had a higher content of phenols in head and stem than Roja (first harvest), probably due to stage of development; this confirms that the late harvesting has a positive effect on the quality of the product, as reported by Šamec et al. [28] and Soengas [25]. Our study confirms that total phenolic content is influenced by genotype and plant developmental stage, in accordance with previous studies [26,29].

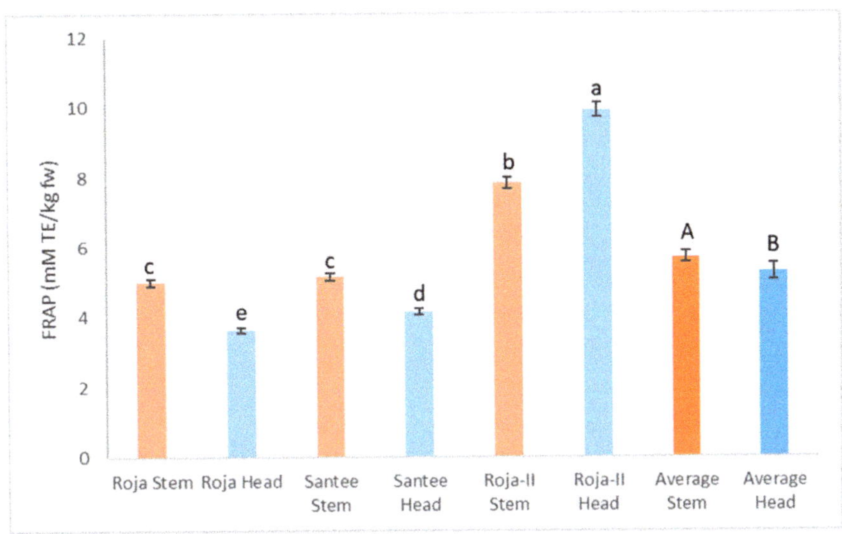

**Figure 1.** Average values and standard deviation of total antioxidant capacity measured by Ferric Reducing Antioxidant Power (FRAP) of Broccoli analyzed: Roja, Santee and Roja-II harvest divided into head and stem, and their average. Values with the same lowercase letter were not significantly different (Test SNK $p \leq 0.05$, $n = 3$). Lowercase letters refer to Roja, Santee and Roja-II harvest divided into head and stem. Values with the same uppercase letter were not significantly different (Test SNK $p \leq 0.05$, $n = 9$). Uppercase letters refer to head and stem average values.

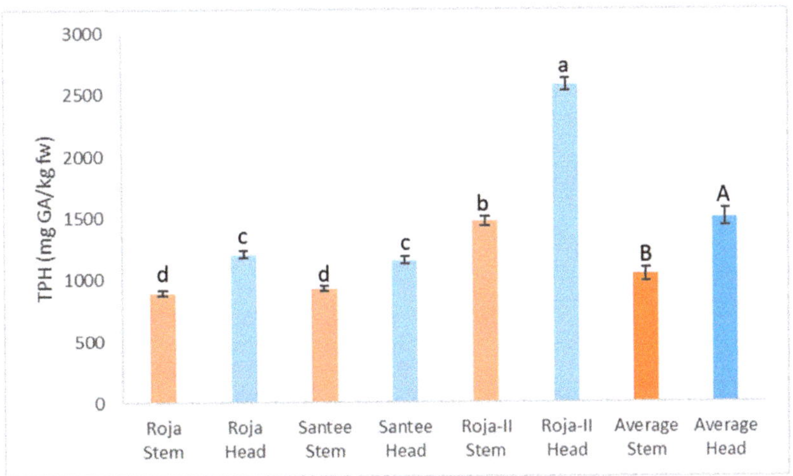

**Figure 2.** Average values of Total Phenol Content (TPH) with standard deviation of different cultivars and type of tissues of Broccoli: Roja, Santee and Roja-II harvest divided into head and stem, and their average. Values with the same lowercase letter were not significantly different (Test SNK $p \leq 0.05$, $n = 3$). Lowercase letters refer to Roja, Santee and Roja II harvest divided into head and stem. Values with the same uppercase letter were not significantly different (Test SNK $p \leq 0.05$, $n = 9$). Uppercase letters refer to head and stem average values.

Phenolic compounds have a great incidence on the total antioxidant capacity of a food matrix: for this reason, TPH and FRAP results usually have the same trend. In our case, the higher TPH average content of head did not correspond to a higher average FRAP value of the same tissue; this result was probably due to the capacity of FRAP method to detect other antioxidant compounds over phenols, such as ascorbic acid, carotenoids, vitamins, and others, although the phenolics content represented 80% of the total antioxidant capacity as reported by Podsedek [30]. However, in Roja-II, TPH values of head was higher than stem, as well as FRAP value was higher in Roja-II head than stem.

The highest value of anthocyanins content was detected in broccoli heads (Figure 3), as suggested by their slightly violaceous coloration. The highest concentration of anthocyanins has been registered for Roja-II broccoli head, probably due to an increase in purple color intensity during plant development. However, the less colored part of the plant (Stem) also showed a statistically higher ACY content in Roja-II than in Roja (first harvest), suggesting an increased accumulation of these compounds during development, likewise in absence of purple coloration.

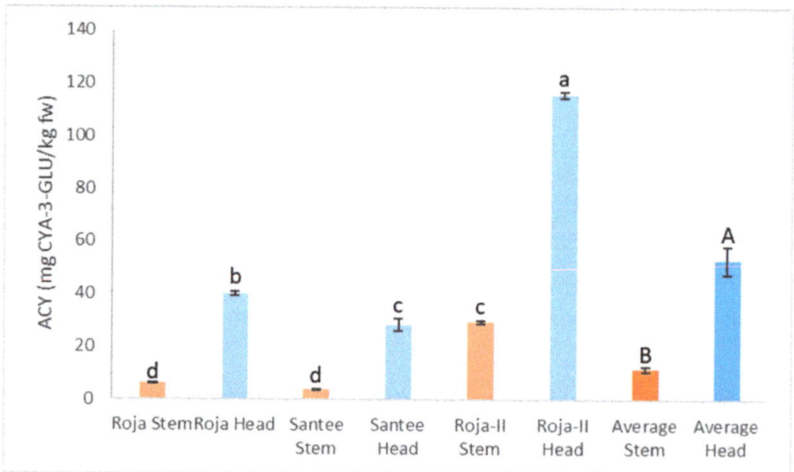


**Figure 3.** Average of total anthocyanin content (ACY) with standard deviation of different cultivars and type of tissues of Broccoli: Roja, Santee and Roja-II harvest divided into head and stem, and their average. Values with the same lowercase letter were not significantly different (Test SNK $p \leq 0.05$, $n = 3$). Lowercase letters refer to Roja, Santee and Roja-II harvest divided into head and stem. Values with the same uppercase letter were not significantly different (Test SNK $p \leq 0.05$, $n = 9$). Uppercase letters refer to head and stem average values.

Stem tissue had 78% lower accumulation of anthocyanins than head tissue. Sotelo et al. [31] found similar values in broccoli (2.2–6.3 mg CYA/kg fw), while Rodriguez-Hernandez et al. [32] made a comparison between inflorescences and leaves in purple sprouting broccoli, finding higher values than the present study.

## 4. Conclusions

The aim of the study was to analyze the nutritional properties of fresh *Brassica* plants to identify new high-quality products for the consumer, with richest content of phytochemical compounds with health benefits for the consumer. Indeed, the study on the variation of quality in different plant portion and developmental stage resulted interesting for the creation of some possible new products for the consumers. Considering the plant portion, it was evident that in black cabbage, seeds possess higher content of phytochemical compounds than leaves. These results are not surprising if we consider that the function and role of seeds is to protect from oxidation their lipids, which are very important during germination when demand of oxygen is high [9,33]. Related to the concerns of broccoli, heads had a higher phytochemicals content than stems, but the FRAP method revealed that the total antioxidant capacity followed the opposite trend. More deep analyses could reveal the factors that determine the different values of antioxidant capacity among the two developmental stages. Regarding TPH and ACY values, the Roja and Santee tissues harvested at the same time showed the same content of these antioxidant compound, not resulting in a significant difference. The big differences highlighted between Roja-II and Roja (first harvest) samples were probably due to developmental stage and harvesting time, because Roja-II was harvested approximately 20 days later than Roja (first harvest). The developmental stage and the harvest time brought some phytochemical compounds content differences: this evidence deserves to be further investigated, to better understand the behavior of phytochemicals concentration during time.

Therefore, as a conclusion, these results clearly indicate that to identify a new high sensorial and nutritional quality fresh or processed *Brassica* product it is important to identify the most appropriate cultivar, the type of tissue to be used and also the harvesting time.

**Author Contributions:** Conceptualization, F.B. (Francesca Biondi), B.M. and M.V. (Marino Visciglio); Methodology, F.B. (Francesca Biondi), L.M., F.B. (Francesca Balducci), E.M. and M.V. (Massimo Vagnoni); Software, F.C. and L.M.; Validation, F.B. (Francesca Biondi), F.C., B.M. and M.V. (Marino Visciglio); Formal Analysis, F.B. (Francesca Biondi), F.B. (Francesca Balducci) and L.M.; Investigation, F.B. (Francesca Biondi), M.V. (Marino Visciglio) and F.B. (Francesca Balducci); Resources, M.V. (Massimo Vagnoni), M.V. (Marino Visciglio) and B.M.; Data Curation, F.B. (Francesca Biondi), L.M. and F.C.; Writing—Original Draft Preparation, F.B. (Francesca Biondi); Writing—Review and Editing, L.M. and B.M.; Visualization, L.M.; Supervision, B.M., M.V. (Massimo Vagnoni), M.V. (Marino Visciglio) and E.M.; Project Administration, B.M. and M.V. (Marino Visciglio); Funding Acquisition, B.M. and M.V. (Marino Visciglio). All authors have read and agreed to the published version of the manuscript.

**Funding:** This research received no external funding.

**Institutional Review Board Statement:** Not applicable.

**Informed Consent Statement:** Not applicable.

**Data Availability Statement:** Not applicable.

**Conflicts of Interest:** The authors declare no conflict of interest.

# References

1. Kris-Etherton, P.M.; Etherton, T.D.; Carlson, J.; Gardner, C. Recent discoveries in inclusive food-based approaches and dietary patterns for reduction in risk for cardiovascular disease. *Curr. Opin. Lipidol.* **2002**, *13*, 397–407. [CrossRef] [PubMed]
2. Bjorkman, M.; Klingen, I.; Birch, A.; Bones, A.; Bruce, T.; Johansen, T.J.; Meadow, R.; Molmann, J.; Seljasen, R.; Smart, L.E.; et al. Phyto-chemicals of Brassicaceae in plant protection and human health. Influences of climate, environment and agronomic practice. *Phytochemistry* **2011**, *72*, 538–556. [CrossRef] [PubMed]
3. Zhou, K.; Yu, L. Total phenolic contents and antioxidant properties of commonly consumed vegetables grown in Colorado. *LWT—Food Sci. Technol.* **2006**, *39*, 1155–1162. [CrossRef]
4. Biondi, F.; Balducci, F.; Capocasa, F.; Visciglio, M.; Mei, E.; Vagnoni, M.; Mezzetti, B.; Mazzoni, L. Environmental Conditions and Agronomical Factors Influencing the Levels of Phytochemicals in *Brassica* Vegetables Responsible for Nutritional and Sensorial Properties. *Appl. Sci.* **2021**, *11*, 1927. [CrossRef]
5. Pereira, F.M.V.; Rosa, E.; Fahey, J.W.; Stephenson, K.K.; Carvalho, R.; Aires, A. Influence of temperature and ontogeny on the levels of glucosinolates in broccoli (*Brassica oleracea* var. *italica*) sprouts and their effect on the induction of mammalian phase 2 enzymes. *J. Agric. Food Chem.* **2002**, *50*, 6239–6244. [CrossRef]
6. Moreno, D.A.; Perez-Balibrea, S.; Ferreres, F.; Gil-Izquierdo, A.; Garcia-Viguera, C. Acylated anthocyanins in broccoli sprouts. *Food Chem.* **2010**, *123*, 358–363. [CrossRef]
7. Lo Scalzo, R.; Genna, A.; Branca, F.; Chedin, M.; Chassaigne, H. Anthocyanin composition of cauliflower (*Brassica oleracea* L. var. *botrytis*) and cabbage (*B. oleracea* L. var. *capitata*) and its stability in relation to thermal treatments. *Food Chem.* **2008**, *107*, 136–144. [CrossRef]
8. Fernandes, F.; Valentao, P.; Sousa, C.; Pereira, C.; Pereira, J.A.; Seabra, R.M.; Andrade, P.B. Chemical and antioxidative assessment of dietary turnip (*Brassica rapa* var. *rapa* L.). *Food Chem.* **2007**, *105*, 1003–1010. [CrossRef]
9. Ferreres, F.; Fernandes, F.; Sousa, C.; Valentao, P.; Pereira, J.A.; Andrade, P.B. Metabolic and Bioactivity Insights into *Brassica oleracea* var. *acephala*. *J. Agric. Food Chem.* **2009**, *57*, 8884–8892. [CrossRef]
10. Francisco, M.; Moreno, D.A.; Cartea, M.E.; Ferreres, F.; Garcia-Viguera, C.; Velasco, P. Simultaneous identification of glucosinolates and phenolic compounds in a representative collection of vegetable *Brassica rapa*. *J. Chromatogr.* **2009**, *1216*, 6611–6619. [CrossRef]

11. Francisco, M.; Cartea, M.E.; Soengas, P.; Velasco, P. Effect of Genotype and Environmental Conditions on Health-Promoting Compounds in *Brassica rapa*. *J. Agric. Food Chem.* **2011**, *59*, 2421–2431. [CrossRef] [PubMed]

12. Schmidt, S.; Zietz, M.; Schreiner, M.; Rohn, S.; Kroh, L.W.; Krumbein, A. Genotypic and climatic influences on the concentration and composition of flavonoids in kale (*Brassica oleracea* var. *sabellica*). *Food Chem.* **2010**, *119*, 1293–1299. [CrossRef]

13. Vallejo, F.; Tomas-Barberan, F.; Garcia-Viguera, C. Changes in broccoli (*Brassica oleracea* L. var. *italica*) health-promoting compounds with inflorescence development. *J. Agric. Food Chem.* **2003**, *51*, 3776–3782. [CrossRef] [PubMed]

14. Guo, J.T.; Lee, H.L.; Chiang, S.H.; Lin, F.I.; Chang, C.Y. Antioxidant properties of the extracts from different parts of broccoli in Taiwan. *J. Food Drug Anal.* **2001**, *9*, 96–101. [CrossRef]

15. Wang, S.Y.; Lin, H.S. Antioxidant activity in fruits and leaves of blackberry, raspberry, and strawberry varies with cultivar and developmental stage. *J. Agric. Food Chem.* **2000**, *48*, 140–146. [CrossRef]

16. Diamanti, J.; Capocasa, F.; Balducci, F.; Battino, M.; Hancock, J.; Mezzetti, B. Increasing Strawberry Fruit Sensorial and Nutritional Quality Using Wild and Cultivated Germplasm. *PLoS ONE* **2012**, *7*, e46470. [CrossRef]

17. Benzie, I.F.F.; Strain, J.J. The ferric reducing ability of plasma (FRAP) as a measure of "antioxidant power": The FRAP assay. *Anal. Biochem.* **1996**, *239*, 70–76. [CrossRef]

18. Deighton, N.; Brennan, R.; Finn, C.; Davies, H.V. Antioxidant properties of domesticated and wild Rubus species. *Sci. Food Agric.* **2000**, *80*, 1307–1313. [CrossRef]

19. Slinkard, K.; Singleton, V.L. Total phenol analysis: Automation and comparision with manual methods. *Am. J. Enol. Vitic.* **1977**, *28*, 49–55.

20. Giusti, M.; Wrolstad, R.E. Characterization and Measurement of anthocyanins by UV-visible Spectroscopy. *Curr. Protoc. Food Anal. Chem.* **2001**, *1*, F1.2.1–F1.2.13. [CrossRef]

21. Nicoletto, C.; Santagata, S.; Pino, S.; Sambo, P. Antioxidant characterization of different italian broccoli landraces. *Hortic. Bras.* **2016**, *34*, 74–79. [CrossRef]

22. Armesto, J.; Gómez-Limia, L.; Carballo, J.; Martínez, S. Effects of different cooking methods on some chemical and sensory properties of Galega kale. *Int. J. Food Sci. Technol.* **2016**, *51*, 2071–2080. [CrossRef]

23. Armesto, J.; Gómez-Limia, L.; Carballo, J.; Martínez, S. Impact of vacuum cooking and boiling and refrigerated storage on the quality of Galega kale (*Brassica oleracea* var. *acephala* cv. Galega). *LWT—Food Sci. Technol.* **2017**, *79*, 267–277. [CrossRef]

24. Martìnez, S.; Olmos, I.; Carballo, J.; Franco, I. Quality parameters of *Brassica* spp. grown in northwest Spain. *Int. J. Food Sci. Technol.* **2010**, *45*, 776–783. [CrossRef]

25. Soengas, P.; Sotelo, T.; Velasco, P.; Cartea, M.E. Antioxidant Properties of *Brassica* Vegetables. *Funct. Plant Sci. Biotechnol.* **2011**, *5*, 43–55.

26. Kaur, C.; Kumar, K.; Anil, D.; Kapoor, H.C. Variations in antioxidant activity in broccoli (*Brassica oleracea* L.) cultivars. *J. Food Biochem.* **2006**, *31*, 621–638. [CrossRef]

27. Pellegrini, N.; Chiavaro, E.; Gardana, C.; Mazzeo, T.; Continuo, D.; Gallo, M.; Riso, P.; Fogliano, V.; Porrini, M. Effect of different cooking methods on color, phytochemical concentration, and antioxidant capacity of raw and frozen *Brassica* vegetables. *J. Agric. Food Chem.* **2010**, *58*, 4310–4321. [CrossRef]

28. Šamec, D.; Piljac-Žegarac, J.; Bogović, M.; Habjanič, K.; Grùz, J. Antioxidant potency of white (*Brassica oleracea* L. var. *capitata*) and Chinese (*Brassica rapa* L. var. *pekinensis* (Lour.)) cabbage: The influence of development stage, cultivar choice and seed selection. *Sci. Hortic.* **2011**, *128*, 78–83. [CrossRef]

29. Bahorun, T.; Luximon-Ramma, A.; Crozier, A.; Aruoma, O.I. Total phenol, flavonoid, proanthocyanidin and vitamin C levels and antioxidant activities of Mauritian vegetables. *J. Sci. Food Agric.* **2004**, *84*, 1553–1561. [CrossRef]

30. Podsedek, A. Natural antioxidants and antioxidant capacity of *Brassica* vegetables: A review. *LWT—Food Sci. Technol.* **2007**, *40*, 1–11. [CrossRef]

31. Sotelo, T.; Cartea, M.E.; Velasco, P.; Soengas, P. Identification of Antioxidant Capacity—Related QTLs in *Brassica oleracea*. *PLoS ONE* **2014**, *9*, e107290. [CrossRef] [PubMed]

32. Rodriguez-Hernandez, M.; Moreno, D.A.; Carvajal, M.; Garcìa-Viguera, C.; Martínez-Ballesta, M. Natural Antioxidants in Purple Sprouting Broccoli under Mediterranean Climate. *J. Food Sci.* **2012**, *77*, C1058–C1063. [CrossRef] [PubMed]

33. Sousa, C.; Lopes, G.; Pereira, D.M.; Taveira, M.; Valentao, P.; Seabra, R.M.; Pereira, J.A.; Baptista, P.; Ferreres, F.; Andrade, P.B. Screening of antioxidant compounds during sprouting of *Brassica oleracea* L. var. *costata* DC. *Comb. Chem. High Throughput Screen.* **2007**, *10*, 377–386. [CrossRef] [PubMed]

*Article*

# The Impact of the Fermentation Method on the Pigment Content in Pickled Beetroot and Red Bell Pepper Juices and Freeze-Dried Powders


Emilia Janiszewska-Turak [1,*], Kacper Tracz [1], Patrycja Bielińska [1], Katarzyna Rybak [1], Katarzyna Pobiega [2], Małgorzata Gniewosz [2], Łukasz Woźniak [3] and Anna Gramza-Michałowska [4,*]

[1] Department of Food Engineering and Process Management, Institute of Food Sciences, Warsaw University of Life Sciences—SGGW, 02-787 Warsaw, Poland; 193068@sggw.edu.pl (K.T.); b.patrycja60@gmail.com (P.B.); katarzyna_rybak@sggw.edu.pl (K.R.)

[2] Department of Food Biotechnology and Microbiology, Institute of Food Sciences, Warsaw University of Life Sciences—SGGW, 02-787 Warsaw, Poland; katarzyna_pobiega@sggw.edu.pl (K.P.); malgorzata_gniewosz@sggw.edu.pl (M.G.)

[3] Department of Food Safety and Chemical Analysis, Institute of Agricultural and Food Biotechnology, 36 Rakowiecka Street, 02-532 Warsaw, Poland; lukasz.wozniak@ibprs.pl

[4] Department of Gastronomy Science and Functional Foods, Faculty of Food Science and Nutrition, Poznań University of Life Sciences, Wojska Polskiego 31, 60-624 Poznań, Poland

* Correspondence: emilia_janiszewska_turak@sggw.edu.pl (E.J.-T.); anna.gramza@up.poznan.pl (A.G.-M.); Tel.: +48-22-593-7366 (E.J.-T.); +48-61-848-7327 (A.G.-M.)



**Citation:** Janiszewska-Turak, E.; Tracz, K.; Bielińska, P.; Rybak, K.; Pobiega, K.; Gniewosz, M.; Woźniak, Ł.; Gramza-Michałowska, A. The Impact of the Fermentation Method on the Pigment Content in Pickled Beetroot and Red Bell Pepper Juices and Freeze-Dried Powders. *Appl. Sci.* **2022**, *12*, 5766. https://doi.org/10.3390/app12125766

Academic Editors: Luca Mazzoni, Maria Teresa Ariza Fernández and Franco Capocasa

Received: 29 April 2022
Accepted: 26 May 2022
Published: 7 June 2022

**Publisher's Note:** MDPI stays neutral with regard to jurisdictional claims in published maps and institutional affiliations.

**Copyright:** © 2022 by the authors. Licensee MDPI, Basel, Switzerland. This article is an open access article distributed under the terms and conditions of the Creative Commons Attribution (CC BY) license (https://creativecommons.org/licenses/by/4.0/).



**Abstract:** The beetroot and red bell pepper are vegetables rich in active ingredients, and their potential for health benefits are crucial. Both presented raw materials are rich in natural pigments, but are unstable and seasonal; thus, it was decided to take steps to extend their durability. Lactic fermentation has been recognized as a food preservation method, requiring minimal resources. The activities undertaken were also aimed at creating a new product with a coloring and probiotic potential. For this reason, the study aimed to evaluate the impact of the method of fermentation on the content of active compounds (pigments) in pickled juices and freeze-dried powders. The lactic acid fermentation guided in two ways. The second step of the research was to obtain powders in the freeze-drying process. For fermentation, *Levilactobacillus brevis* and *Limosilactobacillus fermentum* were used. In juices and powders, pigments, color, and dry matter were tested. In this research, no differences in fermented juice pigment contents were seen; however, the color coefficient differed in raw juices. The freeze-drying process resulted in lowering the pigment content, and increasing dry matter and good storage conditions (glass transition temperatures 48–66 °C). The selection of vegetable methods suggested the use of fermentation and mixing it with a marinade (higher pigments and lactic acid bacteria content). All powders were stable and can be used as a colorant source, whereas for probiotic properties, a higher number of bacteria is needed.

**Keywords:** beetroot; red bell pepper; pigments; freeze-drying; LAB; color; betalain; carotenoids


## 1. Introduction

Lactic acid fermentation has been recognized for centuries as a safe technique of food preservation which requires minimum of resources. The fermentation, with lactic acid bacteria (LAB) application, leads to a decrease in pH at the end of the process. Processing vegetables, such as by fermentation or heating, increases the bioavailability of nutrients and pigments, most likely by disrupting the plant tissues cell walls. The strains of *Levilactobacillus brevis*, *Limosilactobacillus fermentum*, and *Lactiplantibacillus plantarum* are most often used in the fermentation of vegetables [1–3]. Pickled foods, because of low pH ensuring the long shelf-life of products, can be preserved for many months. However, it depends mainly on having good storage conditions, such as a low temperature and the lack of UV radiation [4–6]. Fermented products have unique taste properties. The taste of silage depends on both the type of raw material and the method and the parameters of the ensilage process [7]. The desired changes in taste, consistency, and color take place over time, ultimately creating products with a completely different taste impression from the initial ingredients. This is mainly due to microorganisms such as lactic acid bacteria, yeast, and filamentous fungi, which contribute to the unique taste of the silage and ensure the safety and required quality of the resulting product.

Many pickled foods have been shown to be a valuable source of protein, carbohydrates, minerals, vitamins, and fiber [8]. As a result of lactic acid bacteria presence, fermented foods are classified as foods that have health-promoting properties. However, to be permanently classified as this type of food, it must have a high (over $1 \times 10^7$ log CFU/g) number of lactic acid bacteria [9]. In addition, vegetables, due to their composition and the increasingly popular trend of a healthy life-style, have become a valuable processed raw material [8]. The combination of the health-promoting properties of vegetables and lactic acid bacteria in fermented products is an excellent replacement for dairy products containing LAB. Natural pigments include betalains, carotenoids, chlorophylls, and anthocyanins. The natural pigments found in vegetables and fruits are now progressively being used as a source for coloring food. They are increasingly replacing the use of artificial pigments, which nowadays are mentioned as "not-safe" and causing undesirable health effects [6]. The disadvantage of using natural pigments is their sensitivity to the influence of the environment through the action of UV rays, oxygen, temperature, pH or the presence of metal ions, enzymes, or sugars [4,5,10,11].

In Western Europe the most popular fermented vegetables are cucumbers, cabbage, beetroot, carrots, and red bell peppers. Of these, the highest amount of pigments are in beetroot, carrots and red bell peppers [12–14]. Moreover, beetroot is rich in carbohydrates, a protein with unsaturated fatty acids [15]. The betalains isolated from the beetroot were vulgaxanthin I, vulgaxanthin II, indicaxanthin, betanin, neobetanin, prebetanin, and isobetanin [3,16]. Betalains are soluble in water N-acylated pigments divided into yellow-orange-colored betaxanthins and red-violet-colored betacyanins. They are stable in a low pH range (3–7), their stability increases with decreasing oxygen concentration, and they are resistant to fermentation temperatures [11,17]. However, the thermal stability of betacyanin and betaxanthins is still unclear, as mentioned in the beetroot pigments analysis made by Lombardelli et al. [4] Despite the process conditions, yellow or red beetroot pigments could be stable.

Red bell peppers are rich in carotenoids and are botanically included in fruits, whereas a carrot is a root vegetable. Carrots have been recognized as an excellent source of flavonoids, quercetin, vitamin C, and carotenoids [18]. Carotenoids are tetraterpenoid compounds with a conjugated double-bond system; they are mainly lipophilic pigments whose structure is based on the number of carbons linked in chains. Their cyclic structure can be modified by hydrogenation, dehydrogenation, cyclization, and oxidation [19].

However, fermented vegetables can be difficult to store, mainly due to their subsequent refrigerated storage and space requirements. Moreover, the lactic acid bacteria, after using the carbon source, is not able to multiply further, which is disadvantageous from the consumer's perspective. Therefore, it was decided to use the lyophilization process in order to protect the active ingredients (pigments) and bacteria (LAB). The freeze-drying process can also be called low-thermal drying. After this process, dried substances are obtained. However, after the completion of the process of freeze-drying vegetable juices, water can be absorbed very quickly from the environment and the powder structure broken down. Freeze-drying can be a solution to the problem of protecting sensitive substances from juices containing high amounts of low molecular weight sugars. This is caused by the low glass transition temperatures of the juice itself. When the glass transition temperature is exceeded, the product becomes compact with a glassy sheen on the surface of the agglomerated powder [20,21]. This phenomenon can be prevented by adding a high molecular weight carrier to the juice prior to the freeze-drying process. Maltodextrin is most often added [22,23].

Both presented raw materials are rich in natural pigments, but are unstable and seasonal; thus, it was decided to take steps to extend their durability. The activities undertaken were also aimed at creating a new potential product with coloring and probiotic potential. For this reason, the study aimed at the determination of the fermentation method influence on the content of active compounds (pigments) in pickled juices and freeze-dried powders.

## 2. Materials and Methods

### 2.1. Materials

Materials were purchased at the Bronisze Market (Warsaw, Poland) Beetroot (*Beta vulgaris*) and red bell pepper (*Capsicum annuum* L.) and cold-stored (4 to 6 °C) before use. As an inoculum for fermentation two bacterial strains: *Levilactobacillus brevis* KKP 804 (LB) and *Limosilactobacillus fermentum* KKP 811 (LF) were applied. The strains originated from the Collection of Industrial Microorganisms (KKP, Warsaw, Poland). Maltodextrin DE10 was supplied by Pepees S.A. (Łomża, Poland).

### 2.2. Technological Treatment

#### 2.2.1. Fermentation Process

Two different kinds of fermentation were used. The first type was the fermentation of juices obtained from the beetroot and red s bell pepper. The second type was the fermentation of sliced vegetables/fruits in 200 mL jars. In the first type of juice fermentation, NaCl was directly added to the juice with 2% of juice volume. In the second type of fermentation, NaCl was dissolved in water in a concentration of 2% and then was added to the jars with vegetable slices. Inoculum of 1% to the water/juice volume, which matched to the $1 \times 10^7$ CFU/mL of bacterial content, was added. For spontaneous fermentation (SF) no inoculum was added. For anaerobic conditions, jars were closed and kept in an incubator at a stable temperature of 28 °C. Fermentation process was carried out for 7 days after addition of the inoculum. All experiments were carried out in duplicate and in parallel.

#### 2.2.2. Juice Pressing

The juice was obtained from raw and fermented vegetables. The process was carried out with an NS-621CES juicer model (Kuvings, Daegu, Korea). Separately, juice and pomace were collected. Juice was used in this research.

In the second method, juice was obtained from fermented vegetables. This juice was mixed in a proportion 1:1 with brine (S). In previous research, only pressed juices were tested separately from the post-fermentation solution [3,9] for comparison with the post-fermentation solution, and as a result a high amount of LAB was observed. This is the reason why we mixed it in this research.

Before freeze-drying, the addition of a carrier material was needed. In the presented research, 15% $v/v$ maltodextrin with a low dextrose equivalent (DE = 10) was added to the juices. Fermented juices without a carrier addition were highly hygroscopic, and after being taken from the freeze-dryer their structure collapsed.

#### 2.2.3. Freeze-Drying

The freeze-dried protocol was used in accordance to methodology presented by [24] with some changes. Obtained from both types of fermentation were frozen at −40 °C (Shock Freezer HCM 51.20, Irinox, Treviso, Italy) for 10 h on a petri dish. Freeze-drying was carried out in an ALPHA 1–4 freeze-dryer (Christ, Osterode, Germany) for 24 h at a heating shelf temperature of 30 °C and the constant pressure of 63 Pa; a safety pressure was set up at 103 Pa. The experiments were carried out in duplicate.

### 2.3. Analytical Method

#### 2.3.1. Dry Matter

Gravimetrical method was used to dry matter determination. For juice aproximatelly 0.6–1 g was placedat filter paper in a dish, while 1 g was used for powders.Drying was made in vacuum-dryer (Memmert VO400, Schwabach, Germany) under the pressure of 10 mPa at 75 °C for 24 h until constant weight. Measurements were performed in triplicate.

#### 2.3.2. Total Acidity

For the measurement of total acidity, the titration method with 0.1 M NaOH was used. A known mass of sample juice/powder ($m$) was taken and diluted with distilled water to a volume of 50 mL ($V_1$), then 25 mL of diluted juice ($V_2$) was taken for testing. Titration with the NaOH solution ended when the pH reached 8.1, and the amount of NaOH solution used ($V_{NaOH}$) was used for calculations. The measurement was conducted in triplicate for each sample.

$$Total\ acidity = \frac{(V_{NaOH}\cdot 0.1 \cdot V_1 \cdot 0.09 \cdot 100)}{(V_2 \cdot m)}\ (g\ lactic\ \frac{acid}{100g} product) \tag{1}$$

where 0.1 is the molarity of NaOH and 0.09 is the index for correction for lactic acid in the sample.

#### 2.3.3. Color Parameters

The color of powders was analysed with the colorimeter CR-5 (Konica Minolta Sensing Inc., Osaka, Japan) in the CIE L*a*b* system. Parameters used: Illuminant D65, an angle of 2°, and calibration with white plate. All measurements were made in 3 repetitions. In addition, the ΔE* factor of the color differences between juices before (parameters L*a*b* with index raw) and after fermentation was calculated from formula [5,25]:

$$\Delta E = \sqrt{\left(L^* * - L_{raw}^*\right)^2 + \left(a^* * - a_{raw}^*\right)^2 + \left(b^* * - b_{raw}^*\right)^2} \tag{2}$$

#### 2.3.4. Thermal Properties

The weight loss upon heating was measured using a TGA/DSC 3+ thermogravimeter (Mettler-Toledo, Greifensee, Switzerland). Ground material in amounts of 5–8 mg were placed in 70 μL alumina crucibles and pyrolyzed from 30 to 600 °C with a heating rate of 5 °C/min under nitrogen atmosphere (50 mL/min). The maximum temperatures of the energy effects were determined from the DTG curves [26]. Two measurements of each sample were made.

A differential scanning calorimeter (DSC 3+ STAR, Mettler-Toledo, Greifensee, Switzerland) with liquid nitrogen cooling was used to determine the glass transition temperature (Tg). Prior to analysis, the samples were dried in a vacuum oven (30 °C, 10 mb, 48 h) and stored over anhydrous $P_2O_5$. An amount of 3–6 mg of the sample was weighed in 40 μL aluminum pans and subjected to a three-step analysis. In the first stage, the sample was cooled from room temperature to −50 °C, then kept at this temperature for 5 min and heated to 150 °C at the rate of 5 °C/min with a nitrogen flow of 50 mL/min. The obtained DSC curves were analyzed using dedicated STARe software v.16.0.

#### 2.3.5. ATR-FTIR Spectroscopic Analysis

Infrared absorption spectra of the dried samples were recorded using a Cary 630 spectrometer (Agilent Technologies Inc., Santa Clara, CA, USA) with a diamond crystal ATR system. The analysis was carried out in the wavenumber range of 4000–650 cm$^{-1}$ for 32 scans with a resolution of 4 cm$^{-1}$ [27]. A background measurement was performed before each sample. Each sample was scanned in triplicate.

### 2.3.6. Determination of the Number of Lactic Acid Bacteria

For enumeration of viable cells a total count by plate method was made. Juices and freeze-dried juice samples were serially diluted using sterile saline (0.85% NaCl, Biomaxima, Lublin, Poland). On the de Man, Rogosa, and Sharpe agar plates samples were placed (MRS, Biomaxima, Lublin, Poland) and incubated at $28 \pm 1$ °C for $48 \pm 4$ h. The number of grown colonies was counted and recorded as log CFU per g d.m by ProtoCOL 3, (Synbiosis, Frederick, MD, USA). The samples were analyzed in triplicates.

### 2.3.7. Betalain Content

Betalain content was measured by two methods. For the calculation of the amount, the spectrophotometric method was used, and for the identification of betalain, the HPLC method was used.

(A)   HPLC method

The HPLC method is based on research of Janiszewska-Turak et al. [3]. For freeze-dried samples, approximately 1 g was extracted with 20 mL of a mixture of 0.2% formic acid and acetonitrile (4:1, $v/v$). The samples were stirred for 5 min, centrifuged at $5000 \times g$ for 5 min. The supernatants were filtered through 0.45 μm syringe PTFE filters (Macherey-Nagel, Duren, Germany) to chromatographic vials.

The analyses were conducted using an equipment from Waters (Milford, MA, USA): the 2695 Separations Module connected with 2995 PDA detector, and 2475 Multi-Wavelength Fluorescence Detector (Waters, Milford, MA, USA). The samples of 10 μL were injected on a Sunfire C8 column (5 μm, 4.6 × 250 mm, Waters) with a precolumn with identical chemistry, which were kept at the temperature of 30 °C and rinsed at 1.0 mL/min with a gradient of 0.2% formic acid and acetonitrile as described by Janiszewska-Turak et al. [3]. Betacyanins were quantified at a wavelength of 538 nm, while betaxanthins at 480 nm. Two independent analyses were performed for each sample.

The analytes were identified using their retention times, UV/VIS spectra, and previous results from our team [16,28–31].

(B)   Spectrophotometric method

The spectrophotometric method was based on the methodology presented by Janiszewska-Turak et al. [32]. The measurement was conducted on the Evolution 220 (Evolution 220, Thermo Fisher Scientific Inc., Waltham, MA, USA). The 0.5 g of sample (powder or juice) was mixed with phosphate buffer (5000 mg). Betanin (mg betanin/100 g of dm) and vulgaxanthin I (mg vulgaxanthin I/100 g of dm) determination was based on the red and yellow pigments, respectively. The samples were analyzed in triplicates.

### 2.3.8. Carotenoids Analysis

The total carotenoids content (TCC) was measured according to a methodology [33,34] based on spectrophotometric measurements. The absorbance of the colored solutions was determined at 450 nm (Spectronic 200; Thermo Fisher Scientific Inc., Waltham, MA, USA). The analysis was conducted in triplicate.

### *2.4. Statistical Treatment*

The results obtained were statistically analysed with use of Statistica 13 software (StatSoft, Warsaw, Poland). A one-way ANOVA (analysis of variance) with the identification of homogenous groups with Tukey's HSD test at a significance level of $\alpha = 0.05$ was made. Medium and standard deviation parameters, were determined using MS Excel 16.

## 3. Results and Discussion

The paper presents the results of this research on juices obtained due to lactic acid fermentation with the use of dedicated fermentation strains (*Levilactobacillus brevis* KKP 804 (LB) and *Limosilactobacillus fermentum* KKP 811 (LF)) and spontaneous fermentation. Two raw materials significantly different from each other were used for the research: red beetroot (B), which is a root vegetable with a compact structure, and red bell pepper (RBP), which is botanically classified as a fruit and has a different structure of tissues than beetroot.

The fermented juices obtained in two ways were tested for the content of active compounds, which include betalains, carotenoids, and the presence of lactic acid bacteria. Moreover, the physicochemical properties of juices such as dry weight, viscosity, total acidity, and color were determined. The obtained fermented beetroot and red pepper juices were then subjected to the freeze-drying process in order to extend their shelf-life and create a new product with coloring and probiotic potential. In the obtained powders, apart from the determinations made for juices, the physical transformation temperatures (TGA, DSC) were tested in order to determine the storage conditions.

### 3.1. Juices

The process of lactic acid fermentation of the juices alone increased the viscosity of the juices obtained regardless of the type of raw material, with the only exception being beetroot juice with the use of *Limosilactobacillus fermentum* (Table 1). In the second method of fermentation, the juice was obtained from fermented raw materials and then mixed with a marinade into which water-soluble substances released in the fermentation product could pass during the fermentation process, with no statistically significant changes in the lightness value of the final juices. The lightness of non-fermented juices differed between the analyzed raw materials; higher values were observed for beetroot juice, which may be related to the composition of the vegetable itself and the presence of pigments dissolved in water, which could pass into the juice during the pressing process (Table 1). However, the analysis of the fermented juices did not show any difference in the lightness value, despite of the type of raw material and the fermentation.

**Table 1.** Selected physicochemical and microbiological properties of vegetable juices.

| Sample | Viscosity (mPas) | Dry Matter (g/g) | Total Acidity (g Lactic Acid/100 g Product) | Color Coefficients | | | $\Delta E$ | Number of Lactic Acid Bacteria log CFU/g d.m. |
|---|---|---|---|---|---|---|---|---|
| | | | | L* | a* | b* | | |
| B_raw | 1.51 ± 0.09 [ab] | 0.054 ± 0.008 [b] | 0.82 ± 0.13 [b] | 6.33 ± 0.06 [c] | 14.46 ± 0.16 [e] | 1.91 ± 0.08 [g] | - | 2.06 ± 0.03 [a] |
| B_J_LF | 1.15 ± 0.10 [a] | 0.070 ± 0.006 [c] | 0.73 ± 0.02 [b] | 6.47 ± 0.21 [c] | 9.85 ± 0.44 [d] | 0.35 ± 0.06 [f] | 4.73 ± 0.40 [a] | 6.64 ± 0.16 [c] |
| B_J_LB | 1.63 ± 0.27 [b] | 0.072 ± 0.001 [c] | 0.68 ± 0.04 [b] | 5.68 ± 0.05 [b] | 5.37 ± 0.01 [a] | −0.31 ± 0.13 [e] | 9.23 ± 0.04 [d] | 7.94 ± 0.01 [e] |
| B_J_SF | 1.62 ± 0.07 [b] | 0.052 ± 0.002 [b] | 0.35 ± 0.03 [a] | 4.15 ± 0.05 [a] | 9.08 ± 0.16 [c] | −1.20 ± 0.12 [d] | 6.42 ± 0.10 [b] | 6.98 ± 0.06 [d] |
| B_PJ+S_LF | 1.21 ± 0.11 [a] | 0.033 ± 0.000 [a] | 0.61 ± 0.01 [b] | 6.84 ± 0.05 [d] | 4.91 ± 0.08 [a] | −2.57 ± 0.08 [b] | 10.42 ± 0.08 [e] | 6.27 ± 0.05 [b] |
| B_PJ+S_LB | 1.22 ± 0.11 [a] | 0.025 ± 0.001 [a] | 0.70 ± 0.00 [b] | 10.23 ± 0.05 [f] | 5.02 ± 0.07 [a] | −1.72 ± 0.06 [c] | 10.73 ± 0.06 [e] | 8.08 ± 0.09 [e] |
| B_PJ+S_SF | 1.40 ± 0.13 [ab] | 0.032 ± 0.001 [a] | 0.70 ± 0.01 [b] | 9.68 ± 0.06 [e] | 8.25 ± 0.11 [b] | −2.81 ± 0.02 [a] | 8.38 ± 0.06 [c] | 6.62 ± 0.26 [c] |
| RBP_raw | 1.25 ± 0.05 [A] | 0.018 ± 0.003 [A] | 0.73 ± 0.03 [AB] | 14.96 ± 0.03 [A] | 14.21 ± 0.05 [C] | 14.16 ± 0.06 [C] | - | 2.21 ± 0.09 [A] |
| RBP_J_LF | 1.51 ± 0.14 [C] | 0.047 ± 0.006 [C] | 1.96 ± 0.08 [D] | 24.39 ± 0.01 [F] | 18.98 ± 0.02 [D] | 31.40 ± 0.09 [E] | 20.19 ± 0.07 [C] | 7.91 ± 0.04 [D] |
| RBP_J_LB | 1.31 ± 0.08 [AB] | 0.035 ± 0.000 [B] | 1.60 ± 0.16 [C] | 24.00 ± 0.01 [F] | 19.17 ± 0.01 [E] | 28.28 ± 0.04 [D] | 17.45 ± 0.03 [B] | 7.06 ± 0.03 [B] |
| RBP_J_SF | 1.46 ± 0.01 [B] | 0.038 ± 0.002 [BC] | 1.33 ± 0.10 [C] | 21.96 ± 0.01 [E] | 15.38 ± 0.02 [D] | 31.38 ± 0.12 [E] | 20.14 ± 0.09 [C] | 8.37 ± 0.09 [E] |
| RBP_PJ+S_LF | 1.27 ± 0.09 [A] | 0.013 ± 0.002 [A] | 0.94 ± 0.04 [B] | 17.18 ± 0.03 [D] | 8.26 ± 0.07 [A] | 10.74 ± 0.04 [A] | 7.26 ± 0.07 [A] | 7.71 ± 0.07 [C] |
| RBP_PJ+S_LB | 1.36 ± 0.06 [AB] | 0.015 ± 0.000 [A] | 0.77 ± 0.04 [AB] | 16.97 ± 0.01 [B] | 8.45 ± 0.04 [B] | 11.05 ± 0.15 [B] | 6.65 ± 0.07 [A] | 7.15 ± 0.06 [B] |
| RBP_PJ+S_SF | 1.33 ± 0.01 [AB] | 0.076 ± 0.001 [D] | 0.63 ± 0.04 [A] | 17.06 ± 0.00 [C] | 8.26 ± 0.08 [A] | 10.66 ± 0.04 [A] | 7.26 ± 0.09 [A] | 8.33 ± 0.06 [E] |

B—beetroot; RBP—red bell pepper; LF—*Limosilactobacillus fermentum* KKP 811; LB—*Levilactobacillus brevis* KKP 804; SF—spontaneous fermentation; J—juice pressed before fermentation; PJ+S—juice pressed from fermented vegetable and mixed with post-fermentation solution. [a], [b], [c] and specific letters—different indexes for the beetroot columns mean statistically significant differences for given values at the level of $p < 0.05$; [A], [B], [C] and specific letters—different indexes for red bell pepper columns (RBP) mean statistically significant differences for given values at the level of $p < 0.05$.

The viscosity of the juices depends on the content of the solute. This is confirmed by the dry matter obtained from the juices. The higher the dry matter content and, therefore, the higher the content of soluble and insoluble substances in the juice, the higher the observed viscosity value (Table 1).

Higher values of dry matter were observed for beet juice, regardless of the fermentation method used and the starter cultures used. For both raw materials, the second fermentation method resulted in the achievement of lower values of dry matter, which is strictly related to the addition of a 1:1 water marinade to the juice. The exception was fermented red bell pepper with the addition of a water marinade, for which the highest dry matter values were obtained as compared to other pepper juices.

Similar viscosity was observed for beetroot juices (1.56 mPas) by Janiszewska [35]; however, higher values were observed for red bell pepper juices (1.99 mPas) by Rybak et al. [36]. Differences in viscosity and dry matter in presented juices are connected to the ingredients that are present in raw materials; in the beetroot water 87.6 g/100 g product, carbohydrates accounted for approximately 9.5–10 g/100 g of the product, protein 1.6–1.7 g/100 g of the product, and fat 0.17–0.18 g/100 g of the product [12,37], whereas for the red bell pepper water 92.2 g/100 g product, carbohydrates account for approximately 6.03 g/100 g of the product, protein 0.99 g/100 g of the product, and fat 0.3 g/100 g of the product [13]. In the research of Hallmann et al. [38], the dry matter for fermented beetroot was 0.0814–0.0747 g/g, which was a little higher than in the presented research for beetroot juices obtained through the first type of fermentation; however, it could be related to the used cultivar.

The total acidity for both juices tested was similar, as it was 0.82 for beetroot juice and 0.73 for red pepper juice. It was observed that for beet juice after the fermentation process, regardless of the method of fermentation, no changes in acidity values were visible, which is related to the low pH of the beetroot juice itself (pH about 5) before the fermentation process. Similar results were presented by Czyżowska et al. [39] for the lactic acid content in fermented beetroot juices (0.5 g/100 mL).

In the case of fermented red pepper juice, the total acidity was twice as high as for the juice before fermentation and for the juice obtained by squeezing and mixing with the marinade.

The content of lactic acid bacteria is an substantial parameter that determines the quality of fermented juices. The presence of LAB at the level of two log cycles was demonstrated in raw juices. The presence of 6.5–8.0 log CFU/g d.m. of LAB was found in fermented beetroot juices, whereas in juices squeezed from beetroot after fermentation and combined with marinade, the presence of LAB was demonstrated at a comparable level for the respective types of fermentation. Similar dependencies were shown in the case of pepper juices, but when it comes to spontaneous fermentation, the number of lactic acid bacteria after fermentation was higher than in the case of fermentation with the use of one type of bacteria. Both *L. brevis* and *L. fermentum* grew in red bell pepper and beetroot juices.

The color of juices differed depending on the raw material tested. The red beet juices were redder and darker compared to the red bell pepper juices, which is related to the content of specific pigments in those raw materials (Table 1).

It was observed that the color components decreased after the fermentation process of the red beet juice, which means that the juices after the fermentation process were darker (L* decrease), less red (a* decrease), and much greener (b* decrease). The color components of the red bell pepper juice after fermentation increased, which means that the juices after the fermentation process were lighter (L* increase), blacker (a* increase), and definitely more yellow (b* increase).

Changing the fermentation method into vegetable/fruit fermentation and then squeezing it and mixing it with water brine resulted in a statistically significant reduction in the value of redness (+a*) and yellowness/greenness (b*) for both juices, in addition to an increase in brightness (+L*).

The color difference factor (ΔE*) is a combination of all color coefficients related to the reference color coefficients. A comparison was made in the presented research with the raw juices, thus the differences after the fermentation process could be observed. It is stated that values below 5 mean there is no difference in color as seen with the human eye, values between 5 and 12 give information about minor variations of the color, and values above 12 mean that the color is totally different than the reference sample [5,25,40]. For both juice vegetables, the calculated factor of color differences (ΔE*) was higher than 5. For fermented beetroot juices, the color difference factor was in the range from 4.7 to 10.7, which means that slight differences were seen and the juices were similar to the raw beetroot juice, however, the difference can be detected by the human eye. For red bell pepper, slight differences were observed for fermented juices (values from 6.6–7.3), whereas juices obtained from the fermented vegetable and mixed with marinade differed at first sight (values above 17) (Table 1).

Analysis of the content of color pigments in red beetroot and red bell pepper showed the influence of the fermentation method on the pigment content (Figures 1 and 2). In beetroot juice, an increase in betacyanin content was observed for juices obtained by pressing fermented vegetables and using of marinade after fermentation (approximately from 316 for raw juice to 720–950 mg/100 g dm) and in both pigments (betacyanin and betaxanthin) obtained by spontaneous fermentation of beet juice (680 for betacyanin, 530 mg/100 g dm for betaxanthin) (Figure 1). The increase in the content of betalain pigments in the second method of lactic fermentation may be related to the transition of betalain pigments from the inside of the fermented beetroot tissue to the brine. Betalain pigments are water-soluble compounds, and during the fermentation process they can migrate to the marinade [41,42]. NaCl in fermented vegetables can perform several functions: It is a preservative, i.e., it protects against the development of unfavorable microflora and increases the osmotic pressure that causes the extraction of vegetable juice. Increasing the osmotic pressure during the fermentation process of vegetables in brine causes the migration of water from the inside of the vegetable to the brine [43]. In addition to the migrating water, soluble substances, including betalain pigments, also enter the brine.

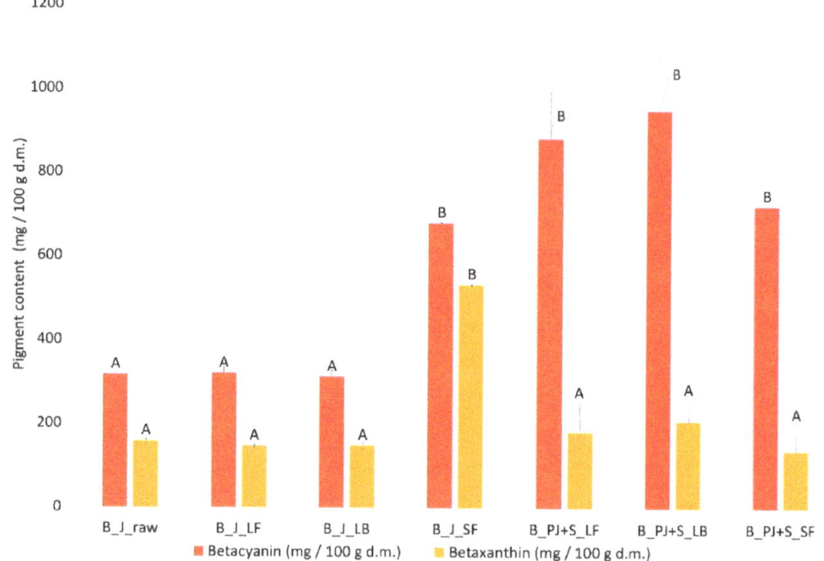


**Figure 1.** Pigment content in beetroot juices, betacyanin and betaxanthin; A, B—different indexes for series mean statistically significant differences for given values at the level of $p < 0.05$.

There was no influence of the type of bacterial addition to the beetroot juice/vegetable fermentation process on the beetroot juice/vegetables.

Analysis of the carotenoid content in fermented red bell pepper juices showed the influence of this process on the pigment content (Figure 2). In fermented juices, a decrease in the carotenoid content was observed, regardless of the type of lactic acid bacteria used. Changing the fermentation method to the fermentation of the raw material caused only a slight decrease in the content of carotenoids compared to juice without fermentation. However, it is not possible to unequivocally establish a relationship between the way the lactic acid fermentation process is carried out and the amount of carotenoids. The highest values achieved in the spontaneous fermentation of the juice (about 130 mg/100 g d.m.)

have not been confirmed in the second fermentation method, of which the carotenoid content was about 10 mg/100 g d.m.

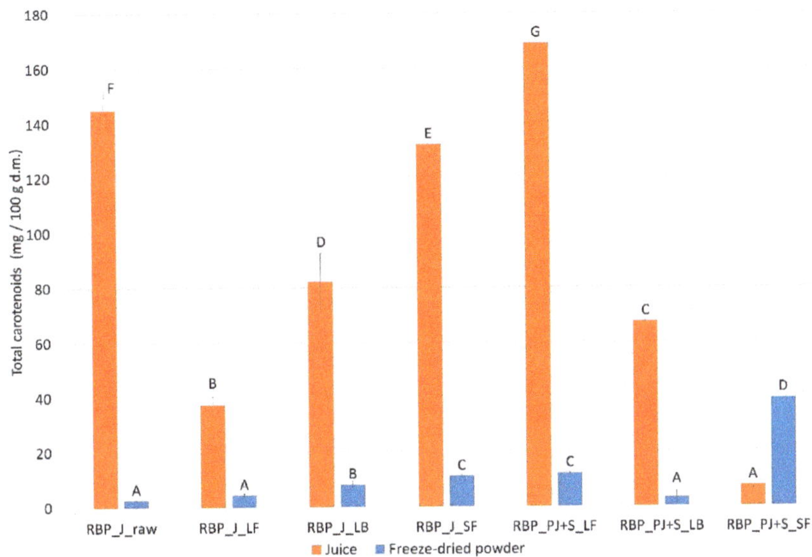


**Figure 2.** Carotenoids content in juice and its powders; A, B, C, and other letters—different indexes for series mean statistically significant differences for given values at the level of $p < 0.05$.

On the other hand, the use of LF strains for whole pieces of red bell pepper resulted in obtaining the highest values of carotenoids (170 mg/100 g d.m.) in juices mixed with post-fermentation solution. It could be caused by tissue disturbance during the fermentation process, resulting in the release of larger amounts of carotenoids from the interior of pepper particles during pressing [44].

In their research, Hallmann et al. [38] observed cis-β-carotene, α -carotene, capsorubin, α- and β-cryptoxanthin, lutein, and zeaxanthin in fermented red bell peppers. They observed that the highest values were for cis-β-carotene and zeaxanthin (0.38 mg/100 g fw) [38].

### 3.2. Freeze-Dried Powders

To each juice before freeze-drying, 15% $v/v$ of maltodextrin DE = 10 was added. Dried beetroot and red bell pepper juices without the addition of a carrier agent were very hygroscopic, and after a few minutes the FD powder's structure collapsed and the glass transition process took place.

Obtained after freeze-drying the fermented juices, beetroot powders were character-ized by a high dry matter content (84–98%); beetroot powders had a higher total acidity (TA), approximately 3 g lactic acid/100 g, higher lightness (L*) and redness (a*), and lower yellowness/greenness (b*) than in juices (Table 2). For red bell pepper powders, almost the same correlation was observed as for beetroot powders. The smallest differences were observed for the color coefficients a* and b * with respect to the juices (Tables 1 and 2).

The higher total acidity of red bell pepper compared to beetroot powders could be related to the concentration of salt in the marinade and lactic acid concentration in the powder.

Freeze-drying causes a decrease in microbial survival, but it is less than while spray-drying or convection-drying. The number of lactic acid bacteria in the powders obtained as a result of lyophilization is presented in Table 2. A decrease in approximately one to three cycles of the powders' log number of LAB in relation to juices was observed. The influence of the lactic acid bacteria used and the fermentation method (fermented juices or juices obtained as a result of squeezed fermented vegetables) on the LAB number was not observed.

**Table 2.** Selected physicochemical and microbiological properties of fermented freeze-dried powders.

| Sample | Dry Matter (g/g) | Total Acidity (g Lactic Acid/100 g Product) | Color Coefficients | | | Tg (°C) | Number of Lactic Acid Bacteria log CFU/g d.m. |
|---|---|---|---|---|---|---|---|
| | | | L* | a* | b* | | |
| B_J_LF | 0.840 ± 0.201 [a] | 2.78 ± 0.21 [a] | 37.92 ± 0.35 [cd] | 25.51 ± 0.26 [ab] | −10.38 ± 0.17 [b] | 61.51 ± 0.49 [b] | 5.80 ± 0.01 [d] |
| B_J_LB | 0.947 ± 0.006 [a] | 2.86 ± 0.09 [a] | 36.09 ± 1.41 [bc] | 26.81 ± 3.24 [bc] | −10.83 ± 1.17 [b] | 58.68 ± 0.86 [a] | 5.89 ± 0.01 [d] |
| B_J_SF | 0.982 ± 0.000 [a] | 2.96 ± 0.39 [a] | 35.31 ± 1.35 [b] | 21.82 ± 0.70 [a] | −9.62 ± 0.66 [b] | 59.60 ± 0.23 [a] | 5.37 ± 0.05 [bc] |
| B_PJ+S_LF | 0.940 ± 0.003 [a] | 2.72 ± 0.00 [a] | 36.05 ± 0.01 [bc] | 36.53 ± 0.01 [e] | −13.31 ± 0.02 [a] | 66.00 ± 0.49 [d] | 5.25 ± 0.01 [b] |
| B_PJ+S_LB | 0.931 ± 0.038 [a] | 2.57 ± 0.06 [a] | 32.85 ± 0.00 [a] | 33.03 ± 0.01 [de] | −0.29 ± 0.01 [c] | 62.55 ± 0.49 [b] | 5.15 ± 0.10 [b] |
| B_PJ+S_SF | 0.960 ± 0.000 [a] | 2.57 ± 0.06 [a] | 39.37 ± 0.01 [d] | 29.60 ± 0.04 [cd] | −10.70 ± 0.02 [b] | 64.12 ± 0.33 [c] | 4.53 ± 0.16 [a] |
| RBP_J_LF | 0.840 ± 0.016 [A] | 6.18 ± 0.76 [BC] | 66.42 ± 0.00 [B] | 27.07 ± 0.01 [E] | 37.28 ± 0.01 [D] | 48.56 ± 0.23 [C] | 5.81 ± 0.01 [C] |
| RBP_J_LB | 0.971 ± 0.001 [C] | 7.50 ± 1.41 [C] | 62.48 ± 0.01 [A] | 28.99 ± 0.01 [F] | 37.88 ± 0.01 [E] | 61.17 ± 0.35 [E] | 5.92 ± 0.06 [D] |
| RBP_J_SF | 0.923 ± 0.004 [B] | 3.49 ± 0.04 [AB] | 75.78 ± 0.15 [E] | 14.47 ± 0.03 [B] | 25.81 ± 0.03 [B] | 56.43 ± 0.15 [D] | 5.54 ± 0.11 [B] |
| RBP_PJ+S_LF | 0.964 ± 0.010 [C] | 4.17 ± 0.73 [AB] | 71.58 ± 0.01 [C] | 22.51 ± 0.01 [D] | 31.21 ± 0.01 [C] | 52.86 ± 0.41 [D] | 5.61 ± 0.03 [B] |
| RBP_PJ+S_LB | 0.950 ± 0.013 [BC] | 3.89 ± 0.05 [AB] | 73.56 ± 0.01 [D] | 18.44 ± 0.01 [C] | 26.17 ± 0.01 [B] | 49.55 ± 0.41 [B] | 5.25 ± 0.20 [A] |
| RBP_PJ+S_SF | 0.824 ± 0.000 [A] | 2.20 ± 0.01 [A] | 83.58 ± 0.38 [F] | 8.50 ± 0.36 [A] | 16.48 ± 0.50 [A] | 46.52 ± 0.39 [A] | 5.11 ± 0.27 [A] |

B—beetroot; RBP—red bell pepper; LF—*Limosilactobacillus fermentum* KKP 811; LB—*Levilactobacillus brevis* KKP 804; SF—spontaneous fermentation; J—juice pressed before fermentation; PJ+S—juice pressed from fermented vegetable and mixed with post-fermentation solution, [a], [b], [c] and specific letters—different indexes for the beetroot columns mean statistically significant differences for given values at the level of $p < 0.05$, [A], [B], [C] and specific letters—different indexes for red bell pepper columns (RBP) mean statistically significant differences for given values at the level of $p < 0.05$.

Storage conditions for powders from fermented juices obtained by freeze-drying can be predicted on the basis of thermal analysis methods. The differential scanning calorimetry (DSC) method determines the glass transition temperature (Tg) below which materials should be stored. In the powders obtained, the temperatures ranged from 59 to 66 °C for beetroot and 46 to 61 for red bell pepper (Table 2). The obtained values allow that the obtained powders can be stored at room temperature, as the literature suggests values lower than the Tg by about 20 °C are safe [45–47]. Similar values for dried beetroot powders with maltodextrin were obtained by Flores-Mancha et al. [48] and reached 61 °C. Those values for powders can be related to the addition of maltodextrin to juices before freeze-drying. The glass transition is related to the molecular weight of the compound and water content in the sample. It was measured that for maltodextrin, the Tg is above 140 °C, for sucrose 60 °C, and for glucose 31 °C [49,50]. The mixture of 15% MD with vegetable juices that contained approximately 4–7 g/100 g of sugars could result in a decrease in the Tg to the level mentioned in Table 2.

In addition to the DSC method, thermal analysis predicts changes during storage. The TGA method can show phase transitions or chemical reactions in powders, which can occur when temperature increases. Thermogravimetry shows "changes in the mass of a substance during heating or cooling as a function of time and temperature" [51]. In the TGA analysis, specific regions can be selected by using steps in measurement. In the analysis provided in this research, three steps were used: In the first step, the temperature ranged from 30 to 140, the second step was 140–420, and the third step was 420–600 °C. The first step is associated with the loss of moisture, whereas the second step, above 120 °C, shows decomposition processes of, e.g., proteins and carbohydrates [52–54]. In analysis samples the highest mass loss (41–75%) was determined in the second region (step 2) (Table 3), which can be related to the carbohydrates from vegetables (6–9 g/100 g product) [12,13] as well as maltodextrin (a polysaccharide with glucose parts linked with each other) present in all samples. The temperature of maltodextrin decomposition is approximately 130–160 °C and is connected to the water content in the powder sample [55]. The highest water evaporation in the first step is related to the highest water content in that sample (Table 3). No specific differences between fermentation type, bacteria type, and material used were observed.

**Table 3.** Thermal properties of fermented freeze-dried powders.

| | Step 1 | | Step 2 | | Step 3 | | Sum [%] | Decomposition Temperature [°C] | | | |
|---|---|---|---|---|---|---|---|---|---|---|---|
| | Temp. Range [°C] | Mass Loss [%] | Temp. Range [°C] | Mass Loss [%] | Temp. Range [°C] | Mass Loss [%] | | 1 | | 2 | |
| B_J_LF | 30–140 | 2.0 | 140–420 | 41.8 | 420–600 | 8.7 | 52.5 | 78.1 | 162.0 | 197.7 | 268.2 |
| B_J_LB | 30–140 | 2.2 | 140–420 | 45.3 | 420–600 | 5.2 | 52.8 | 79.3 | 161.3 | 195.7 | 267.6 |
| B_J_SF | 30–120 | 2.4 | 120–420 | 52.3 | 420–600 | 2.2 | 56.9 | 73.3 | 152.0 | | 286.2 |
| B_PJ+S_LF | 30–140 | 0.4 | 140–420 | 44.1 | 420–600 | 3.8 | 48.3 | 84.5 | 173.7 | | 289.3 |
| B_PJ+S_LB | 30–140 | 7.2 | 140–420 | 46.4 | 420–600 | 3.4 | 57.0 | | 170.0 | | 290.6 |
| B_PJ+S_SF | 30–140 | 5.4 | 140–420 | 55.5 | 420–600 | 2.2 | 63.2 | | 169.4 | | 291.8 |
| RBP_J_LF | 30–130 | 4.4 | 130–420 | 43.4 | 420–600 | 2.7 | 50.5 | 85.3 | 155.0 | | 284.3 |
| RBP_J_LB | 30–130 | 3.7 | 130–420 | 46.2 | 420–600 | 3.5 | 53.5 | 80.3 | 154.0 | | 285.9 |
| RBP_J_SF | 30–120 | 1.6 | 120–420 | 53.4 | 420–600 | 6.6 | 61.6 | | 153.2 | | 287.2 |
| RBP_PJ+S_LF | 30–140 | 2.6 | 140–420 | 47.8 | 420–600 | 3.9 | 54.3 | 92.2 | 168.6 | | 293.3 |
| RBP_PJ+S_LB | 30–140 | 0.6 | 140–420 | 47.1 | 420–600 | 6.8 | 54.5 | 75.9 | | | 293.0 |
| RBP_PJ+S_SF | 30–140 | 1.4 | 140–420 | 75.1 | 420–600 | 0.5 | 77.0 | 67.0 | | | 290.5 |

B—beetroot; RBP—red bell pepper; LF—*Limosilactobacillus fermentum* KKP 811; LB—*Levilactobacillus brevis* KKP 804; SF—spontaneous fermentation; J—juice pressed before fermentation; PJ+S—juice pressed from fermented vegetable and mixed with post-fermentation solution.

FTIR analysis showed that similar results were observed for both types of powders (beetroot and red bell pepper) (Figure 3). No differences in region size were observed, independently of the type of fermentation or starter cultures used. The most visible regions were: the region of O-H hydrogen bonds seen at wavenumbers of 3250–3050 cm$^{-1}$, region of C-H bonds at 2926 cm$^{-1}$, spectral region of C=O bonds at 1500–2000 cm$^{-1}$, region of C=N hydroxyl compounds at 1620 cm$^{-1}$, region of C-O stretch vibrations where alcohols arises, e.g., from xanthophyll 1144 C-O; C-C; and C-O-C, and the region of the presence of C-O phenols at 1050 cm$^{-1}$ and at 800–600 cm$^{-1}$. In the FTIR figures, fingerprint regions (600–1500 cm$^{-1}$) of functional groups can be detected. For beetroot powders at the 1314, 1350-methylene C-H bend, between 1450 and 1650 cm$^{-1}$, the presence of nitrogen was seen. Similar results and conclusions have been observed for beetroot juices and beetroot samples [56–58]. The region linked to β-carotene is at wavenumbers from 1550 to 1600 cm$^{-1}$, which is related to stretching vibrations of the C=C bonds and the region seen at 960 cm$^{-1}$ (Figure 3b). Comparable results were observed by Quijano-Ortega et al. [59] for dried pumpkin species and also for dried red pepper by Castañeda-Pérez et al. [60].

In the presented research two methods were used for the identification of betalain and its content. First, a spectrophotometric method was used in juices and in freeze-dried powders to test the total betaxanthin and betacyanin content in the samples. A second method, HPLC analysis, was used to obtain the betalain profile in freeze-dried samples, as well as its concentration too. The results of the HPLC method are presented in Table 4, whereas the summary of these two methods is compared in Table 5.

**Table 4.** HPLC results.

| Sample | Betaxanthin (mg/100 g dm) Vulgaxanthin I | A | Betanin | B | Betacyanin (mg/100 g dm) Isobetanin | Betanidin | C | Neobetanin |
|---|---|---|---|---|---|---|---|---|
| B_J_LF | 1.22 ± 0.08 [b] | 0 [a] | 60.22 ± 1.27 [d] | 0.29 ± 0.04 [a] | 5.77 ± 0.25 [b] | 6.38 ± 0.38 [c] | 0 [a] | 0 [a] |
| B_J_LB | 1.70 ± 0.08 [c] | 0.21 ± 0.05 [b] | 72.83 ± 0.77 [e] | 0.17 ± 0.02 [a] | 7.93 ± 0.32 [c] | 6.80 ± 0.08 [c] | 0.43 ± 0.04 [b] | 0 [a] |
| B_J_SF | 2.19 ± 0.00 [d] | 0 [a] | 41.96 ± 0.69 [ab] | 0.51 ± 0.03 [b] | 3.67 ± 0.32 [a] | 16.35 ± 0.90 [d] | 0.88 ± 0.08 [c] | 0 [a] |
| B_PJ+S_LF | 0.15 ± 0.00 [a] | 0.85 ± 0.06 [c] | 48.36 ± 1.87 [c] | 0.80 ± 0.08 [c] | 14.69 ± 0.40 [d] | 0.38 ± 0.07 [a] | 0 [a] | 0.54 ± 0.07 [b] |
| B_PJ+S_LB | 0.09 ± 0.00 [a] | 1.24 ± 0.01 [d] | 60.40 ± 4.73 [d] | 1.09 ± 0.04 [d] | 16.41 ± 0.32 [e] | 0 [a] | 0 [a] | 0.83 ± 0.03 [c] |
| B_PJ+S_SF | 1.71 ± 0.11 [c] | 0 [ac] | 36.90 ± 0.01 [a] | 0.54 ± 0.08 [b] | 7.08 ± 0.11 [c] | 4.15 ± 0.08 [b] | 0.73 ± 0.04 [c] | 0.60 ± 0.04 [b] |

B—beetroot; LF—*Limosilactobacillus fermentum* KKP 811; LB—*Levilactobacillus brevis* KKP 804; SF—spontaneous fermentation; J—juice pressed before fermentation; PJ+S—juice pressed from fermented vegetable and mixed with post-fermentation solution; **A**—possible decarboxylated derivative of betanin, **B**, **C**—other betacyanins, [a], [b], [c], [d], [e]—different indexes for the columns mean statistically significant differences for given values at the level of $p < 0.05$.

**Table 5.** Pigment content in beetroot samples (comparison of two methods).

| Sample | Betalain Results from HPLC Method | | Betalain Results from the Spectrophotometric Method | |
|---|---|---|---|---|
| | Betacyanin (mg/100 g dm) | Betaxanthin (mg/100 g dm) | Betacyanin (mg/100 g dm) | Betaxanthin (mg/100 g dm) |
| B_J_LF | 72.67 ± 1.94 [bc] | 1.22 ± 0.08 [b] | 94.89 ± 1.96 [a] | 50.72 ± 2.00 [c] |
| B_J_LB | 88.38 ± 1.28 [d] | 1.70 ± 0.08 [c] | 99.85 ± 0.14 [ab] | 54.82 ± 0.28 [cd] |
| B_J_SF | 63.37 ± 2.02 [ab] | 2.19 ± 0.00 [d] | 104.61 ± 7.55 [ab] | 56.14 ± 1.66 [d] |
| B_PJ+S_LF | 65.62 ± 2.54 [bc] | 0.15 ± 0.00 [a] | 98.39 ± 2.09 [ab] | 7.96 ± 0.59 [a] |
| B_PJ+S_LB | 79.96 ± 5.14 [cd] | 0.09 ± 0.00 [a] | 122.50 ± 0.79 [c] | 17.02 ± 1.88 [b] |
| B_PJ+S_SF | 49.99 ± 0.37 [a] | 1.71 ± 0.11 [c] | 111.53 ± 2.79 [bc] | 15.52 ± 0.24 [b] |

B—beetroot; LF—*Limosilactobacillus fermentum* KKP 811; LB—*Levilactobacillus brevis* KKP 804; SF—spontaneous fermentation; J—juice pressed before fermentation; PJ+S—juice pressed from fermented vegetable and mixed with post-fermentation solution; [a], [b], [c], [d]—different indexes for the columns mean statistically significant differences for given values at the level of $p < 0.05$.

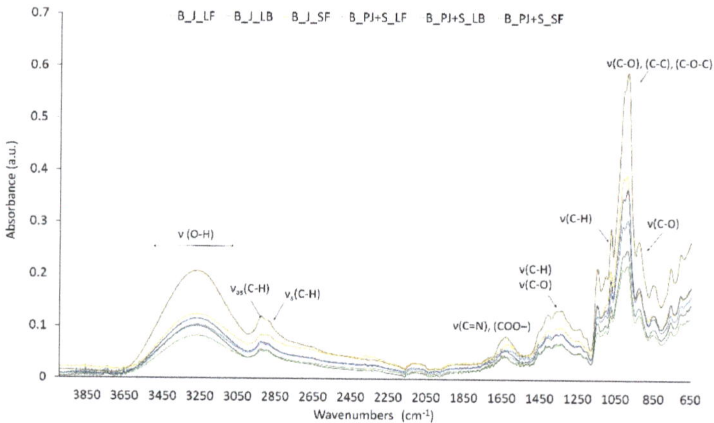

(**a**)

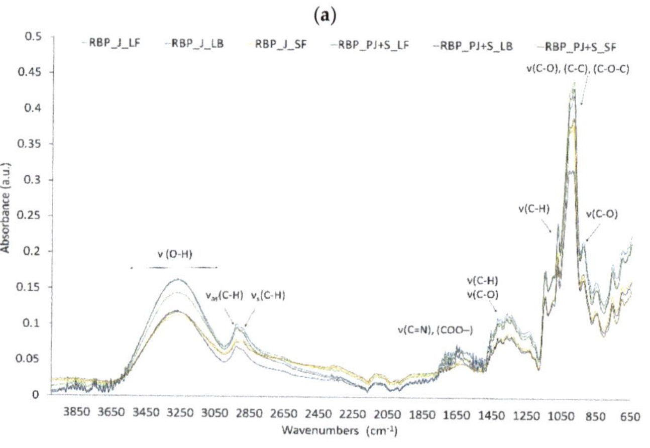

(**b**)

**Figure 3.** FTIR results (**a**) for beetroot, (**b**) for red bell pepper. B—beetroot; RBP—red bell pepper; LF—*Limosilactobacillus fermentum* KKP 811; LB—*Levilactobacillus brevis* KKP 804; SF—spontaneous fermentation; J—juice pressed before fermentation; PJ+S—juice pressed from fermented vegetable and mixed with post-fermentation solution.

As mentioned by other authors [16,39], the betalains identified in beetroot pigments were: vulgaxanthin I, vulgaxanthin II, indicaxanthin, betanin, neobetanin, prebetanin, and isobetanin. In the presented research, identification showed that in fermented freeze-dried beetroot powders only vulgaxanthin I was observed from the betaxanthin group. Of the betacyanins in all samples, betanin and isobetanin were seen in higher amounts than betanidin, neobetanin, and other betacyanins which were not identified in this research. This lower amount of detected betaxanthins could be connected to the lower pH of the juice, as these compounds are more sensitive to pH changes than betacyanins [41,61,62]. In fermented beetroot juices, other authors have identified similar betacyanin and betaxanthin compounds [3,63]. High degradation of both betacyanins and betaxanthins from fresh beetroot juice during fermentation could be resulted by the low pH of the process, as well as to the active peroxidase enzyme present in beetroot cell walls. After slicing the beetroot tissue this enzyme comes in contact with pigments that can fasten its degradation, as was acknowledged by Czyżowska et al. [64]. In addition, the fermentation process temperature used in this study was 26 °C; when betacyanins are more stable than betaxanthins [63,64].

The content of individual compounds given by other authors and their identification in beetroot show that after the extraction process degradation takes place, and the fermentation and freeze-drying process itself may contribute to the reduction of dye content as a result of the influence of the environmental pH and the drying process, which can influence the degradation of the beads [41,42,63,64].

Analysis of the betalain content in the powders obtained as a result of freeze-drying showed a higher betacyanin content than betaxanthins, regardless of the fermentation method and the type of bacterial strain used (Table 5). The different values for the determination of betaxanthins in the HPLC method and the spectrophotometric method may have resulted from the measurement methods themselves. In the spectrophotometric method, waves of 476, 538, and 600 are used, whereas at wave 450 nm absorption of carotenoids has its maximum peak. That is the reason why the possibility of absorption at wave 476 of carotenoids is present in beetroots. The retention time helps the HPLC method in the separation of those compounds [65,66].

## 4. Conclusions

Both raw materials, beetroot and red bell pepper, are rich in natural pigments but are unstable and seasonal; thus, it was decided to check the influence of two processes, which can prolong storage time, on the stability of the pigments in beetroot and red bell pepper. The analysis of two methods of fermentation and bacterial strains did not show differences in bacterial number; however, there was an increase in pigment content in fermentation of raw material as compared to juice. Thermal analysis of freeze-dried powders showed that they can be stored under normal temperature conditions (25 °C).

The results showed that the conducted process of lactic acid fermentation carried out should be verified, so that the amounts of lactic acid bacteria, both after the fermentation process and after the process of freeze-drying, constitute the basis for stating that this is a product with potential probiotic properties. Further analysis of the process may include standardization or enrichment of the juice obtained with ingredients that can serve as a carbon source necessary for bacterial growth.

The content of colored compounds in the obtained juices was at a high level; however, the use of juices as a coloring substance is very difficult. On the other hand, the obtained powders, which had high stability, did not contain the appropriate amount of dyes that would allow them to be used in food industry products such as yoghurts or loose products.

**Author Contributions:** Conceptualization, E.J.-T.; methodology, E.J.-T., K.R., K.P. and Ł.W.; software, E.J.-T., K.R., K.P., Ł.W. and A.G.-M.; validation, E.J.-T., K.R., K.P., M.G., Ł.W. and A.G.-M.; formal analysis, E.J.-T., K.T., P.B., K.R., K.P. and Ł.W.; investigation, E.J.-T., K.T., P.B., K.R., K.P. and Ł.W.; data curation, E.J.-T., K.R., K.P., Ł.W. and A.G.-M.; writing—original draft preparation, E.J.-T., K.R., K.P., M.G., Ł.W. and A.G.-M.; writing—review and editing, E.J.-T., K.R., K.P., M.G., Ł.W. and A.G.-M.; visualization, E.J.-T., K.R., K.P., Ł.W. and A.G.-M.; supervision, E.J.-T. All authors have read and agreed to the published version of the manuscript.

**Funding:** Research equipment was purchased as part of the "Food and Nutrition Centre—modernisation of the WULS campus to create a Food and Nutrition Research and Development Centre (CŻiŻ)" co-financed by the European Union from the European Regional Development Fund under the Regional Operational Programme of the Mazowieckie Voivodeship for 2014–2020 (Project No. RPMA.01.01.00-14-8276/17).

**Institutional Review Board Statement:** Not applicable.

**Informed Consent Statement:** Not applicable.

**Data Availability Statement:** Data available from the corresponding author.

**Conflicts of Interest:** The authors declare no conflict of interest.

# References


1. Khubber, S.; Marti-Quijal, F.J.; Tomasevic, I.; Remize, F.; Barba, F.J. Lactic acid fermentation as a useful strategy to recover antimicrobial and antioxidant compounds from food and by-products. *Curr. Opin. Food Sci.* **2022**, *43*, 189–198. [CrossRef]
2. Bontsidis, C.; Mallouchos, A.; Terpou, A.; Nikolaou, A.; Batra, G.; Mantzourani, I.; Alexopoulos, A.; Plessas, S. Microbiological and Chemical Properties of Chokeberry Juice Fermented by Novel Lactic Acid Bacteria with Potential Probiotic Properties during Fermentation at 4 degrees C for 4 Weeks. *Foods* **2021**, *10*, 768. [CrossRef] [PubMed]
3. Janiszewska-Turak, E.; Walczak, M.; Rybak, K.; Pobiega, K.; Gniewosz, M.; Woźniak, Ł.; Witrowa-Rajchert, D. Influence of Fermentation Beetroot Juice Process on the Physico-Chemical Properties of Spray Dried Powder. *Molecules* **2022**, *27*, 1008. [CrossRef] [PubMed]
4. Lombardelli, C.; Benucci, I.; Mazzocchi, C.; Esti, M. Betalain Extracts from Beetroot as Food Colorants: Effect of Temperature and UV-Light on Storability. *Plant Foods Hum. Nutr.* **2021**, *76*, 347–353. [CrossRef] [PubMed]
5. Lombardelli, C.; Benucci, I.; Esti, M. Novel food colorants from tomatoes: Stability of carotenoid-containing chromoplasts under different storage conditions. *LWT* **2021**, *140*, 110725. [CrossRef]
6. Lombardelli, C.; Liburdi, K.; Benucci, I.; Esti, M. Tailored and synergistic enzyme-assisted extraction of carotenoid-containing chromoplasts from tomatoes. *Food Bioprod. Process.* **2020**, *121*, 43–53. [CrossRef]
7. Tang, Y.; Chen, G.; Wang, D.; Hu, R.; Li, H.; Liu, S.; Zhang, Q.; Ming, J.; Chi, Y. Effects of dry-salting and brine-pickling processes on the physicochemical properties, nonvolatile flavour profiles and bacterial community during the fermentation of Chinese salted radishes. *LWT* **2022**, *157*, 113084. [CrossRef]
8. Behera, S.S.; El Sheikha, A.F.; Hammami, R.; Kumar, A. Traditionally fermented pickles: How the microbial diversity associated with their nutritional and health benefits? *J. Funct. Foods* **2020**, *70*, 103971. [CrossRef]
9. Janiszewska-Turak, E.; Hornowska, Ł.; Pobiega, K.; Gniewosz, M.; Witrowa-Rajchert, D. The influence of Lactobacillus bacteria type and kind of carrier on the properties of spray-dried microencapsules of fermented beetroot powders. *Int. J. Food Sci. Technol.* **2021**, *56*, 2166–2174. [CrossRef]
10. Rodriguez-Amaya, D.B. Update on natural food pigments—A mini-review on carotenoids, anthocyanins, and betalains. *Food Res. Int.* **2019**, *124*, 200–205. [CrossRef]
11. Benucci, I.; Lombardelli, C.; Mazzocchi, C.; Esti, M. Natural colorants from vegetable food waste: Recovery, regulatory aspects, and stability—A review. *Compr. Rev. Food Sci. Food Saf.* **2022**, *21*, 2715–2737. [CrossRef] [PubMed]
12. USDA1. Beet Raw, FDC ID: 169145. Available online: https://fdc.nal.usda.gov/fdc-app.html#/food-details/169145/nutrients (accessed on 25 April 2022).
13. USDA2. Peppers, Sweet, Red, Raw FDC ID: 170108. Available online: https://fdc.nal.usda.gov/fdc-app.html#/food-details/170108/nutrients (accessed on 25 April 2022).
14. USDA3; Carrots Raw FDC ID: 170393. Available online: https://fdc.nal.usda.gov/fdc-app.html#/food-details/170393/nutrients (accessed on 25 April 2022).
15. Chhikara, N.; Kushwaha, K.; Sharma, P.; Gat, Y.; Panghal, A. Bioactive compounds of beetroot and utilization in food processing industry: A critical review. *Food Chem.* **2019**, *272*, 192–200. [CrossRef] [PubMed]
16. Nemzer, B.; Pietrzkowski, Z.; Spórna, A.; Stalica, P.; Thresher, W.; Michałowski, T.; Wybraniec, S. Betalainic and nutritional profiles of pigment-enriched red beet root (*Beta vulgaris* L.) dried extracts. *Food Chem.* **2011**, *127*, 42–53. [CrossRef]

17. Nowacka, M.; Dadan, M.; Janowicz, M.; Wiktor, A.; Witrowa-Rajchert, D.; Mandal, R.; Pratap-Singh, A.; Janiszewska-Turak, E. Effect of nonthermal treatments on selected natural food pigments and color changes in plant material. *Compr. Rev. Food Sci. Food Saf.* **2021**, *20*, 5097–5144. [CrossRef] [PubMed]

18. Arimboor, R.; Natarajan, R.B.; Menon, K.R.; Chandrasekhar, L.P.; Moorkoth, V. Red pepper (*Capsicum annuum*) carotenoids as a source of natural food colors: Analysis and stability-a review. *J. Food Sci. Technol.* **2015**, *52*, 1258–1271. [CrossRef] [PubMed]

19. Petito, N.D.L.; Devens, J.M.; Falcão, D.Q.; Dantas, F.M.L.; Passos, T.S.; Araujo, K.G.D.L. Nanoencapsulation of Red Bell Pepper Carotenoids: Comparison of Encapsulating Agents in an Emulsion Based System. *Colorants* **2022**, *1*, 132–148. [CrossRef]

20. Moayyedi, M.; Eskandari, M.H.; Rad, A.H.E.; Ziaee, E.; Khodaparast, M.H.H.; Golmakani, M.-T. Effect of drying methods (electrospraying, freeze drying and spray drying) on survival and viability of microencapsulated *Lactobacillus rhamnosus* ATCC 7469. *J. Funct. Foods* **2018**, *40*, 391–399. [CrossRef]

21. Krzykowski, A.; Dziki, D.; Rudy, S.; Gawlik-Dziki, U.; Polak, R.; Biernacka, B. Effect of pre-treatment conditions and freeze-drying temperature on the process kinetics and physicochemical properties of pepper. *LWT* **2018**, *98*, 25–30. [CrossRef]

22. Jakubczyk, E.; Jaskulska, A. The Effect of Freeze-Drying on the Properties of Polish Vegetable Soups. *Appl. Sci.* **2021**, *11*, 654. [CrossRef]

23. Liu, Y.; Zhang, Z.; Hu, L. High efficient freeze-drying technology in food industry. *Crit. Rev. Food Sci. Nutr.* **2021**, *62*, 1–19. [CrossRef]

24. Janiszewska-Turak, E.; Kołakowska, W.; Pobiega, K.; Gramza-Michałowska, A. Influence of Drying Type of Selected Fermented Vegetables Pomace on the Natural Colorants and Concentration of Lactic Acid Bacteria. *Appl. Sci.* **2021**, *11*, 7864. [CrossRef]

25. Cierach, M.; Niedźwiedź, J. Effects of three lighting intensities during display on discolouration of beef semitendinosus muscle. *Eur. Food Res. Technol.* **2014**, *239*, 377–383. [CrossRef]

26. Mikus, M.; Galus, S.; Ciurzyńska, A.; Janowicz, M. Development and Characterization of Novel Composite Films Based on Soy Protein Isolate and Oilseed Flours. *Molecules* **2021**, *26*, 3738. [CrossRef]

27. Fernandes, G.; Bastos, M.C.; Mondamert, L.; Labanowski, J.; Burrow, R.A.; Rheinheimer, D.D.S. Organic composition of epilithic biofilms from agricultural and urban watershed in South Brazil. *Environ. Sci. Pollut. Res.* **2021**, *28*, 28808–28824. [CrossRef]

28. Sokołowska, B.; Woźniak, Ł.; Skąpska, S.; Porębska, I.; Nasiłowska, J.; Rzoska, S.J. Evaluation of quality changes of beetroot juice after high hydrostatic pressure processing. *High Press. Res.* **2017**, *37*, 214–222. [CrossRef]

29. Khan, M.I.; Giridhar, P. Plant betalains: Chemistry and biochemistry. *Phytochemistry* **2015**, *117*, 267–295. [CrossRef] [PubMed]

30. Miguel, M. Betalains in Some Species of the Amaranthaceae Family: A Review. *Antioxidants* **2018**, *7*, 53. [CrossRef] [PubMed]

31. Gandía-Herrero, F.; Escribano, J.; García-Carmona, F. Betaxanthins as pigments responsible for visible fluorescence in flowers. *Planta* **2005**, *222*, 586–593. [CrossRef]

32. Janiszewska-Turak, E.; Rybak, K.; Grzybowska, E.; Konopka, E.; Witrowa-Rajchert, D. The Influence of Different Pretreatment Methods on Color and Pigment Change in Beetroot Products. *Molecules* **2021**, *26*, 3683. [CrossRef]

33. Janiszewska-Turak, E.; Witrowa-Rajchert, D. The influence of carrot pretreatment, type of carrier and disc speed on the physical and chemical properties of spray-dried carrot juice microcapsules. *Dry. Technol.* **2021**, *39*, 439–449. [CrossRef]

34. Rybak, K.; Wiktor, A.; Pobiega, K.; Witrowa-Rajchert, D.; Nowacka, M. Impact of pulsed light treatment on the quality properties and microbiological aspects of red bell pepper fresh-cuts. *LWT* **2021**, *149*, 111906. [CrossRef]

35. Janiszewska, E. Microencapsulated beetroot juice as a potential source of betalain. *Powder Technol.* **2014**, *264*, 190–196. [CrossRef]

36. Rybak, K.; Samborska, K.; Jedlinska, A.; Parniakov, O.; Nowacka, M.; Witrowa-Rajchert, D.; Wiktor, A. The impact of pulsed electric field pretreatment of bell pepper on the selected properties of spray dried juice. *Innov. Food Sci. Emerg. Technol.* **2020**, *65*, 102446. [CrossRef]

37. Dhiman, A.; Suhag, R.; Chauhan, D.S.; Thakur, D.; Chhikara, S.; Prabhakar, P.K. Status of beetroot processing and processed products: Thermal and emerging technologies intervention. *Trends Food Sci. Technol.* **2021**, *114*, 443–458. [CrossRef]

38. Hallmann, E.; Marszałek, K.; Lipowski, J.; Jasińska, U.; Kazimierczak, R.; Średnicka-Tober, D.; Rembiałkowska, E. Polyphenols and carotenoids in pickled bell pepper from organic and conventional production. *Food Chem.* **2019**, *278*, 254–260. [CrossRef] [PubMed]

39. Czyżowska, A.; Siemianowska, K.; Śniadowska, M.; Nowak, A. Bioactive Compounds and Microbial Quality of Stored Fermented Red Beetroots and Red Beetroot Juice. *Pol. J. Food Nutr. Sci.* **2020**, *70*, 35–44. [CrossRef]

40. Massimo, C.; Marina, C.; Riccardo, M.; Danilo, M.; Roberto, M. Effects of controlled atmospheres and low temperature on storability of chestnuts manually and mechanically harvested. *Postharvest Biol. Technol.* **2011**, *61*, 131–136. [CrossRef]

41. Fu, Y.; Shi, J.; Xie, S.Y.; Zhang, T.Y.; Soladoye, O.P.; Aluko, R.E. Red Beetroot Betalains: Perspectives on Extraction, Processing, and Potential Health Benefits. *J. Agric. Food Chem.* **2020**, *68*, 11595–11611. [CrossRef]

42. Hadipour, E.; Taleghani, A.; Tayarani-Najaran, N.; Tayarani-Najaran, Z. Biological effects of red beetroot and betalains: A review. *Phytother. Res.* **2020**, *34*, 1847–1867. [CrossRef]

43. Wolkers-Rooijackers, J.C.M.; Thomas, S.M.; Nout, M.J.R. Effects of sodium reduction scenarios on fermentation and quality of sauerkraut. *LWT-Food Sci. Technol.* **2013**, *54*, 383–388. [CrossRef]

44. Mapelli-Brahm, P.; Barba, F.J.; Remize, F.; Garcia, C.; Fessard, A.; Mousavi Khaneghah, A.; Sant'Ana, A.S.; Lorenzo, J.M.; Montesano, D.; Meléndez-Martínez, A.J. The impact of fermentation processes on the production, retention and bioavailability of carotenoids: An overview. *Trends Food Sci. Technol.* **2020**, *99*, 389–401. [CrossRef]

45. Shrestha, A.K.; Ua-arak, T.; Adhikari, B.P.; Howes, T.; Bhandari, B.R. Glass Transition Behavior of Spray Dried Orange Juice Powder Measured by Differential Scanning Calorimetry (DSC) and Thermal Mechanical Compression Test (TMCT). *Int. J. Food Prop.* **2007**, *10*, 661–673. [CrossRef]
46. Otalora, M.C.; Carriazo, J.G.; Iturriaga, L.; Nazareno, M.A.; Osorio, C. Microencapsulation of betalains obtained from cactus fruit (*Opuntia ficus-indica*) by spray drying using cactus cladode mucilage and maltodextrin as encapsulating agents. *Food Chem.* **2015**, *187*, 174–181. [CrossRef] [PubMed]
47. Tan, S.; Zhong, C.; Langrish, T. Encapsulation of caffeine in spray-dried micro-eggs for controlled release: The effect of spray-drying (cooking) temperature. *Food Hydrocoll.* **2020**, *108*, 105979. [CrossRef]
48. Flores-Mancha, M.A.; Ruiz-Gutierrez, M.G.; Sanchez-Vega, R.; Santellano-Estrada, E.; Chavez-Martinez, A. Characterization of Beet Root Extract (*Beta vulgaris*) Encapsulated with Maltodextrin and Inulin. *Molecules* **2020**, *25*, 5498. [CrossRef] [PubMed]
49. Foster, K.D.; Bronlund, J.E.; Paterson, A.H.J. Glass transition related cohesion of amorphous sugar powders. *J. Food Eng.* **2006**, *77*, 997–1006. [CrossRef]
50. Avaltroni, F.; Bouquerand, P.; Normand, V. Maltodextrin molecular weight distribution influence on the glass transition temperature and viscosity in aqueous solutions. *Carbohydr. Polym.* **2004**, *58*, 323–334. [CrossRef]
51. Małecka, B. Metody analizy termicznej połączone z analizą produktów gazowych (TG-DSC-MS). *LAB Lab. Apar. Bad.* **2012**, *17*, 6–16.
52. Fritzen-Freire, C.B.; Prudêncio, E.S.; Amboni, R.D.M.C.; Pinto, S.S.; Negrão-Murakami, A.N.; Murakami, F.S. Microencapsulation of bifidobacteria by spray drying in the presence of prebiotics. *Food Res. Int.* **2012**, *45*, 306–312. [CrossRef]
53. Macêdo, R.; de Moura, O.; de Souza, A.; Macêdo, A. Comparative studies on some analytical methods: Thermal decomposition of powder milk. *J. Therm. Anal. Calorim.* **1997**, *49*, 857–862. [CrossRef]
54. Carmo, E.L.D.; Teodoro, R.A.R.; Felix, P.H.C.; Fernandes, R.V.B.; Oliveira, E.R.; Veiga, T.; Borges, S.V.; Botrel, D.A. Stability of spray-dried beetroot extract using oligosaccharides and whey proteins. *Food Chem.* **2018**, *249*, 51–59. [CrossRef] [PubMed]
55. Siemons, I.; Politiek, R.; Boom, R.; van der Sman, R.; Schutyser, M. Dextrose equivalence of maltodextrins determines particle morphology development during single sessile droplet drying. *Food Res. Int.* **2020**, *131*, 108988. [CrossRef] [PubMed]
56. Aztatzi-Rugerio, L.; Granados-Balbuena, S.Y.; Zainos-Cuapio, Y.; Ocaranza-Sánchez, E.; Rojas-López, M. Analysis of the degradation of betanin obtained from beetroot using Fourier transform infrared spectroscopy. *J. Food Sci. Technol.* **2019**, *56*, 3677–3686. [CrossRef]
57. Mocanu, G.-D.; Chirilă, A.C.; Vasile, A.M.; Andronoiu, D.G.; Nistor, O.-V.; Barbu, V.; Stănciuc, N. Tailoring the functional potential of red beet purées by inoculation with lactic acid bacteria and drying. *Foods* **2020**, *9*, 1611. [CrossRef]
58. Devadiga, D.; Ahipa, T.N. *Betanin: A Red-Violet Pigment-Chemistry and Applications*; IntechOpen: London, UK, 2020.
59. Quijano-Ortega, N.; Fuenmayor, C.A.; Zuluaga-Dominguez, C.; Diaz-Moreno, C.; Ortiz-Grisales, S.; García-Mahecha, M.; Grassi, S. FTIR-ATR Spectroscopy Combined with Multivariate Regression Modeling as a Preliminary Approach for Carotenoids Determination in *Cucurbita* spp. *Appl. Sci.* **2020**, *10*, 3722. [CrossRef]
60. Castañeda-Pérez, E.; Osorio-Revilla, G.I.; Gallardo-Velázquez, T.; Proal-Nájera, J. Uso de FTIR-HATR y análisis multivariable para el seguimiento de la degradación de compuestos bioactivos durante el secado de pimiento rojo. *Rev. Mex. Ing. Química* **2013**, *12*, 193–204.
61. Khan, M.I. Stabilization of betalains: A review. *Food Chem.* **2016**, *197 Pt B*, 1280–1285. [CrossRef]
62. Coy-Barrera, E. Chapter 17-Analysis of betalains (betacyanins and betaxanthins). In *Recent Advances in Natural Products Analysis*; Sanches Silva, A., Nabavi, S.F., Saeedi, M., Nabavi, S.M., Eds.; Elsevier: Amsterdam, The Netherlands, 2020; pp. 593–619.
63. Wilkowska, A.; Ambroziak, W.; Czyżowska, A.; Adamiec, J. Effect of Microencapsulation by Spray Drying and Freeze Drying Technique on the Antioxidant Properties of Blueberry (*Vaccinium myrtillus*) Juice Polyphenolic Compounds. *Pol. J. Food Nutr. Sci.* **2016**, *66*, 11–16. [CrossRef]
64. Czyżowska, A.; Klewicka, E.; Libudzisz, Z. The influence of lactic acid fermentation process of red beet juice on the stability of biologically active colorants. *Eur. Food Res. Technol.* **2006**, *223*, 110–116. [CrossRef]
65. Wrolstad, R.E.; Acree, T.E.; Decker, E.A.; Penner, M.H.; Reid, D.S.; Schwartz, S.J.; Shoemaker, C.F.; Smith, D.; Sporns, P. Carotenoids. In *Handbook of Food Analytical Chemistry*; John Wiley & Sons, Inc.: London, UK, 2005; pp. 71–119.
66. Giusti, M.; Wrolstad, R. *Characterization and measurement with UV-visible spectroscopy. Unit F2. 2, Ch. 2 In Current Protocols in Food Analytical Chemistry*; King, S., Gates, M., Scalettar, L., Eds.; John Wiley & Sons: Hoboken, NJ, USA, 2000.

*applied sciences*

*Article*

# Antidepressant-Like Effect of Traditional Medicinal Plant Carthamus Tinctorius in Mice Model through Neuro-Behavioral Tests and Transcriptomic Approach

Mohamed H. Alegiry [1], Abdelfatteh El Omri [1,2], Ahmed Atef Bayoumi [1], Mohammed Y. Alomar [1], Irfan A. Rather [1,3,*] and Jamal S. M. Sabir [1,3,*]

[1] Department of Biological Sciences, Faculty of Science, King Abdulaziz University (KAU), Jeddah 21589, Saudi Arabia; alegirym@gmail.com (M.H.A.); omriabdel@gmail.com (A.E.O.); ahmad.atefaig2@yahoo.com (A.A.B.); mohadlomar.88@gmail.com (M.Y.A.)
[2] Surgery Research Section, Department of Surgery, Hamad Medical Corporation, Doha 3050, Qatar
[3] Center of Excellence in Bionanoscience Research, King Abdulaziz University, Jeddah 21589, Saudi Arabia
* Correspondence: ammm@kau.edu.sa (I.A.R.); jsabir@kau.edu.sa (J.S.M.S.)

**Citation:** Alegiry, M.H.; El Omri, A.; Bayoumi, A.A.; Alomar, M.Y.; Rather, I.A.; Sabir, J.S.M. Antidepressant-Like Effect of Traditional Medicinal Plant Carthamus Tinctorius in Mice Model through Neuro-Behavioral Tests and Transcriptomic Approach. *Appl. Sci.* **2022**, *12*, 5594. https://doi.org/10.3390/app12115594

Academic Editors: Luca Mazzoni, Maria Teresa Ariza Fernández and Franco Capocasa

Received: 14 May 2022
Accepted: 30 May 2022
Published: 31 May 2022

**Publisher's Note:** MDPI stays neutral with regard to jurisdictional claims in published maps and institutional affiliations.

**Copyright:** © 2022 by the authors. Licensee MDPI, Basel, Switzerland. This article is an open access article distributed under the terms and conditions of the Creative Commons Attribution (CC BY) license (https://creativecommons.org/licenses/by/4.0/).

**Abstract:** Major depression disorder (MDD) has become a common life-threatening disorder. Despite the number of studies and the introduced antidepressants, MDD remains a major global health issue. *Carthamus tinctorius* (safflower) is traditionally used for food and medical purposes. This study investigated the chemical profile and the antidepressant-like effect of the Carthamus tincto-rius hot water extract in male mice and its mechanism using a transcriptomic analysis. The antidepressant effect of hot water extract (50 mg/kg and 150 mg/kg) was investigated in mice versus the untreated group (saline) and positive control group (fluoxetine 10 mg/kg). Hippocampus transcriptome changes were investigated to understand the *Carthamus tinctorius* mechanism of action. The GC-MS analysis of *Carthamus tinctorius* showed that hot water extract yielded the highest amount of oleamide as the most active ingredient. Neuro-behavioral tests demonstrated that the safflower treatment significantly reduced immobility time in TST and FST and improved performance in the YMSAT compared to the control group. RNA-seq analysis revealed a significant differential gene expression pattern in several genes such as Ube2j2, Ncor1, Tuba1c, Grik1, Msmo1, and Casp9 related to MDD regulation in 50 mg/kg safflower treatment as compared to untreated and fluoxetine-treated groups. Our findings demonstrated the antidepressant-like effect of safflower hot water extract and its bioactive ingredient oleamide on mice, validated by a significantly shortened immobility time in TST and FST and an increase in the percentage of spontaneous alternation.

**Keywords:** safflower; oleamide; depression; transcriptome; neuro-behavioral test; grik1; casp9

## 1. Introduction

Major depression disorder (MMD) is one of the most common diseases that affect the quality of life of millions worldwide [1]. According to WHO, 264 million people of all ages suffer from depression, and around 850,000 suicides annually [2]. In countries such as the United States, Canada, and China, the lifetime prevalence of MMD varies between 16.2%, 11.3%, and 3.4%, respectively [3]. Besides the modern lifestyle contributing to stress and MDD, the outbreak of COVID-19 had a huge impact on the mental health of many individuals. During the pandemic, the fear of being sick, loss of beloved ones, social distancing, self-isolation, remote education, remote working, loss of job, and dramatic change in daily life activities, among others, negatively impacted our mental health and led to severe psychological distress including depression [4]. In Saudi Arabia, it has been reported that at the beginning of the pandemic, nearly one-fourth of the population experienced psychological impacts that ranged from moderate to severe [5].

MDD is a mental and mood disorder, and it is considered a complex disease. Patients suffering from depression usually suffer from low self-esteem, loss of pleasure, loss of interest, lack of arousal, sleep disturbance, loss of appetite, loss of libido, delusion, hallucinations, and suicidal tentative. Moreover, depression episodes may severely impact a patient's life, social relations, and general health, leading in some cases to life-long disabilities [2,6]. With different causative factors involved, such as heritability and environmental factors, the mechanism of MDD is still unclear, and no mechanism could explain it [7]. Several antidepressants have been developed to treat MDD, including tricyclic antidepressants, monoamine oxidase inhibitors, selective serotonin reuptake inhibitors (SSRIs), Atypical antidepressants, and serotonin-norepinephrine reuptake inhibitors (SNRIs). However, the effect of these drugs seems to be questionable in terms of benefits versus side effects [8]. These therapeutics have not provided the needed relief for those who suffer.

On the contrary, there is a common risk of relapse/recurrence of depressive symptoms after successful acute MDD treatment [1]. Recently, there has been much scientific evidence that complementary and alternative medicine (CAM), which is not considered to be part of conventional medicine, is promising as an alternative to treat and manage mental distress. There are over 120 CAM formulations meant to alleviate and or treat MDD symptoms [9]. A growing body of evidence supports that dietary phytochemicals such as polyphenols (a special class of natural active ingredients) could provide a suitable option to improve mental health alone or integrated with allopathic medicine with minimum side effects [10]. However, more scientific evidence is needed to confirm their activity and understand their molecular mechanisms. In our previous research, we conducted a survey among Saudi people about the ethnobotanical preparations to alleviate stress and depression. We could identify safflower as one of the most cited herbal preparations [11]. Traditionally, safflower petals are soaked in water for around 2 h at room temperature. The filtrate is ingested orally during sorrow, distress, and panic events as a sedative and hypnotic.

Safflower (*Carthamus tinctorius*) is a part of the Composite or Asteraceae family, traditionally used for culinary and medicinal purposes [12]. It was demonstrated to exert numerous medical properties such as antioxidant, analgesic, anti-inflammatory, and antidiabetic [13]. Moreover, safflower has shown anticoagulant, vasodilating, antihypertensive, neuroprotective, and immunosuppressive activities [12].

The present study was conducted based on the ethnopharmacological background to evaluate the antidepressant-like effect of safflower dried petals in mice using different neuro-behavioral tests, optimize the best extraction technique, fingerprint the main active entities, and understand the molecular mechanism using transcriptomic profiling of mice hippocampus.

## 2. Materials and Methods

### 2.1. Preparation of Safflower (Carthamus tinctorius)

Safflower dried petals (SFP) were crushed in a mortar and extracted at 10 g/100 mL using different techniques: (i) hydro-alcoholic maceration (SFP soaked in 70% ethanol for 24 h, at 10 g/100 mL with continuous shaking), (ii) hot water extraction (extraction at 115 °C for 1 min and allowing the maceration to cool down in the autoclave), and (iii) the traditional method used in the Arabic peninsula (Kingdom of Saudi Arabia) which consists of soaking SFP in distilled water for 2, 12, and 24 h at room temperature. The obtained macerate was filtered using the Whatman filter (1001-150, 11 μm particle retention). The filtrate was placed in glass Petri dishes and dried at 37 °C in the incubator. Dried resin with dark orange to brown color was then collected in Eppendorf tubes and stored at −80 °C for further experiments. For convenience, SFP extracts are abbreviated as follows: 70% ethanol extract: (SFPE), hot water extract: SFPWH, and water extract at room temperature for 2 h (SFPW2), 12 h (SFPW12), and 24 h (SFPW24).

### 2.2. Safflower Extract Chemical Profiling Using GC-MS

The GC-MS system (Agilent Technologies, Santa Clara, CA, USA) used in the current study was equipped with a gas chromatograph (7890B) and mass spectrometer detector (5977A) at Central Laboratories Network, National Research Centre, Cairo, Egypt. The GC was equipped with an HP-5MS column (30 m × 0.25 mm internal diameter and 0.25 μm film thickness). Analyses were carried out using helium as a gas carrier at a flow rate of 1 mL/min at a splitless mode, injection volume of 1 μL, and the following temperature program: 60 °C for 2 min; rising at 10 °C/min to 280 °C and held for 10 min. The injector and detector were held at 250 and 300 °C, respectively. Mass spectra were obtained by electron ionization (EI) at 70 eV and using a spectral range of $m/z$ 50–550 and solvent delay of 3 min. Identification of different constituents was determined by comparing the spectrum fragmentation pattern with those stored in Wiley and NIST Mass Spectral Library data.

### 2.3. Animals

The animal experiment was conducted in the animal laboratory of the Department of Biological Sciences, Faculty of Sciences at King Abdulaziz University, Jeddah, Saudi Arabia. The study was approved by the ethical committee of the university (201060288). Thirty-two (32) male albino mice were 8 weeks old, with an average body weight of 30.53 g. Animals were housed individually under a 12/12 h light/dark cycle at 25 °C, with free access to food and water. Animals were allowed to acclimatize to the laboratory conditions for 7 days prior to the experimentation. All experiments were carried out between 9:00 and 16:00. On the last day of the experiment, mice were sacrificed, and the brain tissue was collected.

### 2.4. Carthamus tinctorius Administration in Mice

Fluoxetine and SFPWH were freshly prepared daily in saline prior to the experiments. Drug and extracts were administered by oral gavage using a 1 mL syringe and 20 G × 25 mm (2 mm tip diameter)/straight gavage needle (Petsurgical (AFN2425S)) daily for 15 consecutive days at 9:00 AM, and all mice were administered the same equivalent volumes. Mice were randomly divided into 4 groups (n = 8 each). Control group: administered 200 μL of normal saline per day; positive control group: fluoxetine 10 mg/kg per day; treatment groups 50 and 150 mg/kg SFPWH extract per day.

### 2.5. Neuro-Behavioral Tests

Neuro-behavioral tests were conducted in mice 3 h after treatment. The tail suspension test (TST) and forced swimming test (FST) were conducted 7 days each. On the last day of the experiment (day 15), alternation in the Y maze test (YMT) was conducted for each mouse, and animals were sacrificed by spinal cord dislocation, the brain was removed in PBS on ice, and the hippocampus was collected and immediately frozen in liquid nitrogen, then stored at −80 °C for further experiments.

The tail suspension test (TST) has been widely used to evaluate and assess antidepressant activity in mice [14]. Treatment was given 3 h before the beginning of the experiment. Mice from each group were suspended simultaneously, using tape placed at 1 cm from the tip of the tail and hooked to the roof, inside a 4-chamber nonreflective black plexiglass box for 6 min. Animal behavior was recorded using a video camera, and the last 4 min of the experiment were considered to assess TST in SFPWH extract in mice. Recorded videos were analyzed manually, and scores were assigned to each mouse based on our previous study [15].

### 2.6. Forced Swimming Test (FST) in Mice

Mice from each group were individually and simultaneously forced to swim in 4 open cylindrical containers (diameter 20 cm, height 30 cm) containing 15 cm of water at 25 ± 1 °C. FST sessions duration was 6 min, and the immobility time was recorded during the last

4 min of each experiment. Mice are considered immobile when they remain floating motionless in the water, making only necessary limb movements to keep floating above water [16]. Scores were assigned based on recorded videos.

### 2.7. Y Maze Test in Mice

Mice were individually placed in a starting point with a gate at arm A. Exploring time was not allowed. The duration of each test was 10 min, which took place on the last day of the experiment (day 15). The alternation occurred when the mouse moved from one arm and entered entirely into one of the other two arms with its four limbs. The number of alternations between 3 arms was calculated and reported as a percentage of the total number of entries (ABC, ACB, BAC, BCA, CAB, CBA) with no re-entry counted. The way maze was made from black nonreflective plexiglass. It consists of 3 equal arms (length, width, height: $35 \times 5 \times 10$ cm) in a Y shape. Alternation was tracked and scored visually.

### 2.8. Transcriptomic Analysis (RNA-Seq)

The hippocampus was dissected from the brain tissue. The total RNA of three mice of each group was extracted using PureLink™ RNA Mini Kit (ThermoFisher Scientific, Waltham, MA, USA) according to the manufacturer's protocol. The RNA was incubated for 1 h at 37 °C with DNase I (ThermoFisher Scientific, Waltham, MA, USA) before heating to 90 °C for 30 min. Quantity/quality and RNA integrity were assessed by running the RNA sample on an Agilent Bioanalyzer® RNA 6000 Nano/Pico Chip. mRNA was extracted using NEBNext Poly(A) mRNA Magnetic Isolation beads following the manufacturer's protocol. cDNA libraries were constructed using the NEBNext Ultra II Directional RNA-Seq library. Sequencing was carried out using the Illumina NovaSeq 6000 SP v1.5 Lane (150PE) libraries at the Earlham Institute Genomics core facility. The raw data were processed by the Trinity RNA-seq assembly package [17]. Briefly, the Trimmomatic tool was used in order to remove reads containing adapters. Then, the reads were quantified and mapped to the reference transcript of GRCm39 (Ensembl, Cambridge, UK) by Bowtie v0.12.1. software [18], RSEM v1.1.6 [19] had been used for transcription quantification.

Differential expression analysis of expected read counts was performed by EdgeR (version 3.0.0, R version 2.1.5) [20]. Quantification of transcript expression levels was presented by FPKM (fragments per kilobase of exon per million fragments mapped). The 3rd replicate of SFPWH (150 mg/kg) was excluded due to the lack of consistency.

## 3. Results

### 3.1. Preliminary Phytochemical Screening

Gas chromatography-mass spectrometry was applied to study the chemical profile of *Carthamus tinctorius* extracts to profile the bioactive ingredients in each extract, including 70% ethanol extraction, water/heat extraction, and traditional extraction. GC-MS analysis showed the presence of oleamide (9-Octadecenamide C18H35NO), as shown in Table 1, which is a fatty amide derived from oleic acid and is considered a potential treatment for mood and sleep disorders [21]. The detailed GC/MS profiling is shown in Table S1. As shown in Figure 1a, hot water extraction yielded the maximum amount of oleamide with 46.52% compared to the other extracts, where 70% ethanol extract yielded 11.15% of oleamide. The traditional extraction method for 2, 12, and 24 h have yielded 34.22%, 32.28%, and 34.53%, respectively. Our results indicate that the maceration time at room temperature did not improve the yield of oleamide. However, hot water extraction at 115 °C for 1 min increased the oleamide level by almost 1.5-fold.

### 3.2. Effect of Safflower Hot Water Extract on Mice Body Weight

The body weight of each mouse was measured daily before the oral administration session. The data show no significant change in the body weight during the 15-day experiment at $p < 0.05$ (Figure 1b).

**Table 1.** Major chemical constituents present in *Carthamus tinctorius* hot water extract (detailed profiling refer to Table S1).

| Name | Structure |
| --- | --- |
| Undecane |  |
| Hexadecanoic acid methyl ester | 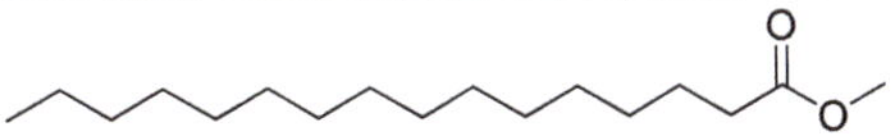 |
| 9,12-Octadecadienoicacid(Z,Z)-methylester | 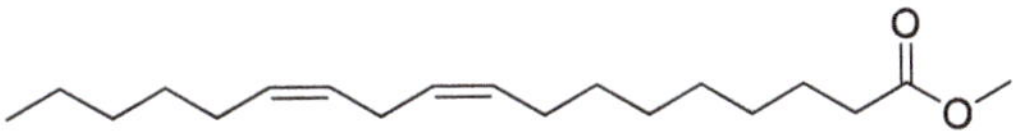 |
| 9-Octadecenoicacid-(Z)-methyl ester | 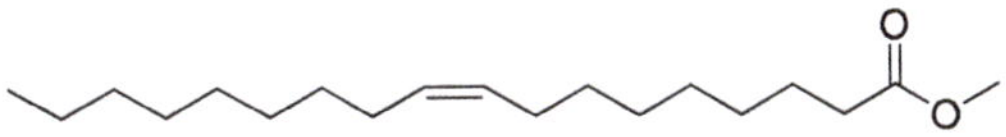 |
| Octadecanoicacid methylester | 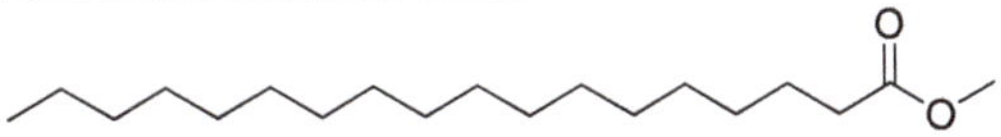 |
| Hexadecanamide |  |
| 9-Octadecenamide-(Z) |  |
| Octadecanamide |  |
| 13-Docosenamide-(Z) |  |

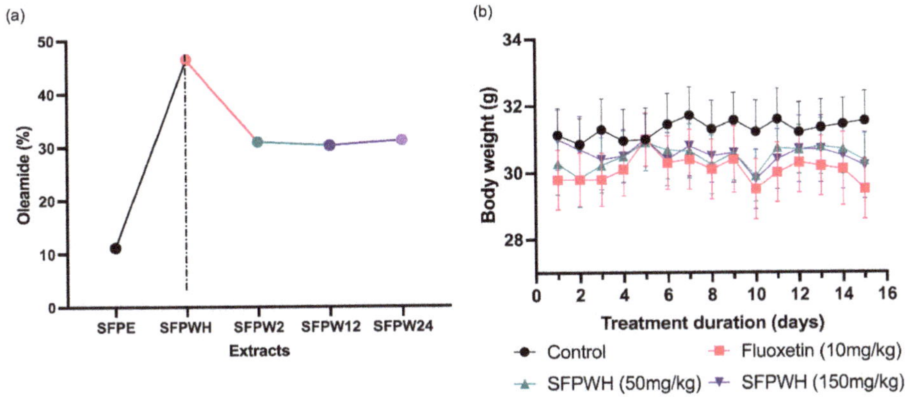

**Figure 1.** Characterization and efficacy of extracts. (**a**) Oleamide ($CH_3(CH_2)_7CH=CH(CH_2)_7CONH_2$ (CID 5283387)) yield in *Carthamus tinctorius* extracts. (SFPE) 70% ethanol for 24 h, (SFPWH) hot water extraction, and (SFPW) the traditional method using maceration in water at room temperature for 2, 12, and 24 h. Oleamide (%) is reported as the percentage of the sum area in each extract. (**b**) Daily effect of safflower hot water extract treatment on body weight in depression-induced mice. Data represent the mean ± SEM (n = 8 mice).

### 3.3. Tail Suspension Test (TST)

The antidepressant-like effect of *Carthamus tinctorius* hot water extract was evaluated every other day using a tail suspension test in mice fed with the extract for 14 days. As shown in Figure 2a, a significant reduction in the immobility time during the last 4 min of the total 6 min period tail suspension was noticed ($p < 0.05$) in fluoxetine 10 mg/kg (47.921 $\pm$ 8.725 s) and safflower hot water extract treatments at 50 mg/kg (45.398 $\pm$ 5.564 s) and 150 mg/kg (51.940 $\pm$ 3.506 s) as compared to the control group (84.375 $\pm$ 10.513 s).

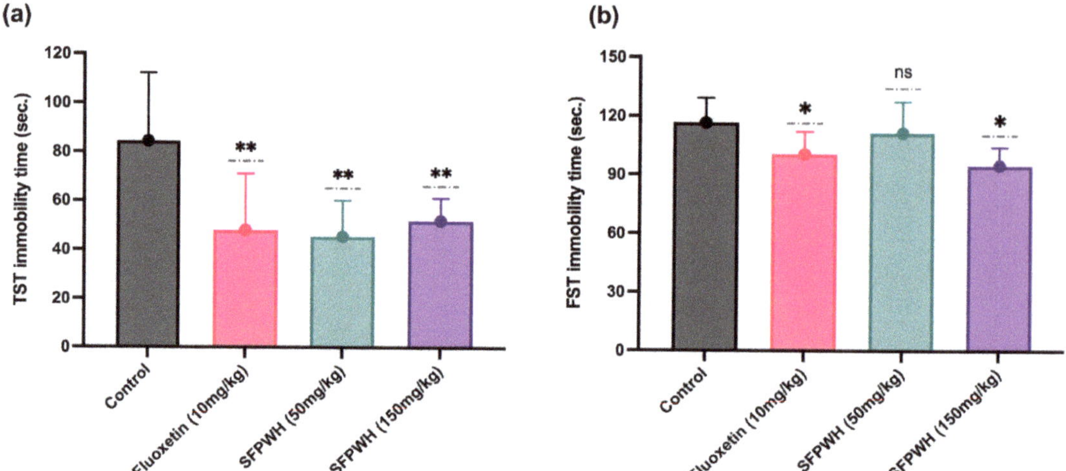


**Figure 2.** Effect of safflower hot water extract on immobility time in TST. (**a**) Effect on overall immobility time in TST-induced stress in mice. (**b**). Effect on overall immobility time in FST-induced stress in mice. ns = non-significant, * and ** indicates significantly different at $p < 0.05$ and $p < 0.01$, respectively versus untreated group, Student t-test.

The immobility time in all mice groups was wavy with a general trend of increase. The TST immobility time increase was the highest in the untreated mice (the control group), where it fluctuated in the interval [60 s–107 s]. In contrast, safflower-treated groups (50 and 150 mg/kg) along with positive control (Fluoxetine 10 mg/kg) group have shown a significant reduction in the immobility times at the same period fluctuating the following intervals [31 s–56 s], [12 s–64 s], and [19 s–54 s], respectively.

### 3.4. Forced Swimming Test (FST)

Safflower hot water extract-treated mice were subjected to FST compared to the untreated group (vehicle) and fluoxetine-treated mice as a positive control group. The antidepressant effect was assessed based on the last 4 min of 6 min session immobility time in FST. Mice were considered immobile when they were motionless, floating above the water. As indicated in Figure 2b, safflower and fluoxetine showed a mild effect on immobility time in FST. The maximum reduction in the immobility time was observed in SFPWH (150 mg/kg) with 80.601 $\pm$ 13.52 s compared to the control group 101.949 $\pm$ 19.317 s and fluoxetine (10 mg/kg) group 100.501 $\pm$ 16.556 s.

### 3.5. Y Maze Spontaneous Alternation Test (YMSAT)

YMSAT is a behavioral test used to evaluate the exploratory behavior in mice or rodents in general. Typically, rodents in normal conditions prefer to explore a new arm instead of re-visiting the same arm they explored before. This exercise involves several brain parts, including the hippocampus and prefrontal cortex [22]. YMSAT is commonly used to assess and quantify cognitive deficits in rodents exposed to novel chemical entities [23].

As indicated in Figure 3, mice treated with fluoxetine and safflower hot water extract 50 mg/kg and 150 mg/kg had their exploratory behavior significantly improved and scored 68.41 ± 3.16%, 67.12 ± 1.39%, and 72.82 ± 2.08%, respectively in the percentage of spontaneous alternation as compared to stressed and untreated mice (vehicle group) (60.41 ± 0.98%). Results indicated better performance in safflower-treated mice than in the fluoxetine-treated group.

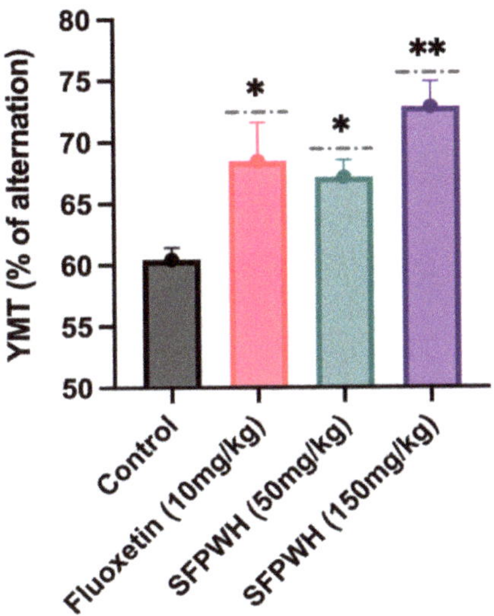


**Figure 3.** Effect of safflower water extract treatment on spontaneous alternation in YMT. Each mouse was allowed to explore the Y maze in an 8 min session. The number of alternations between 3 arms was calculated and reported as a percentage of the total number of entries. Each bar represents the mean ± SEM (n = 8). * and ** indicates significantly different at $p < 0.05$ versus untreated group, Student *t*-test.

### 3.6. Transcriptomic Analysis (RNA-Seq)

Using next-generation sequencing, RNA-seq is a powerful tool capable of detecting the quantitative measurement of RNA abundance or gene expression [24]. RNA-seq analysis of the hippocampus tissue was used to compare the differential gene expression pattern of the vehicle group compared to SFPWH-treated and fluoxetine-treated groups. As shown in Figure 4a, 72 detected subclusters and 82 transcripts. Gene ontology enrichment analysis (Figure 4b) was performed by ShinyGO v0.741 [25], revealing the top pathways for the detected genes. Data analysis showed a significant differential gene expression pattern in 22 genes, which might be potentially related to mood disorders (Figure 4c).

The expression pattern of the mean SFPWH (50 mg/kg) group replicates revealed a significantly different gene expression compared to the control group. Where 11 transcripts were downregulated in genes (Ube2j2, Dctn6, Ncor1, Acp2, Tuba1c, Gm14150, Actr3-ps, Grik1, Commd5, and Gm3226). In addition, 7 transcripts were upregulated in genes (Msmo1, Casp9, Hspe1-rs1, Tmem44, Rps13-ps2, Rps3a2,and Dctn6). In comparison to the control group. At the same time, fluoxetine-treated group showed a significantly different gene expression in six genes where two transcripts were downregulated in genes Cacna1h and Dync1i2. Moreover, five transcripts were upregulated in genes (Rbl2, Mas1, Hyi, Morf4l2, and Dync1i2). SFPWH (150 mg/kg) group has not shown any significant gene expression compared to the control group (Table 2).

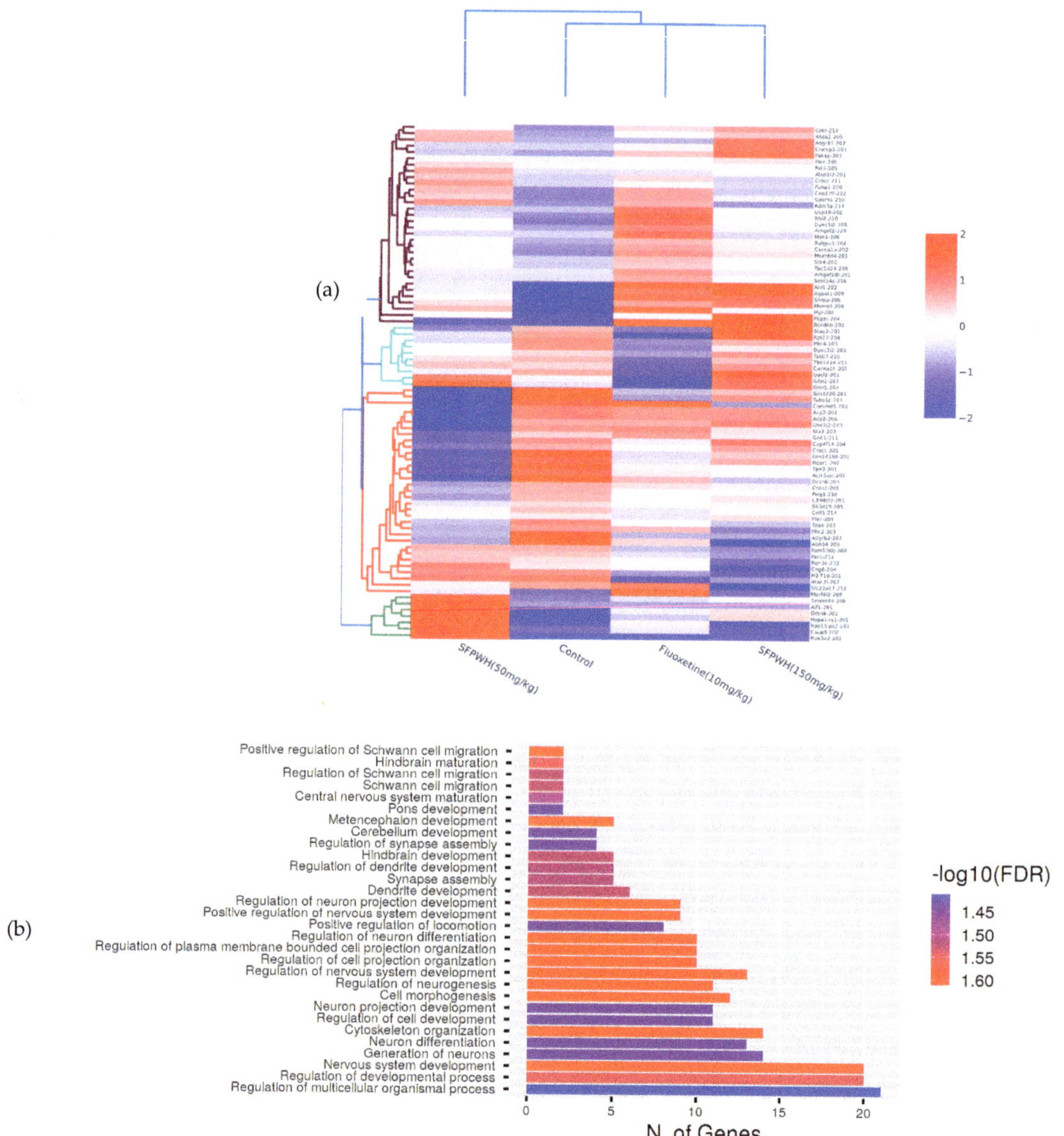



**Figure 4.** *Cont.*

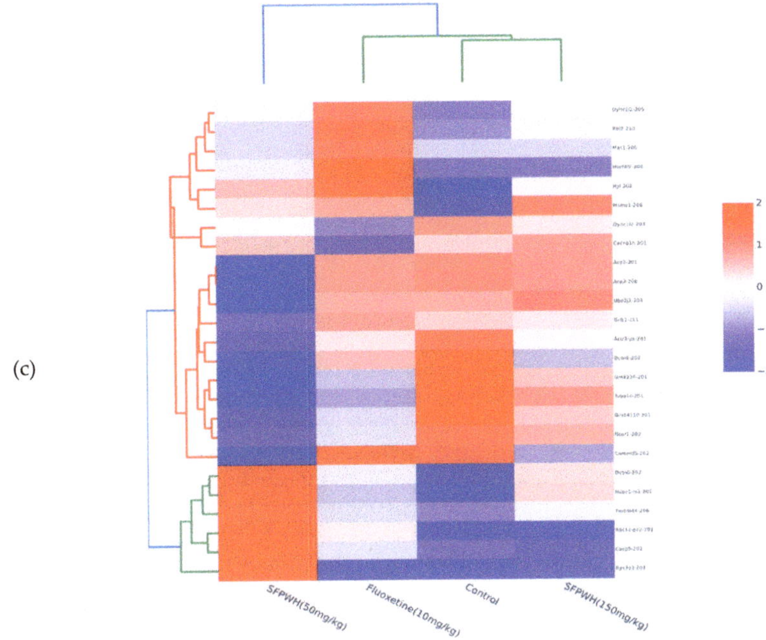



**Figure 4.** Transcriptome differences detected between the mean of the control group and treatment groups. (**a**). Hierarchical cluster analysis of gene expression based on log-ratio FPKM data for the mean of transcripts replicates illustrating the impact of the administration of safflower hot water extract and fluoxetine treatment in comparison to control where red indicates upregulation pattern while blue color indicates downregulation pattern. (**b**). Heatmap showing significantly different gene expression patterns (log2 fold). (**c**). The bar charts for the gene ontology enrichment analysis show the top pathways for the significant enrichment genes for all groups. *p*-value cutoff (FDR) 0.05.

**Table 2.** List of the significant differential expressed genes of safflower hot water extract SFPWH (50 mg/kg) and fluoxetine-treated group compared to the control group.

| Gene Symbol | NCBI ID | log2 Fold | Gene Symbol | NCBI ID | log2 Fold |
|---|---|---|---|---|---|
| SFPWH 50 (mg/kg) | | | Fluoxetine (10 mg/kg) Groups | | |
| *Ube2j2* | 140,499 | −2.06 | *Cacna1h* | 58,226 | −1.42 |
| *Dctn6-206* | 22,428 | −1.90 | *Dync1i2-203* | 13,427 | −0.97 |
| *Ncor1* | 20,185 | −1.43 | *Rbl2* | 19,651 | 1.37 |
| *Acp2* | 11,432 | −2.42 | *Mas1* | 17,171 | 1.24 |
| *Tuba1c* | 22,146 | −1.87 | *Dync1i2-205* | 13,427 | 1.22 |
| *Gm14150* | 100,043,840 | −1.56 | *Hyi* | 68,180 | 1.47 |
| *Actr3-ps* | 667,799 | −1.46 | *Morf4l2* | 56,397 | 2.23 |
| *Grik1* | 14,805 | −1.29 | | | |
| *Commd5* | 66,398 | −2.68 | | | |
| *Gm3226* | 100,041,240 | −2.14 | | | |
| *Msmo1* | 66,234 | 0.32 | | | |
| *Casp9* | 12,371 | 2.67 | | | |
| *Hspe1-rs1* | 628,438 | 2.35 | | | |
| *Tmem44* | 224,090 | 1.51 | | | |
| *Rps13-ps2* | 100,039,924 | 2.83 | | | |
| *Rps3a2* | 100,043,780 | 4.56 | | | |
| *Dctn6-202* | 22,428 | 1.84 | | | |

## 4. Discussion

In recent years, dietary phytochemicals and botanicals received the scientific community's attention regarding their potential use as a complementary and alternative therapy to either prevent, manage, or treat several diseases, including stress and depressive symptoms, with minimum side effects, in comparison to allopathic treatments, known to exert several adverse effects such as gastrointestinal problems, sleep disturbances, sexual dysfunction, congenital disabilities, seizures, and even death, among others [8]. Dietary phytochemicals are a part of our daily diet, easy to find, and cheap. Additionally, they have a safe "green image" that may potentially help to prevent and alleviate MDD symptoms. In this respect, many studies and experiments demonstrated that several polyphenols might produce an antidepressant-like effect [26]. The Chinese Health Ministry has recorded safflower formulations as a drug against cardiovascular diseases [27]. Safflower is one of the plants that has been used in traditional and modern medicine for years in several countries. Many studies illustrated the protective effect and other health benefits on musculoskeletal and cardiovascular systems and reproductive organs [12]. Safflower is known for its bioactivities such as anti-inflammatory, antioxidant, and antitumor effects [28]. Precisely, the water extract of safflower is considered an anticoagulant, vasodilating, antihypertensive, antioxidative, neuroprotective, immunosuppressive, and anticancer agent [12]. Our study demonstrated that safflower dried petals hot water extract exerts an antidepressant effect in mice subjected to TST and FST-induced stress, as shown by the improvement in the percentage of spontaneous alternation in YMSAT. The tail suspension test is an inexpensive and easy-to-use behavioral test used to detect depression-related behaviors and measure the potentiality of antidepressants on mice since the 1980s. TST has advantages over other behavioral tests in overcoming motor dysfunction of the mice and hypothermia issues [29].

A forced swimming test is also another widely used, easy, inexpensive, and fast behavioral test that could be used along with TST. TST and FST both use immobility as a measurement that represents depressive behavior. FST contributes to depression research, especially in genetic analysis, where immobility in FST is determined by heritable traits [30]. The Y maze test could easily assess the memory and learning patterns of mice, which allows continuous spontaneous alternation. During this test, the hippocampus is hugely involved in retrieving essential information for building a mental map [31]. Fluoxetine ($C_{17}H_{18}F_3NO$) is a high CNS penetration, orally administrated, and widely used drug, which is a selective serotonin reuptake inhibitor (SSRI) antidepressant [32].

Chromatographic analysis showed that oleamide (9-octadecenamide) would be the main active ingredient extracted in high yield in water than in hydro-alcoholic solvent. Moreover, we report for the first time the optimum procedure of extraction that yielded 46% oleamide-rich tea by soaking safflower in water at 115 °C for 1 min. Olamide is known as a hypnotic and sedative chemical entity that naturally accumulates in human and mammals cerebrospinal fluid during sleeping deprivation as a natural modulator to induce sleeping and/or to reduce psychological excitement through the interaction with the cannabinoid receptor [33]. Additionally, oleamide was reported to interact with several neurotransmitter-receptor systems, including serotonin and GABA [34]. Moreover, oleamide provides anti-inflammatory activity along with other pharmacological properties such as human monoamine oxidase B enzyme inhibitor and TRPV1 vanilloid receptors activator [35].

Our study reports for the first-time safflower extract-administered mice 'hippocampus transcriptomic profiling. Transcriptome analysis, including highly reproducible and significant target genes ($p < 0.001$), resulted in a pool of 22 genes. Interestingly, some of the genes are highly relevant to MDD, anxiety, and neurobehavior. A study on rats has revealed that NcoR (nuclear receptor corepressor) expression reduction within the developing amygdala is associated with a significant increase in anxiety-like behavior during the juvenile period in both genders [36]. On the contrary, hippocampal NcoR1 was upregulated in the control group with an average of 1.33 log2 for the three replicates, while a significant reduction in expression level was detected in SFPWH (50 mg/kg) group with −1.4 log2 fold. However, no significant difference was detected in SFPWH (150 mg/kg) and fluoxetine (10 mg/kg) groups. The gene ontology enrichment analysis showed that NcoR1 involves in the thyroid hormone signaling pathway (33.15-fold enrichment, FDR 0.0030). Noting that an adequate level of thyroid hormone is essential for the brain to function normally [37].

Apoptosis plays an important role in the development of the brain and the peripheral nervous system as well [38]. There is a link between apoptosis and major depressive disorder and other neurodegenerative diseases such as Parkinson's and Alzheimer's disease [38]. Casp9 is one of the differentially expressed genes in the SFPWH (50 mg/kg)-treated group and the enrichment analysis revealed involvement in several pathways, including apoptosis and neurodegeneration. Interestingly, casp9 was highly expressed in SFPWH (50 mg/kg) group (2.67 log2 fold) in comparison with the control group (−1.27 log2 fold).

Glutamate is a neurotransmitter that has an excitatory effect [39]. Glutamate system abnormalities and behavior that lead to several disorders, including MDD, were linked [39]. In our study, Grik1 (glutamate receptor, ionotropic, kainate 1) was significantly downregulated (−1.29 log2 fold) in SFPWH (50 mg/kg) group. It has been reported that female patients with MDD showed a significantly higher expression level of the Grik1 gene in the dorsolateral prefrontal cortex [39].

Ube2j2 gene is predicted to be involved in protein polyubiquitination and ubiquitin-dependent endoplasmic reticulum-associated protein degradation pathway [40]. Endoplasmic reticulum stress has a major role in the pathophysiology of depression, where an abnormal function is linked with triggering apoptosis signals [41]. In our study, Ube2j2 was significantly downregulated in SFPWH (50 mg/kg) group (−2.06 log2 fold) compared to the control group.

Transcriptomic analysis revealed two transcripts of the Dctn6 gene that were expressed differently. However, Dctn6-202 is the transcript that encodes the protein Dynactin subunit 6. The function of Dctn6 involves enabling dynein complex binding activity [42]. It has been reported that chronic stress reduces the dynein motor protein expression in the rat hippocampus [43]. In SFPWH (50 mg/kg) group, the Dctn6-206 transcript was downregulated (−1.90 log2 fold) while the transcript Dctn6-202 was upregulated (1.84 log2 fold). Tuba1c is another significantly downregulated gene detected in SFPWH (50 mg/kg) group. It is predicted that Tuba1c, which encodes Tubulin alpha-1C chain protein, enables GTP binding activity and structural constituent of the cytoskeleton [44]. Cytoskeletal (microtubules) post-translation modification is linked with several neuropsychiatric diseases [45]. In our study, the data analysis showed that Tuba1c is involved with several pathways, including neurodegeneration, Parkinson's disease, Huntington's disease, amyotrophic lateral sclerosis, and apoptosis pathways. Cacna1h gene was significantly downregulated in the fluoxetine-treated group with an average of −1.42 log2 fold. It has been reported that mice deficient in CaV3.2 channels, which is encoded by the Cacna1h gene, showed an increase in anxiety-like behavior [46], where safflower hot water extract-treated and control groups were not significantly different.

## 5. Conclusions

Our results represent the first scientific evidence supporting the ethnopharmacological use of safflower and its main active ingredient, oleamide, as a hypnotic and sedative. These properties are mainly related to a large amount of oleamide in the hot water extract. To the best of our knowledge, our study is the first study to elucidate the mechanism of action of safflower through the regulation of several genes related to MDD using a transcriptomic approach. Further experiments are needed to validate the RNA-seq outcomes at translational and post-translational levels. Our study may provide insight into the mechanism of action of safflower extract that may interfere with the apoptosis, glutamate, and endoplasmic reticulum-associated protein degradation pathways playing essential roles in protection against depression. Further studies on this issue at the post-translational level using the transcriptomic approach and immuno-histochemistry targeting different brain parts will provide more information to understand the mechanism of action of safflower and its main active ingredient, oleamide, in the prevention of depression and stress.

**Supplementary Materials:** The following are available online at https://www.mdpi.com/article/10.3390/app12115594/s1, Table S1: Chemical profile of *Carthamus tinctorius* extract.

**Author Contributions:** Conceptualization, M.H.A., A.E.O. and J.S.M.S.; methodology, M.H.A., A.E.O., A.A.B. and M.Y.A.; software, I.A.R.; validation, A.E.O., A.A.B., M.Y.A., I.A.R. and J.S.M.S.; formal analysis, all authors; investigation, M.H.A.; resources, J.S.M.S.; data curation, all authors; writing—original draft preparation, M.H.A.; writing—review and editing, all authors; visualization, all authors; supervision, J.S.M.S.; project administration, A.E.O. and J.S.M.S.; funding acquisition, J.S.M.S. All authors have read and agreed to the published version of the manuscript.

**Funding:** This research received no external funding.

**Institutional Review Board Statement:** The animal study was conducted in accordance with the Declaration of Helsinki, and the protocol was approved by the Ethics Committees, Department of Biological Sciences, King Abdulaziz University, Saudi Arabia (201060288).

**Informed Consent Statement:** Not applicable.

**Data Availability Statement:** The date generated is cited in the manuscript and provided as Supplement File.

**Conflicts of Interest:** The authors declare no conflict of interest.

# References

1. Kato, M.; Hori, H.; Inoue, T.; Iga, J.; Iwata, M.; Inagaki, T.; Shinohara, K.; Imai, H.; Murata, A.; Mishima, K.; et al. Discontinuation of antidepressants after remission with antidepressant medication in major depressive disorder: A systematic review and meta-analysis. *Mol. Psychiatry* **2021**, *26*, 118–133. [CrossRef] [PubMed]
2. Depression. Available online: https://www.who.int/news-room/fact-sheets/detail/depression (accessed on 9 July 2020).
3. Duan, L.; Gao, Y.; Shao, X.; Tian, C.; Fu, C.; Zhu, G. Research on the development of theme trends and changes of knowledge structures of drug therapy studies on major depressive disorder since the 21st century: A bibliometric analysis. *Front. Psychiatry* **2020**, *11*, 647. [CrossRef]
4. Xiong, J.; Lipsitz, O.; Nasri, F.; Lui, L.; Gill, H.; Phan, L.; Chen-Li, D.; Iacobucci, M.; Ho, R.; Majeed, A.; et al. Impact of COVID-19 pandemic on mental health in the general population: A systematic review. *J. Affect. Disord.* **2020**, *277*, 55–64. [CrossRef] [PubMed]
5. Alkhamees, A.A.; Alrashed, S.A.; Alzunaydi, A.A.; Almohimeed, A.S.; Aljohani, M.S. The psychological impact of COVID-19 pandemic on the general population of Saudi Arabia. *Compr. Psychiatry* **2020**, *102*, 152192. [CrossRef] [PubMed]
6. Depression. Available online: https://www.nimh.nih.gov/health/topics/depression#part_2255 (accessed on 27 February 2022).
7. Otte, C.; Gold, S.; Penninx, B.; Pariante, C.; Etkin, A.; Fava, M.; Mohr, D.; Schatzberg, A. Major depressive disorder. *Nat. Rev. Dis. Primers* **2016**, *2*, 16065. [CrossRef] [PubMed]
8. Jakobsen, J.C.; Gluud, C.; Kirsch, I. Should antidepressants be used for major depressive disorder? *BMJ Evid.-Based Med.* **2020**, *25*, 130. [CrossRef]
9. Ng, J.; Nazir, Z.; Nault, H. Complementary and alternative medicine recommendations for depression: A systematic review and assessment of clinical practice guidelines. *BMC Complementary Med. Ther.* **2020**, *20*, 299. [CrossRef]
10. Godos, J.; Castellano, S.; Ray, S.; Grosso, G.; Galvano, F. Dietary Polyphenol Intake and Depression: Results from the Mediterranean Healthy Eating, Lifestyle and Aging (MEAL) Study. *Molecules* **2018**, *23*, 999. [CrossRef]
11. Alegiry, M.H.; Hajrah, N.H.; Alzahrani, N.; Shawki, H.H.; Khan, M.; Zrelli, H.; Atef, A.; Kim, Y.; Alsafari, I.A.; Arfaoui, L.; et al. Attitudes Toward Psychological Disorders and Alternative Medicine in Saudi Participants. *Front. Psychiatry* **2021**, *12*, 577103. [CrossRef]
12. Delshad, E.; Yousefi, M.; Sasannezhad, P.; Rakhshandeh, H.; Ayati, Z. Medical uses of *Carthamus tinctorius* L. (Safflower): A comprehensive review from Traditional Medicine to Modern Medicine". *Electron. Phys.* **2018**, *10*, 6672–6681. [CrossRef]
13. Asgarpanah, J.; Kazemivash, N. Phytochemistry, pharmacology and medicinal properties of *Carthamus tinctorius* L. *Chin. J. Integr. Med.* **2013**, *19*, 153–159. [CrossRef] [PubMed]
14. Can, A.; Dao, D.T.; Terrillion, C.E.; Piantadosi, S.C.; Bhat, S.; Gould, T.D. The tail suspension test. *J. Vis. Exp.* **2012**, *59*, e3769. [CrossRef] [PubMed]
15. Kondo, S.; El Omri, A.; Han, J.; Isoda, H. Antidepressant-like effects of rosmarinic acid through mitogen-activated protein kinase phosphatase-1 and brain-derived neurotrophic factor modulation. *J. Funct. Foods* **2015**, *14*, 758–766. [CrossRef]
16. Can, A.; Dao, D.; Arad, M.; Terrillion, C.; Piantadosi, S.; Gould, T. The mouse forced swim test. *J. Vis. Exp.* **2012**, *59*, e3638. [CrossRef]

17. Grabherr, M.; Haas, B.; Yassour, M.; Levin, J.; Thompson, D.; Amit, I.; Adiconis, X.; Fan, L.; Raychowdhury, R.; Zeng, Q.; et al. Full-length transcriptome assembly from RNA-seq data without a reference genome. *Nat. Biotechnol.* **2011**, *29*, 644–652. [CrossRef] [PubMed]
18. Langmead, B.; Trapnell, C.; Pop, M.; Salzberg, S. Ultrafast and memory-efficient alignment of short DNA sequences to the human genome. *Genome Biol.* **2009**, *10*, R25. [CrossRef]
19. Li, B.; Dewey, C.N. RSEM: Accurate transcript quantification from RNA-Seq data with or without a reference genome. *BMC Bioinform.* **2011**, *12*, 323. [CrossRef]
20. Robinson, M.; McCarthy, D.; Smyth, G. edgeR: A Bioconductor package for differential expression analysis of digital gene expression data. *Bioinformatics* **2010**, *26*, 139–140. [CrossRef]
21. National Center for Biotechnology Information. "PubChem Compound Summary for CID 5283387, Oleamide" PubChem. Available online: https://pubchem.ncbi.nlm.nih.gov/compound/Oleamide (accessed on 18 June 2021).
22. Kraeuter, A.; Guest, P.; Sarnyai, Z. The Y-Maze for Assessment of Spatial Working and Reference Memory in Mice. *Methods Mol. Biol.* **2019**, *1916*, 105–111. [CrossRef]
23. Hölter, S.; Garrett, L.; Einicke, J.; Sperling, B.; Dirscherl, P.; Zimprich, A.; Fuchs, H.; Gailus-Durner, V.; Hrabě de Angelis, M.; Wurst, W. Assessing Cognition in Mice. *Curr. Protoc. Mouse Biol.* **2015**, *5*, 331–358. [CrossRef]
24. Le, T.; Savitz, J.; Suzuki, H.; Misaki, M.; Teague, T.; White, B.; Marino, J.; Wiley, G.; Gaffney, P.; Drevets, W.; et al. Identification and replication of RNA-Seq gene network modules associated with depression severity. *Transl. Psychiatry* **2018**, *8*, 180. [CrossRef]
25. ShinyGO v0.741. Available online: Bioinformatics.sdstate.edu/go (accessed on 28 January 2022).
26. Trebatická, J.; Ďuračková, Z. Psychiatric disorders and polyphenols: Can they be helpful in therapy? *Oxidative Med. Cell. Longev.* **2015**, *2015*, 248529. [CrossRef] [PubMed]
27. Li, S.; Li, T.; Jin, Y.; Qin, X.; Tian, J.; Zhang, L. Antidepressant-like effects of coumaroylspermidine extract from safflower injection residues. *Front. Pharmacol.* **2020**, *11*, 713. [CrossRef]
28. Wu, X.; Cai, X.; Ai, J.; Zhang, C.; Liu, N.; Gao, W. Extraction, structures, bioactivities and structure-function analysis of the polysaccharides from safflower (*Carthamus tinctorius* L.). *Front. Pharmacol.* **2021**, *12*, 767947. [CrossRef] [PubMed]
29. Yan, H.; Cao, X.; Das, M.; Zhu, X.; Gao, T. Behavioral animal models of depression. *Neurosci. Bull.* **2010**, *26*, 327–337. [CrossRef] [PubMed]
30. Yankelevitch-Yahav, R.; Franko, M.; Huly, A.; Doron, R. The forced swim test as a model of depressive-like behavior. *J. Vis. Exp.* **2015**, *97*, 52587. [CrossRef]
31. Famitafreshi, H.; Karimian, M. Assessment of improvement in oxidative stress indices with resocialization in memory retrieval in Y-Maze in male rats. *J. Exp. Neurosci.* **2018**, *12*, 1179069518820323. [CrossRef] [PubMed]
32. Szoke-Kovacs, Z.; More, C.; Szoke-Kovacs, R.; Mathe, E.; Frecska, E. Selective inhibition of the serotonin transporter in the treatment of depression: Sertraline, fluoxetine and citalopram. *Neuropsychopharmacol. Hung. Magy. Pszichofarmakologiai Egyes. Lapja=Off. J. Hung. Assoc. Psychopharmacol.* **2020**, *22*, 4–15.
33. Mendelson, W.B.; Basile, A.A. The hypnotic actions of the fatty acid amide, oleamide. *Neuropsychopharmacol. Off. Publ. Am. Coll. Neuropsychopharmacol.* **2001**, *25*, S36–S39. [CrossRef]
34. Fedorova, I.; Hashimoto, A.; Fecik, R.A.; Hedrick, M.P.; Hanus, L.O.; Boger, D.L.; Rice, K.C.; Basile, A.S. Behavioral evidence for the interaction of oleamide with multiple neurotransmitter systems. *J. Pharmacol. Exp. Ther.* **2001**, *299*, 332–342.
35. Jug, U.; Naumoska, K.; Metličar, V.; Schink, A.; Makuc, D.; Vovk, I.; Plavec, J.; Lucas, K. Interference of oleamide with analytical and bioassay results. *Sci. Rep.* **2020**, *10*, 2163. [CrossRef] [PubMed]
36. Jessen, H.M.; Kolodkin, M.H.; Bychowski, M.E.; Auger, C.J.; Auger, A.P. The nuclear receptor corepressor has organizational effects within the developing amygdala on juvenile social play and anxiety-like behavior. *Endocrinology* **2010**, *151*, 1212–1220. [CrossRef] [PubMed]
37. Liu, Y.Y.; Brent, G.A. The role of thyroid hormone in neuronal protection. *Compr. Physiol.* **2021**, *11*, 2075–2095. [CrossRef] [PubMed]
38. McKernan, D.P.; Dinan, T.G.; Cryan, J.F. "Killing the Blues": A role for cellular suicide (apoptosis) in depression and the antidepressant response? *Prog. Neurobiol.* **2009**, *88*, 246–263. [CrossRef] [PubMed]
39. Gray, A.L.; Hyde, T.M.; Deep-Soboslay, A.; Kleinman, J.E.; Sodhi, M.S. Sex differences in glutamate receptor gene expression in major depression and suicide. *Mol. Psychiatry* **2015**, *20*, 1057–1068. [CrossRef]
40. Ube2j2 Ubiquitin-Conjugating Enzyme E2J 2 [Mus Musculus (House Mouse)]-Gene-NCBI. Available online: https://www.ncbi.nlm.nih.gov/gene/140499 (accessed on 22 January 2022).
41. Mao, J.; Hu, Y.; Ruan, L.; Ji, Y.; Lou, Z. Role of endoplasmic reticulum stress in depression (Review). *Mol. Med. Rep.* **2019**, *20*, 4774–4780. [CrossRef] [PubMed]
42. Dctn6 Dynactin 6 [Mus Musculus (House Mouse)]-Gene-NCBI. Available online: https://www.ncbi.nlm.nih.gov/gene/22428 (accessed on 22 January 2022).
43. Zavvari, F.; Nahavandi, A. Fluoxetine increases hippocampal neural survival by improving axonal transport in stress-induced model of depression male rats. *Physiol. Behav.* **2020**, *227*, 113140. [CrossRef]
44. Tuba1c Tubulin, Alpha 1C [Mus Musculus (House Mouse)]-Gene-NCBI. Available online: https://www.ncbi.nlm.nih.gov/gene/22146 (accessed on 22 January 2022).

45. Singh, H.; Chmura, J.; Bhaumik, R.; Pandey, G.N.; Rasenick, M.M. Membrane-associated $\alpha$-tubulin is less acetylated in postmortem prefrontal cortex from depressed subjects relative to controls: Cytoskeletal dynamics, HDAC6, and depression. *J. Neurosci. Off. J. Soc. Neurosci.* **2020**, *40*, 4033–4041. [CrossRef]

46. Andrade, A.; Brennecke, A.; Mallat, S.; Brown, J.; Gomez-Rivadeneira, J.; Czepiel, N.; Londrigan, L. Genetic associations between voltage-gated calcium channels and psychiatric disorders. *Int. J. Mol. Sci.* **2019**, *20*, 3537. [CrossRef]

*Article*

# A Pilot Study of whether or Not Vegetable and Fruit Juice Containing *Lactobacillus paracasei* Lowers Blood Lipid Levels and Oxidative Stress Markers in Thai Patients with Dyslipidemia: A Randomized Controlled Clinical Trial

Pattharaparn Siripun [1], Chaiyavat Chaiyasut [1,2,*], Narissara Lailerd [3], Netnapa Makhamrueang [1], Ekkachai Kaewarsar [1] and Sasithorn Sirilun [1,2,*]

1   Department of Pharmaceutical Sciences, Faculty of Pharmacy, Chiang Mai University, Chiang Mai 50200, Thailand; pattharaparn@gmail.com (P.S.); netnapa_ma@cmu.ac.th (N.M.); ekkachai_kaew@cmu.ac.th (E.K.)
2   Innovation Center for Holistic Health, Nutraceuticals and Cosmeceuticals, Faculty of Pharmacy, Chiang Mai University, Chiang Mai 50200, Thailand
3   Department of Physiology, Faculty of Medicine, Chiang Mai University, Chiang Mai 50200, Thailand; narissara.lailerd@cmu.ac.th
*   Correspondence: chaiyavat@gmail.com (C.C.); sasithorn.s@cmu.ac.th (S.S.); Tel.: +66-53-94-4375 (S.S.)

**Citation:** Siripun, P.; Chaiyasut, C.; Lailerd, N.; Makhamrueang, N.; Kaewarsar, E.; Sirilun, S. A Pilot Study of whether or Not Vegetable and Fruit Juice Containing *Lactobacillus paracasei* Lowers Blood Lipid Levels and Oxidative Stress Markers in Thai Patients with Dyslipidemia: A Randomized Controlled Clinical Trial. *Appl. Sci.* **2022**, *12*, 4913. https://doi.org/10.3390/app12104913

Academic Editors: Luca Mazzoni, Maria Teresa Ariza Fernández and Franco Capocasa

Received: 11 April 2022
Accepted: 10 May 2022
Published: 12 May 2022

**Publisher's Note:** MDPI stays neutral with regard to jurisdictional claims in published maps and institutional affiliations.

**Copyright:** © 2022 by the authors. Licensee MDPI, Basel, Switzerland. This article is an open access article distributed under the terms and conditions of the Creative Commons Attribution (CC BY) license (https://creativecommons.org/licenses/by/4.0/).

**Abstract:** Dyslipidemia is one of the risk factors of cardiovascular disease, which is the main cause of mortality worldwide. Meanwhile, lipid-lowering drug side-effects may also occur. Thus, consumption of vegetables and fruits containing probiotics is a good alternative to influence the lipid profile in plasma. This study investigated the effect of consuming vegetable and fruit juice (VFJ) with (probiotic group) and without probiotic *Lactobacillus paracasei* (placebo group), on the body weight, body mass index, waist circumference, lipid profile, lipid peroxidation, oxidative stress enzymes, and bile acid level in dyslipidemic patients (n = 20) at Bhumibol Adulyadej Hospital for 30 days. The levels of total cholesterol, low-density lipoprotein cholesterol, triglyceride (TG), and TG/high-density lipoprotein cholesterol (HDL-C) ratio in the probiotic group were significantly lower than those in the placebo group. The HDL-C concentration in the probiotic group was higher than that in the placebo group. The probiotic group showed significantly decreased malondialdehyde levels; increased oxidative stress enzymes, catalase and glutathione peroxidase in the plasma; and increased bile acid (BA) levels in the feces. Therefore, the findings of this study demonstrate that VFJ enriched with probiotic *L. paracasei* may represent an alternative method for the prevention of dyslipidemia during the primary intervention stage for patients who are not yet taking other medication.

**Keywords:** clinical trial; dyslipidemia; *Lactobacillus paracasei*; probiotics; vegetable and fruit juice

## 1. Introduction

Dyslipidemia is a disorder of the lipid and lipoprotein metabolism characterized by too-high or too-low blood lipid levels [1]. Abnormal serum lipid levels increase the risk of cardiovascular diseases and can be caused by elevated total cholesterol (TC), low-density lipoprotein cholesterol (LDL-C), or triglyceride (TG) levels, or low high-density lipoprotein cholesterol (HDL-C) [2]. Globally, cardiovascular diseases currently represent the primary cause of mortalities, including within Thailand [3]. In fact, the World Health Organization (WHO) [4] estimates that cardiovascular diseases will claim the lives of approximately 23.6 million people by the end of 2030. Regarding Thailand specifically, the Medical Record Section, Service Division, of the Bhumibol Adulyadej Hospital reported that the number of patients with dyslipidemia continues to increase steadily each year. Currently, the recommended treatment options for dyslipidemia include lifestyle modifications and drug therapy [5,6]; however, lipid-lowering drugs can induce adverse side effects. Hence, dietary changes, including reduced fat diets, and increased vegetable and fruit intake, which influence the plasma lipid profile [7], may represent effective alternatives to drug therapy.

Vegetables and fruits contain dietary phytochemicals, including flavonoids, carotenoids, vitamins, minerals, and dietary fiber, which decrease dyslipidemia and cardiovascular risk [8–11]. Indeed, their consumption is inversely correlated with plasma TC and LDL-C levels [11]. Suwimol et al. [7] investigated the effect of vegetable and fruit intake on plasma lipid profiles and oxidative status and reported that the consumption of eight servings of vegetables and fruits per day significantly reduces LDL-C and lipid peroxidation via malondialdehyde (MDA) levels. However, the sensory aspects of fruits and vegetables (i.e., color, smell, flavor, texture) can dissuade patients from consuming adequate amounts. Therefore, altering these characteristics by blending or mixing may improve uptake. Moreover, the combination of vegetables and fruits with various forms of probiotics, such as probiotic powder or probiotic fermented milk, can provide beneficial probiotic and dietary fiber [12].

Probiotics comprise a group of bacteria that are generally recognized as safe (GRAS) [13] for use in food and human health products, including for the treatment of various diseases [14] while also decreasing the risk of dyslipidemia. *L. paracasei* are widely utilized as probiotics or synbiotics to improve clinical outcomes. Chiu et al. [15] reported that, in a hamster model, ingestion of *L. paracasei* NTU 101 fermented milk, along with a high-cholesterol diet, significantly reduced serum cholesterol levels, compared to the control group. Furthermore, Dehkohneh et al. [16] reported significantly reduced serum cholesterol levels in Wistar rats following consumption of *L. paracasei* TD3 with a high-fat diet.

The current 30-day randomized controlled trial therefore aims to investigate the effects of consuming vegetable and fruit juice (VFJ), with probiotic *L. paracasei*, on physical parameters (body weight, BW; body mass index, BMI; and waist circumference, WC) and biological markers (lipid profile, lipid peroxidation, oxidative stress enzymes, and bile acid (BA) level) in dyslipidemia patients at Bhumibol Adulyadej Hospital. The findings will show that the intake of VFJ with *L. paracasei* could provide health benefits by improving lipid profiles, lipid peroxidation, and oxidative stress enzyme activity levels.

## 2. Materials and Methods

### 2.1. Materials

Organic vegetables and fruits from the same plantations and the same farm, i.e., green lettuce (*Lactuca sativa*), Chinese celery (*Apium graveolens* var. *secalinum*), cherry tomato (*Solanum lycopersicum* var. *cerasiforme*), onion (*Alium cepa.* Linn), and lime (*Citrus aurantifolia* (Christm. and Panz.) Swing) were obtained from King Fresh Farm CO., LTD. (contact farming) located in Samut Sakhon, Thailand. Only apple (*Malus pumila*), was not organic, which was obtained from a market in Bangkok, Thailand. The selected raw material requirements were as follows: (i) an edible vegetable or fruit, (ii) of a low price, (iii) available throughout the year, (iv) had rich source of bioactive compounds, and (v) had a favorable taste in amixed juice.

### 2.2. Probiotics Lactobacillus paracasei

The powder form of a probiotic *L. paracasei* that is on the list of notifications of the Ministry of public health, Thailand (2011) was provided by Innovation Center for Holistic Health, Nutraceuticals and Cosmeceuticals, Faculty of Pharmacy, Chiang Mai University, Chiang Mai, Thailand. A total of 10 g of probiotic powder was contained within an aluminum foil sachet. The sachet of probiotic powder was kept in the fridge (4–6 °C) with the cell survival rate $\geq$ 90% for 3–6 months of storage.

### 2.3. Preparation of Vegetable and Fruit Juice (VFJ) with and without Probiotic Lactobacillus paracasei

Organic fresh vegetables and fruits were washed under running tap water and drained in a sieve to remove excess water before creating the juice. Only apples were peeled and immersed in a mixture of 0.5% (*v/v*) vinegar and 0.5% (*w/v*) saline for 15 min. The fruits and vegetables were then sliced, weighed, and prepared according to the recipe shown in Table 1. The composition was blended using a 1500-watt blender (Buono, Model BUO-127799T, Taipei, Taiwan) for 1 min. The resulting juice was then poured into 250 mL plastic bottles and stored at 4–6 °C before consumption on the same day of preparation. For the VFJ with *L. paracasei*, 10 g of the *L. paracasei* probiotic powder was added to the bottle. The final concentration of the probiotic was $10^9$ CFU per bottle.

**Table 1.** Ingredients of vegetable and fruit juice (VFJ).

| Ingredients | Content (% *w/w*) |
| --- | --- |
| Green lettuce | 7 |
| Chinese celery | 0.5 |
| Cherry tomato | 15 |
| Onion | 3 |
| Apple | 20 |
| Lime juice | 3 |
| Honey | 10 |
| Fresh water | 41.5 |

*2.4. Determination of Bioactive Compounds and Nutritions in Vegetable and Fruit Juice (VFJ)*

The antioxidant activity of VFJ was measured using a modified ABTS (2, 20-Azino-bis (3-ethylbenzothiazoline-6-sulfate)) assay, following the method of Saenjum et al. [17]. In brief, the ABTS°+ aqueous solution was prepared by the mixture of 7.0 mM ABTS stock solution (Merck, Darmstadt, Germany) and 2.45 mM potassium persulfate (RCl Labscan, Bangkok, Thailand). The mixture solution was incubated in the dark for 16 h at room temperature. Then, the working solution was created by mixing 1.0 mL of ABTS°+ aqueous with 50 mL deionized water to attain an absorbance of $0.70 \pm 0.05$ at 734 nm using a spectrophotometer. The reaction mixture contained 2000 μL of the ABTS°+ working solution and 100 μL of the sample or positive control using Trolox (Merck, Darmstadt, Germany). Then, the incubation was continued for 3 min at room temperature. The results were shown as μg Trolox equivalent per mL.

Other nutritional information of VFJ was analyzed by the Institute of Nutrition, Mahidol University, Nakhon Pathom, Thailand.

*2.5. Ethical Considerations*

The Ethics Committee of Bhumibol Adulyadej Hospital approved the study protocol (approval number IRB 68/61). In addition, the study was reviewed and approved by the Thai Clinical Trials Registry (TCTR) Committee (TCTR identification number is TCTR20220109001; https://www.thaiclinicaltrials.org/show/TCTR20220109001 (accessed on 9 January 2022). Patients voluntarily decided whether to participate the clinical trial or not. The research team clearly explained the purpose and methodology of this study. All the participants approved of the study procedure and provided their consent before enrollment.

*2.6. Participants and Study Design*

The participant population was recruited from patients of Bhumibol Adulyadej Hospital, Bangkok, Thailand. The inclusion criteria included: $\geq 18$ years of age; an LDL-C level between 130 and 160 mg/dL; and had never undergone medical treatment for dyslipidemia. The exclusion criteria included: family history of dyslipidemia; had undergone gastrointestinal surgery; diagnosed with metabolic disorder, cardiovascular disease, thyroid disorder, kidney disease, or liver disease; regular consumption of prebiotic, probiotic, and/or nutritional supplements; currently taking drugs that may affect lipid metabolism; smoking; alcohol consumption during the trial.

The 20 eligible participants were randomized to the probiotic or placebo groups using random allocation software. The participants were blinded to group assignments. During the 30-day intervention period, the Nutrition Section of Bhumibol Adulyadej Hospital provided both groups with a regular diet in lunch boxes for three meals per day, along with two bottles of VFJ; the probiotic group received VFJ with probiotic *L. paracasei* and the placebo group received VFJ without probiotic *L. paracasei*. Each participant drank the juice daily 30 min before lunch and dinner. A flow diagram is shown as Figure 1.

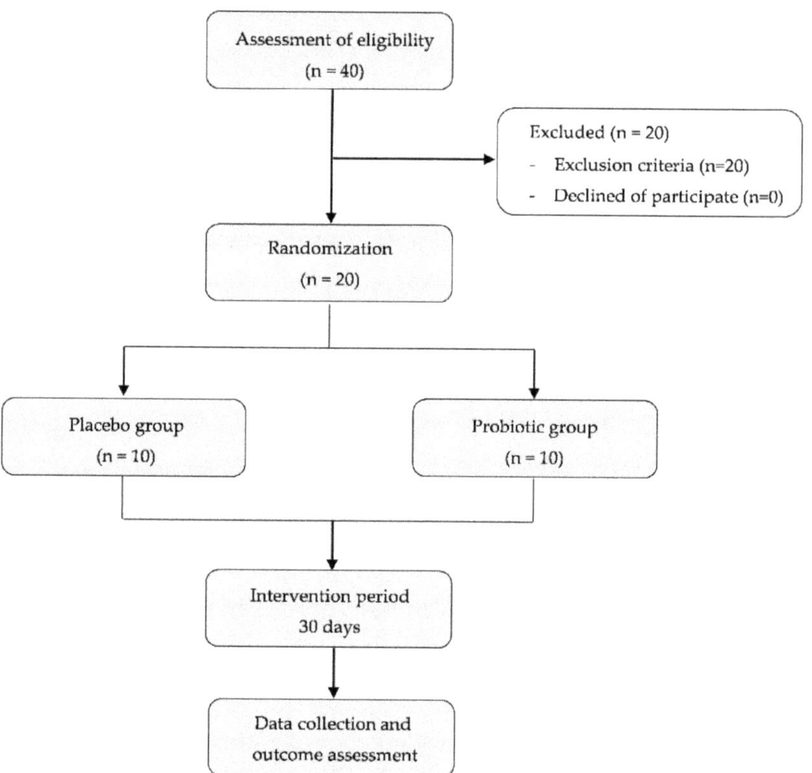


**Figure 1.** The study flow diagram and enrollment.

### 2.7. Outcome Assessment

At baseline and the study endpoint, BW, BMI, and WC physical parameters were measured. Additionally, the lipid profile was assessed based on blood samples collected at the Chemistry and Immunology Laboratory, Department of Pathology, Bhumibol Adulyadej Hospital. Lipid peroxidation and oxidative stress enzymes were measured from blood samples, and bile acid (BA) levels were measured from fecal samples at the Department of Pharmaceutical Sciences, Faculty of Pharmacy, Chiang Mai University (Figure 2). Ten milliliters of blood were collected from each participant after an 8 h overnight fast and stored in an icebox before being transported to the laboratory. A 10 g fecal sample was also collected from each participant and stored in a plastic container at 4–8 °C before being sent to the laboratory the next day. The fecal samples were stored at −20 °C until analysis within a month.

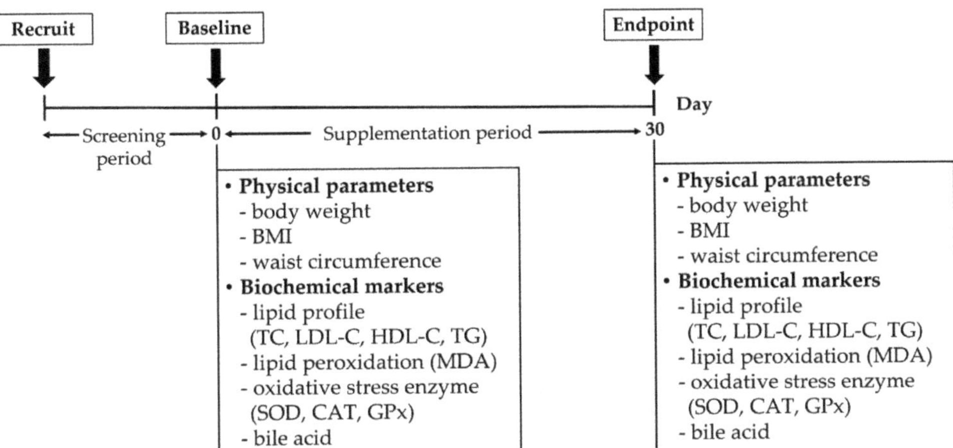


**Figure 2.** The study schedule of the clinical trial.

### 2.7.1. Physical Examination

BW and BMI were measured using a digital weighing scale (Tanita, Model 1584, Tokyo, Japan). WC was measured and recorded using a tape measure (Hoechstmass, Sulzbach, Germany).

### 2.7.2. Lipid Profile

Lipid profile, TC, LDL-C, HDL-C, and TG, levels were measured using an automatic analyzer (Model Cobas®8000, Roche Diagnostics, Mannheim, Germany).

### 2.7.3. Lipid Peroxidation

MDA, the main marker for lipid peroxidation, was determined using a thiobarbituric acid reactive substances (TBARS) assay according to the modified method of Zeb and Ullah [18] and Atasayar et al. [19]. Serum samples (50 µL) were reacted with thiobarbituric acid (TBA) (Sigma-Aldrich, St. Louis, MO, USA) and trichloroacetic acid reagent (TCA; Merck, Darmstadt, Germany) at 100 °C for 30 min and then placed in cool water. The reaction was measured at 532 nm using a multi-mode microplate reader (SpectraMax M3, San Jose, CA, USA).

### 2.7.4. Oxidative Stress Enzymes

Oxidative stress enzyme activities, including superoxide dismutase (SOD) activity, catalase (CAT) activity, and glutathione peroxidase (GPx) activity, were evaluated in the serum samples.

The SOD activity assay was modified from that by Kaya et al. [20] using the inhibition of nitroblue tetrazolium (NBT) reduction by the xanthine/xanthine oxidase system as a superoxide generator. The sample was supplemented with a 0.1 mL mixture of ethanol (RCI Labscan, Bangkok, Thailand) and chloroform (Merck, Darmstadt, Germany) (ratio 5:3, $v/v$) and centrifuged at $4000 \times g$ for 10 min. The ethanol phase of the supernatant was evaluated. One unit of SOD was defined as the amount of enzyme causing 50% inhibition of the NBT reduction rate. SOD activity was inhibited in units per milliliter of the sample.

The CAT activity was determined using a method modified from that by Kaya et al. [20]. A reaction of 750 µL of 0.059 M hydrogen peroxide (Merck, Darmstadt, Germany) and 25 µL of plasma sample or catalase standard (0–700 units/mL) was mixed for 3 min. Kinetic reactions were detected using a spectrophotometer at 240 nm. CAT activity was calculated by comparison with the catalase enzyme standard curve and expressed as units per milliliter.

The modified method of GPx activity was performed according to Rush and Sandiford [21]. The reaction was carried out by mixing β-nicotinamide adenine dinucleotide phosphate (β-NADPH) (Sigma-Aldrich, St. Louis, MO, USA), 1.0 mM sodium azide solution (Sigma-Aldrich), 200 mM glutathione (Sigma-Aldrich, St. Louis, MO, USA), 10 units/mL glutathione reductase (Sigma-Aldrich), and 50 mM phosphate buffer (pH 7.0) with 0.4 mM ethylene diamine tetraacetic acid (Loba Chemie, Mumbai, India). Next, 12.5 μL of plasma was added to the reaction. A mixture without plasma was used as a blank sample. GPx activity was determined using 0.042% hydrogen peroxide as the substrate. The reaction was followed using a spectrophotometer at 340 nm intervals for 15 s for 5 min. The activity of plasma GPx was determined by measuring the rate of oxidation of NADPH, and the result was reported as units per milliliter of plasma.

### 2.7.5. Bile Acid

Bile acid levels were determined in feces using a commercially available IDK® Bile acid kit (Immundiagnostik AG, Bensheim, Germany) according to the manufacturer's protocol and as previously studied [22]. Fifteen milligrams of fecal sample was added to a screw-capped tube containing 1.5 mL of buffer (dilution factor of 1:100). The sample was homogenously shaken and left to stand for 10 min. The mixture sample (10 μL) was added to the reagent in the strip and incubated for 5 min at 25 °C. The absorbance of the sample was measured at 405 nm using a multimode microplate reader.

### 2.8. Statistical Analysis

Data analysis was performed using STATA version 15.1 (StataCorp, College Station, TX, USA) for Windows licensed to the Faculty of Pharmacy, Chiang Mai University. Demographic characteristic data were evaluated using an independent *t*-test. A paired *t*-test was used to assess differences within the treatment groups. Gaussian regression analysis was performed to assess the effects of treatment between groups.

## 3. Results

### 3.1. Nutrition Information of the Vegetable and Fruit Juice (VFJ)

The antioxidant activity in VFJ was reported as 244.68 μg Trolox equivalent/mL, which was expressed as 60.97% of inhibition. Table 2 displays the nutritional values of the VFJ without probiotic *L. paracasei*, such as energy, moisture, protein, total carbohydrate, total dietary fiber, soluble dietary fiber, insoluble dietary fiber, vitamin and mineral, total polyphenol, flavonoid, and carotenoid according to the Association of Official Agricultural Chemists (AOAC) test method. The amount was reported by the Institute of Nutrition, Mahidol University, Nakhon Pathom, Thailand.

**Table 2.** Nutritional value of the vegetable and fruit juice (VFJ).

| Nutrition | Amount (Per 100 mL) |
|---|---|
| Energy | 43.08 kcal |
| Moisture | 91.92 g |
| Protein | 0.43 g |
| Total Carbohydrate | 10.34 g |
| Total dietary fiber | 1.35 g |
| Soluble dietary fiber | 0.71 g |
| Insoluble dietary fiber | 0.65 g |
| Vitamin C | 3.06 mg |
| Calcium | 8.99 mg |
| Phosphorus | 27.97 mg |
| Sodium | 14.73 mg |
| Potassium | 64.94 mg |
| Magnesium | 3.97 mg |
| Iron | 0.51 mg |

**Table 2.** *Cont.*

| Nutrition | Amount (Per 100 mL) |
|---|---|
| Zinc | 0.06 mg |
| Copper | 0.02 mg |
| Chloride | 9.39 mg |
| Pantothenic acid | 0.04 mg |
| Total polyphenol | 41.99 mg eq GA [1] |
| Naringenin | 636.82 µg |
| Quercetin | 1088.34 µg |
| Lutein | 48.76 µg |
| Lycopene | 239.10 µg |

[1] mg eq GA represents milligram equivalent gallic acid.

### 3.2. Participant Characteristics

The participant characteristics are shown in Table 3. Of the 20 participants, 17 were female and 3 were male with ages ranging from 27 to 58 years. All eligible participants were randomized into two groups ($n$ = 10 per group). The placebo group comprised one (10%) man and nine (90%) women, with ages ranging from 27 to 54 years, while two (20%) men and eight (80%) women were assigned to the probiotic group (33–58 years old). No patients withdrew from the study before the endpoint, and no serious adverse events were reported during the intervention period. No statistically significant differences were observed in the demographic characteristics of participants between the two groups at baseline.

**Table 3.** General demographic characteristics of the participants.

| Parameters | Placebo Group | Probiotic Group | $p$-Value |
|---|---|---|---|
| Male, n (%) | 1 (10) | 2 (20) | 1.000 |
| Female, n (%) | 9 (90) | 8 (80) | |
| Age (years) | 44.40 ± 2.40 | 43.80 ± 2.07 | 0.852 |
| Weight (kg) | 62.62 ± 3.20 | 65.01 ± 2.53 | 0.566 |
| Height (cm) | 157.40 ± 1.69 | 157.50 ± 1.87 | 0.969 |
| BMI (kg/m$^2$) | 25.36 ± 1.44 | 26.25 ± 1.06 | 0.625 |
| WC (cm) | 81.50 ± 2.46 | 86.00 ± 3.03 | 0.264 |
| LDL-C (mg/dL) | 145.20 ± 3.93 | 146.60 ± 3.54 | 0.794 |

Data are presented as mean ± standard error. The $p$-value was at 95% confidence interval. The population was analyzed using an exact probability test, and the continuous data were analyzed using a $t$-test. BMI = body mass index, WC = waist circumference, LDL-C = low-density lipoprotein cholesterol.

### 3.3. Effect of the Vegetable and Fruit Juice (VFJ) with and without Probiotic L. paracasei on Physical Parameters

BW, BMI, and WC showed no considerable differences at baseline and endpoint within the placebo group. Moreover, within the probiotic group, no considerable changes were observed in BW or BMI; however, the WC exhibited a decreasing trend ($p$ < 0.10; Table 4). Moreover, no significant differences were observed between the two groups in terms of BW, BMI, or WC, at the endpoint (Table 5).

**Table 4.** Physical examination parameters and biological markers in plasma and feces of the participant at the baseline (Day 0) and endpoint (Day 30).

| Parameters | Placebo Group | | | | Probiotic Group | | | |
|---|---|---|---|---|---|---|---|---|
| | Baseline | Endpoint | % Change | $p$-Value | Baseline | Endpoint | % Change | $p$-Value |
| BW (kg) | 62.62 ± 3.20 | 62.08 ± 3.15 | −0.86 | 0.255 | 65.01 ± 2.53 | 64.57 ± 2.57 | −0.68 | 0.192 |
| BMI (kg/m$^2$) | 25.36 ± 1.43 | 25.16 ± 1.43 | −0.79 | 0.281 | 26.25 ± 1.06 | 26.06 ± 1.04 | −0.72 | 0.163 |
| WC (cm) | 81.50 ± 2.46 | 81.10 ± 2.26 | −0.49 | 0.343 | 86.00 ± 3.02 | 85.20 ± 3.02 | −0.93 | 0.087 * |
| TC (mg/dL) | 202.60 ± 9.46 | 203.00 ± 6.40 | 0.20 | 0.955 | 203.80 ± 5.11 | 192.40 ± 4.53 | −5.59 | 0.041 ** |
| LDL-C (mg/dL) | 145.20 ± 3.93 | 145.80 ± 4.97 | 0.41 | 0.818 | 146.60 ± 3.54 | 137.80 ± 4.04 | −6.00 | 0.003 *** |
| HDL-C (mg/dL) | 51.40 ± 2.87 | 50.80 ± 2.56 | −1.17 | 0.712 | 49.20 ± 2.53 | 53.30 ± 2.36 | 8.33 | 0.009 *** |

**Table 4.** *Cont.*

| Parameters | Placebo Group | | | | Probiotic Group | | | |
|---|---|---|---|---|---|---|---|---|
| | Baseline | Endpoint | % Change | *p*-Value | Baseline | Endpoint | % Change | *p*-Value |
| TG (mg/dL) | 122.40 ± 12.84 | 101.80 ± 8.39 | −16.83 | 0.111 | 121.80 ± 11.59 | 92.40 ± 9.83 | −24.14 | 0.029 ** |
| TG/HDL-C ratio | 2.48 ± 0.32 | 2.06 ± 0.21 | −16.93 | 0.160 | 2.53 ± 0.27 | 1.75 ± 0.18 | −30.83 | 0.016 ** |
| MDA (μM) | 0.18 ± 0.07 | 0.07 ± 0.01 | −61.11 | 0.147 | 0.19 ± 0.07 | 0.06 ± 0.01 | −68.42 | 0.067 * |
| CAT (unit/mL) | 159.08 ± 3.97 | 163.39 ± 5.43 | 2.71 | 0.542 | 153.60 ± 2.89 | 166.00 ± 4.36 | 8.07 | 0.005 *** |
| GPx (unit/mL) | 0.61 ± 0.11 | 0.78 ± 0.08 | 27.87 | 0.196 | 0.63 ± 0.06 | 0.83 ± 0.13 | 31.75 | 0.060 * |
| SOD (unit/mL) | 588.84 ± 62.54 | 643.14 ± 40.22 | 9.22 | 0.350 | 562.41 ± 73.44 | 678.92 ± 51.68 | 20.71 | 0.115 |
| BA (μmol/L) | 30.06 ± 3.65 | 29.29 ± 2.16 | −2.56 | 0.834 | 31.60 ± 2.90 | 37.73 ± 2.48 | 19.40 | 0.054 * |

Data are presented as mean ± standard error. *, **, and *** were significantly different $p < 0.10$, $p < 0.05$, and $p < 0.01$, respectively, within each group at different times of this study. BW = body weight, BMI = body mass index, WC = waist circumference, TC = total cholesterol, LDL-C = low-density lipoprotein cholesterol, HDL-C = high-density lipoprotein cholesterol, TC = triglyceride, MDA = malondialdehyde, CAT = catalase, GPx = glutathione peroxidase, SOD = superoxide dismutase, BA = bile acid.

**Table 5.** Gaussian regression analysis of physical examination parameters and biological markers for the probiotic group compared with the placebo group at the endpoint (Day 30) of the study.

| Parameters | Coefficient | 95% CI | *p*-Value |
|---|---|---|---|
| BW (kg) | 2.49 | −1.24 to 1.31 | 0.953 |
| BMI (kg/m$^2$) | 0.90 | −0.52 to 0.51 | 0.985 |
| WC (cm) | 4.10 | −1.40 to 1.24 | 0.898 |
| TC (mg/dL) | −10.60 | −23.05 to 2.48 | 0.106 |
| LDL-C (mg/dL) | −8.00 | −23.42 to 6.78 | 0.259 |
| HDL-C (mg/dL) | 2.50 | −0.02 to 7.37 | 0.051 * |
| TG (mg/dL) | −9.40 | −35.10 to 15.44 | 0.420 |
| TG/HDL-C ratio | −0.31 | −0.86 to 0.26 | 0.270 |
| MDA (μM) | −0.01 | −0.04 to 0.03 | 0.825 |
| CAT (unit/mL) | 2.61 | −10.82 to 22.17 | 0.475 |
| GPx (unit/mL) | 0.05 | −0.29 to 0.30 | 0.980 |
| SOD (unit/mL) | 35.78 | −101.65 to 160.33 | 0.640 |
| BA (μmol/L) | 8.44 | 1.52 to 15.15 | 0.020 ** |

Data were compared to the placebo group at endpoint (Day 30). The *p*-value was at 95% confidence interval. * and ** were significantly different $p < 0.10$ and $p < 0.05$, respectively. BW = body weight, BMI = body mass index, WC = waist circumference, TC = total cholesterol, LDL-C = low-density lipoprotein cholesterol, HDL-C = high-density lipoprotein cholesterol, TC = triglyceride, MDA = malondialdehyde, CAT = catalase, GPx = glutathione peroxidase, SOD = superoxide dismutase, BA = bile acid.

### 3.4. Effect of the Vegetable and Fruit Juice (VFJ) with and without Probiotic L. paracasei on Lipid Profile

No significant differences were observed in the TC, LDL-C, HCL-C, TG, or TG/HDL-C ratio ($p > 0.05$) at the endpoint in the placebo group compared to the baseline value (Table 4). Meanwhile, significant declines in the probiotic group were observed in TC, LDL-C, TG, and TG/HDL-C ratio, with a significant increase in HDL-C (Table 4). The parameters of TC, LDL-C, TG, and TG/HDL-C ratio significantly declined: ($p < 0.05$), ($p < 0.01$), ($p < 0.05$), and ($p < 0.05$), respectively. In contrast, HDL-C showed a significant increase ($p < 0.01$). Moreover, although the differences in TC, LDL-C, TG, and TG/HDL-C ratio between the probiotic and placebo groups (Table 5) were not statistically significant, HDL-C was significantly higher in the probiotic group at the endpoint ($p < 0.01$).

### 3.5. Effect of the Vegetable and Fruit Juice (VFJ) with and without Probiotic L. paracasei on Lipid Peroxidation

MDA levels in the plasma of the placebo group at the baseline (Day 0) and endpoint (Day 30) were 0.18 ± 0.07 and 0.07 ± 0.01 μM. While the MDA levels of the probiotic group at the baseline (Day 0) and endpoint (Day 30) were 0.19 ± 0.07 and 0.06 ± 0.01 μM, the statistical data showed a significant difference ($p < 0.10$) (Table 4). However, the MDA levels did not differ significantly between the probiotic group and the placebo group at the endpoint (Table 5).

### 3.6. Effect of the Vegetable and Fruit Juice (VFJ) with and without Probiotic L. paracasei on Oxidative Stress Enzyme

Plasma levels of CAT, GPx, and SOD in the placebo group did not differ significantly ($p > 0.05$) between baseline and endpoint (Table 4). Meanwhile, in the probiotic group, CAT levels were significantly increased ($p < 0.01$), whereas GPx exhibited an increased trend ($p < 0.10$). However, the SOD levels were not significantly different after the endpoint compared to the baseline of the study period. Moreover, no significant differences were detected in CAT, GPx, or SOD levels between the groups (Table 5).

### 3.7. Effect of the Vegetable and Fruit Juice (VFJ) with and without Probiotic L. paracasei on Bile Acids Level

Fecal levels of BA did not differ significantly within the placebo group between timepoints and were $30.06 \pm 3.65$ μmol/L at the baseline and $29.29 \pm 2.16$ μmol/L at the endpoint. Meanwhile, a trend toward significantly increased BA levels ($p < 0.10$) was detected in the probiotic group at the endpoint compared to baseline (Table 4). Moreover, BA levels were significantly higher ($p < 0.05$) in the probiotic group compared to the placebo group at the endpoint (Table 5).

## 4. Discussion

In the present study, the selected recipe of VFJ was analyzed in terms of its properties, especially the antioxidant activity, which was related to dyslipidemia and oxidative stress in patients with dyslipidemia [23]. The nutritional value of VFJ, including total dietary fiber, soluble dietary fiber, insoluble dietary fiber, vitamins and minerals, total polyphenols, flavonoids, and carotenoids makes it a potential alternative to conventional medicine for lowering dyslipidemia. Indeed, many reports have verified its ability to reduce dyslipidemia risk by altering the TC, LDL-C, HDL-C, and TG levels. Specifically, soluble dietary fiber has been shown to reduce TC and LDL-C levels [24,25], potentially by increasing BA excretion, thus delaying reabsorption in the intestine [26]. Additionally, the production of short-chain fatty acids by microbial fermentation of dietary fiber in the colon lowers LDL-C levels [27].

Several studies have reported that probiotic bacteria can improve lipid profiles. Fuentes et al. [28] studied the efficacy of *L. plantarum* capsule (containing $1.2 \times 10^9$ CFU of a mixture strain CECT 7527, CECT 7528, and CECT 7529) in hypercholesterolemic patients for 12 weeks. The level of TC in the *L. plantarum* group significantly decreased to 13.6% compared to a placebo group. Ahn et al. [29] suggested that the consumption of a mixed probiotic powder (*L. curvatus* HY7601 and *L. plantarum* KY1032) for 12 weeks decreased 18.3% of TG level compared to a group that consumed the powder without probiotic strains.

Therefore, the VFJ and VFJ containing probiotic bacteria may present useful complementary medicine for patients with dyslipidemia presenting with mild symptoms. In particular, the VFJ offers the benefit of not requiring heat during the preparation process, thereby retaining the full nutrition content of the ingredients. The consumption of VFJ with and without probiotic *L. paracasei* on physical parameters was discussed in term of BW, BMI, and WC. WC was found to significantly decrease in the probiotic group on day 30 compared to baseline. Our results are consistent with those of Zhang et al. [30], who showed that obese participants exhibited a significant reduction in WC after probiotic consumption. Meanwhile, no effect was observed on BM or BMI. Similarly, in a study by Michael et al. [31], obese participants showed reduced WC compared to the placebo group participants after consuming a mixture of probiotic *Lactobacillus* and *Bifidobacterium* (50 billion per day) for six months.

The result of this study showed that the lipid profile of the placebo group (consumption of VFJ without probiotic *L. paracasei*) was not significantly different between baseline and endpoint. However, the lipid profile of the probiotic group (consumption of VFJ with probiotic *L. paracasei*) showed that the level of TC, LDL-C, TG, and TG/HDL-C ratio was significantly declined to 5.59, 6.00, 24.14, and 30.83%, respectively, while the level of

HDL-C was significantly increased to 8.33% at the endpoint, when compared with the baseline. Overall, the VFJ containing probiotic *L. paracasei* could improve lipid profile in the participant with dyslipidemia. This might be associated with bile salt hydrolase (BSH) produced by probiotics, which is the enzymatic deconjugation of bile acids by hydrolysis of conjugated bile acids into deconjugated bile acids [32]. Deconjugated bile acids are less soluble and are reabsorbed in the intestinal lumen less than conjugated bile acids, leading to the excretion of larger amounts of free bile acids in feces [30]. Thus, the deconjugation of bile acids results in a reduction in serum cholesterol by increasing the demand of cholesterol for de novo synthesis of bile acids to replace those lost in feces [33]. The outcomes of lipid profile are related to the previous literature of Mohamadshahi et al. [34]. They studied the consumption of conventional yogurt and probiotic yogurt (containing *L. acidophilus* La-5 and *B. lactis* Bb-12) in diabetic patients. The subjects had a daily intake of 330 g of yogurt for 8 weeks. The result was confirmed that the probiotic yogurt caused a decrease in LDL-C/HDL-C ratio and significantly increased HDL-C levels. Oxidative stress is an imbalance in the cells between the system of reactive oxygen species (or free radicals); production and accumulation lead to the damage of cellular structures such as lipids, proteins, and nucleic acids [35]. Therefore, the harmful effects can respond with dyslipidemia. The lipid peroxidation is one of the possible outcomes of aberrant free radicals, of which MDA is the primary marker. In the current study, the MDA level was significantly decreased to 68.42% in the probiotic group, from $0.19 \pm 0.07$ mg/dL at the baseline to $0.06 \pm 0.01$ mg/dL at the endpoint. These findings agree with the results of a previous study that reported decreased MDA levels in the serum and liver of hyperlipidemic rats following *L. casei* supplementation [36]. Moreover, the ability of an antioxidant defense system is based on main enzymatic components, including CAT, GPx, and SOD. For the probiotic group, the levels of CAT and GPx were significantly increased to 8.07 and 31.75% at Day 30 compared to Day 0. However, the level of SOD was an insignificant difference after the endpoint compared to the baseline of the study period. Similarly, within a clinical trial, Chamari et al. [37] reported that probiotic yogurt decreased oxidative stress in healthy women within a clinical trial. That is, intake of the probiotic yogurt for six weeks significantly increased CAT activity compared to the normal yogurt. Kleniewska et al. [26] found that the supplement containing probiotic *L. casei* ($4 \times 10^8$ CFU) and prebiotic Inulin (400 mg) might be an advantageous influence on the antioxidant properties of healthy human plasma after 7 weeks. Briefly, the symbiotic could be a significant increase in CAT activity; however, the activity of SOD and GPx was an insignificant increase compared to a control group. The result was slightly dissimilar because of the difference in probiotic bacterial strain. These results could show that VFJ with probiotic bacteria affected oxidative stress. Several mechanisms explained the probiotics affecting antioxidant activity and reduced damages caused by oxidation. The probiotics have their antioxidant enzymatic system, especially SOD. Furthermore, the probiotic strain was able to provoke the system of antioxidation in the host and increased the antioxidant enzymatic activity. As a result, they can prevent or decrease the seriousness of intestinal pathology caused by reactive oxygen species [38]. Moreover, the metabolites of antioxidant activity were produced by probiotics. For example, butyrate, as SCFA, was generated by the fermentation of microbiota in the small intestine and/or a final section of the small intestine led to induce antioxidative enzymes [39]. Wang et al. [38] reported that the probiotics *Lactobacillus* and *Bifidobacterium* were able to produce lactic acid, acetic acid, and propionic acid, resulting in a lower intestinal pH, maintaining a balance of the gut microbiota. The probiotics can also regulate the growth of harmful bacteria, which may contribute to reduced oxidative stress.

BAs or bile salts are synthesized from cholesterol in the liver and, subsequently, become conjugated to glycine or taurine, which increases their solubility [40,41]. Approximately 200–600 mg of BA is produced daily in the human liver and excreted in feces [40]. The alteration in the BA level of the probiotic group was significant increased by 19.40%. Thus, based on the marked increase in BAs within the probiotic group of the current study, it could be inferred that the BSH activity of *L. paracasei* reduced the TC, LDL-C, TG, and

TG/HDL-C ratio, resulting in increased fecal BAs [33,42]. The BAs that are lost in the feces are then regenerated from cholesterol, via de novo synthesis, in the liver to maintain a constant BA pool [40,41], thus, reducing serum cholesterol levels [33].

The level of lipid profile, lipid peroxidation, oxidative stress enzyme, and bile acid, especially the TG/HDL-C ratio, can indicate a risk rate of dyslipidemia [43]. These available results could be explained in more details. The treatment period should be extended to more than 30 days and/or the amount of VFJ and probiotic *L. paracasei* should be adjusted to a higher level: more than 500 mL/day and $10^9$ CFU/day, respectively. Further research is warranted to investigate the impact of the consumption of VFJ with probiotics on patients who have higher LDL-C levels or who have been administered lipid-lowering drugs to prove the effectiveness of the juice with probiotics combined with a medical treatment.

## 5. Conclusions

According to the present study, the intake of VFJ with *L. paracasei* at the dosage of $2 \times 10^9$ CFU/day for 30 days could serve as an effective alternative strategy for the primary prevention of dyslipidemia in Bhumibol Adulyadej Hospital patients who had not undergone medical treatment. Specifically, intake of VFJ with *L. paracasei* could provide health benefits by improving lipid profiles, lipid peroxidation, and oxidative stress enzyme activity levels. The level of TC, LDL-C, TG, and TG/HDL-C ratio significantly decreased in the probiotic group (consumption of VFJ with *L. paracasei*) compared to the placebo group (consumption of VFJ without *L. paracasei*). While the HDL-C marker was a high level in the probiotic group. Moreover, the level of MDA in the probiotic group was significantly decreased compared to the placebo group. However, the activity of CAT and GPx significantly increased in the participant group of VFJ with *L. paracasei* consumption. The BA level in feces significantly increased in the probiotic group. Therefore, intake of VFJ with *L. paracasei* could be improved to benefit health in terms of the lipid profile, lipid peroxidation and oxidative stress enzymes.

**Author Contributions:** Conceptualization, P.S., S.S. and C.C.; methodology, P.S., S.S., N.L. and C.C.; formal analysis, P.S.; data curation, P.S.; investigation, P.S., S.S., N.L., C.C., N.M. and E.K.; writing—original draft preparation, P.S.; writing—review and editing, P.S., S.S., N.L., C.C. and N.M.; supervision, S.S., N.L. and C.C.; funding acquisition, P.S., S.S. and C.C.; project administration, S.S. All authors have read and agreed to the published version of the manuscript.

**Funding:** This research was funded by the Research and Invention Center of Bhumibol Adulyadej Hospital, Bangkok, Thailand, and the Faculty of Pharmacy, Chiang Mai University, Chiang Mai, Thailand. Part of this research was supported by the Innovation Center for Holistic Health, Nutraceuticals, and Cosmeceuticals, Faculty of Pharmacy, Chiang Mai University, Chiang Mai, Thailand.

**Institutional Review Board Statement:** Not applicable.

**Informed Consent Statement:** Not applicable.

**Data Availability Statement:** The data presented in this study are available within the article.

**Acknowledgments:** The authors would like to express their gratitude to Bhumibol Adulyadej Hospital, Royal Thai Air Force, Bangkok, Thailand and Chiang Mai University, Chiang Mai, Thailand for research support. This research was partially supported by Chiang Mai University.

**Conflicts of Interest:** The authors declare no conflict of interest.

## References

1. Shenoy, C.; Shenoy, M.M.; Rao, G.K. Dyslipidemia in dermatological disorders. *N. Am. J. Med. Sci.* **2015**, *7*, 421. [CrossRef]
2. Narindrarangkura, P.; Bosl, W.; Rangsin, R.; Hatthachote, P. Prevalence of dyslipidemia associated with complications in diabetic patients: A nationwide study in Thailand. *Lipids Health Dis.* **2019**, *18*, 90. [CrossRef] [PubMed]
3. Lertwanichwattana, T.; Rangsin, R.; Sakboonyarat, B. Prevalence and associated factors of uncontrolled hyperlipidemia among Thai patients with diabetes and clinical atherosclerotic cardiovascular diseases: A cross-sectional study. *BMC Res. Notes* **2021**, *14*, 118. [CrossRef] [PubMed]

4. World Health Organization. *Diet, Nutrition, and the Prevention of Chronic Diseases: Report of a Joint WHO/FAO Expert Consultation (Vol. 916)*; World Health Organization: Geneva, Switzerland, 2003.

5. Reiner, Z.; Catapano, A.L.; De Backer, G.; Graham, I.; Taskinen, M.R.; Wiklund, O.; Agewall, S.; Alegria, E.; Chapman, M.J.; Durrington, P.; et al. ESC/EAS Guidelines for the management of dyslipidaemias: The Task Force for the management of dyslipidaemias of the European Society of Cardiology (ESC) and the European Atherosclerosis Society (EAS). *Eur. Heart J.* **2011**, *32*, 1769–1818. [PubMed]

6. Kelly, R.B. Diet and exercise in the management of hyperlipidemia. *Am. Fam. Physician* **2010**, *81*, 1097–1102. [PubMed]

7. Suwimol, S.; Pimpanit, L.; Aporn, M.; Pichita, S.; Ratiyaporn, S.; Wiroj, J. Impact of fruit and vegetables on oxidative status and lipid profiles in healthy individuals. *Food Public Health* **2012**, *2*, 113–118. [CrossRef]

8. Jideani, A.I.; Silungwe, H.; Takalani, T.; Omolola, A.O.; Udeh, H.O.; Anyasi, T.A. Antioxidant-rich natural fruit and vegetable products and human health. *Int. J. Food Prop.* **2021**, *24*, 41–67. [CrossRef]

9. Van Duyn, M.A.S.; Pivonka, E. Overview of the health benefits of fruit and vegetable consumption for the dietetics professional: Selected literature. *J. Am. Diet. Assoc.* **2000**, *100*, 1511–1521. [CrossRef]

10. Maron, D.J. Flavonoids for reduction of atherosclerotic risk. *Curr. Atheroscler. Rep.* **2004**, *6*, 73–78. [CrossRef]

11. Liu, R.H. Health-promoting components of fruits and vegetables in the diet. *Adv. Nutr.* **2013**, *4*, 384S–392S. [CrossRef]

12. Guan, Q.; Xiong, T.; Xie, M. Influence of probiotic fermented fruit and vegetables on human health and the related industrial development trend. *Engineering* **2021**, *7*, 212–218. [CrossRef]

13. Tsuda, H.; Miyamoto, T. Production of Exopolysaccharide by *Lactobacillus plantarum* and the Prebiotic Activity of the Exopolysaccharide. *Food Sci. Technol. Res.* **2010**, *16*, 87–92. [CrossRef]

14. Markowiak, P.; Śliżewska, K. Effects of probiotics, prebiotics, and synbiotics on human health. *Nutrients* **2017**, *9*, 1021. [CrossRef] [PubMed]

15. Chiu, C.H.; Lu, T.Y.; Tseng, Y.Y.; Pan, T.M. The effects of *Lactobacillus*-fermented milk on lipid metabolism in hamsters fed on high-cholesterol diet. *Appl. Microbiol. Biotechnol.* **2006**, *71*, 238–245. [CrossRef] [PubMed]

16. Dehkohneh, A.; Jafari, P.; Fahimi, H. Effects of probiotic *Lactobacillus paracasei* TD3 on moderation of cholesterol biosynthesis pathway in rats. *Iran. J. Basic Med. Sci.* **2019**, *22*, 1004.

17. Saenjum, C.; Chaiyasut, C.; Chansakaow, S.; Suttajit, M.; Sirithunyalug, B. Antioxidant and anti-inflammatory activities of gamma-oryzanol rich extracts from Thai purple rice bran. *J. Med. Plant Res.* **2012**, *6*, 1070–1077.

18. Zeb, A.; Ullah, F. A simple spectrophotometric method for the determination of thiobarbituric acid reactive substances in fried fast foods. *J. Anal. Methods Chem.* **2016**, *2016*, 9412767. [CrossRef]

19. Atasayar, S.; Orhan, H.; Özgüneş, H. Malondialdehyde quantification in blood plasma of tobacco smokers and non-smokers. *FABAD J. Pharm. Sci.* **2004**, *29*, 15–19.

20. Kaya, S.; Sütçü, R.; Cetin, E.S.; Arıdogan, B.C.; Delibaş, N.; Demirci, M. Lipid peroxidation level and antioxidant enzyme activities in the blood of patients with acute and chronic fascioliasis. *Int. J. Infect. Dis.* **2007**, *11*, 251–255. [CrossRef]

21. Rush, J.W.; Sandiford, S.D. Plasma glutathione peroxidase in healthy young adults: Influence of gender and physical activity. *Clin. Biochem.* **2003**, *36*, 345–351. [CrossRef]

22. Vijayvargiya, P.; Camilleri, M.; Shin, A.; Saenger, A. Methods for diagnosis of bile acid malabsorption in clinical practice. *Clin. Gastroenterol. Hepatol.* **2013**, *11*, 1232–1239. [CrossRef] [PubMed]

23. Khutami, C.; Sumiwi, S.A.; Khairul Ikram, N.K.; Muchtaridi, M. The Effects of Antioxidants from Natural Products on Obesity, Dyslipidemia, Diabetes and Their Molecular Signaling Mechanism. *Int. J. Mol. Sci.* **2022**, *23*, 2056. [CrossRef] [PubMed]

24. Slavin, J. Fiber and prebiotics: Mechanisms and health benefits. *Nutrients* **2013**, *5*, 1417–1435. [CrossRef] [PubMed]

25. Brown, L.; Rosner, B.; Willett, W.W.; Sacks, F.M. Cholesterol-lowering effects of dietary fiber: A meta-analysis. *Am. J. Clin. Nutr.* **1999**, *69*, 30–42. [CrossRef] [PubMed]

26. Kleniewska, P.; Hoffmann, A.; Pniewska, E.; Pawliczak, R. The influence of probiotic *Lactobacillus casei* in combination with prebiotic inulin on the antioxidant capacity of human plasma. *Oxid. Med. Cell. Longev.* **2016**, *2016*, 1340903. [CrossRef] [PubMed]

27. Surampudi, P.; Enkhmaa, B.; Anuurad, E.; Berglund, L. Lipid lowering with soluble dietary fiber. *Curr. Atheroscler. Rep.* **2016**, *18*, 1–13. [CrossRef] [PubMed]

28. Fuentes, M.C.; Lajo, T.; Carrión, J.M.; Cuné, J. Cholesterol-lowering efficacy of *Lactobacillus plantarum* CECT 7527, 7528 and 7529 in hypercholesterolaemic adults. *Br. J. Nutr.* **2013**, *109*, 1866–1872. [CrossRef]

29. Ahn, H.Y.; Kim, M.; Chae, J.S.; Ahn, Y.T.; Sim, J.H.; Choi, I.D.; Lee, S.H.; Lee, J.H. Supplementation with two probiotic strains, *Lactobacillus curvatus* HY7601 and *Lactobacillus plantarum* KY1032, reduces fasting triglycerides and enhances apolipoprotein AV levels in non-diabetic subjects with hypertriglyceridemia. *Atherosclerosis* **2015**, *241*, 649–656. [CrossRef]

30. Zhang, Y.; Yan, T.; Xu, C.; Yang, H.; Zhang, T.; Liu, Y. Probiotics can further reduce waist circumference in adults with morbid obesity after bariatric surgery: A systematic review and meta-analysis of randomized controlled trials. *Evid. Based Complement. Altern. Med.* **2021**, *2021*. [CrossRef]

31. Michael, D.R.; Jack, A.A.; Masetti, G.; Davies, T.S.; Loxley, K.E.; Kerry-Smith, J.; Plummer, J.F.; Marchesi, J.R.; Mullish, B.H.; McDonald, J.A.K.; et al. A randomised controlled study shows supplementation of overweight and obese adults with lactobacilli and bifidobacteria reduces bodyweight and improves well-being. *Sci. Rep.* **2020**, *10*, 4183. [CrossRef]

32. Ooi, L.G.; Liong, M.T. Cholesterol-lowering effects of probiotics and prebiotics: A review of in vivo and in vitro findings. *Int. J. Mol. Sci.* **2010**, *11*, 2499–2522. [CrossRef] [PubMed]

33. Kumar, M.; Nagpal, R.; Kumar, R.; Hemalatha, R.; Verma, V.; Kumar, A.; Chakraborty, C.; Singh, B.; Marotta, F.; Jain, S.; et al. Cholesterol-lowering probiotics as potential biotherapeutics for metabolic diseases. *Exp. Diabetes Res.* **2012**, *2012*, 902917. [CrossRef] [PubMed]
34. Mohamadshahi, M.; Veissi, M.; Haidari, F.; Javid, A.Z.; Mohammadi, F.; Shirbeigi, E. Effects of probiotic yogurt consumption on lipid profile in type 2 diabetic patients: A randomized controlled clinical trial. *J. Res. Med. Sci.* **2014**, *19*, 531. [PubMed]
35. Pizzino, G.; Irrera, N.; Cucinotta, M.; Pallio, G.; Mannino, F.; Arcoraci, V.; Squadrito, F.; Altavilla, D.; Bitto, A. Oxidative stress: Harms and benefits for human health. *Oxid. Med. Cell. Longev.* **2017**, *2017*, 8416763. [CrossRef] [PubMed]
36. Zhang, Y.; Du, R.; Wang, L.; Zhang, H. The antioxidative effects of probiotic *Lactobacillus casei* Zhang on the hyperlipidemic rats. *Eur. Food Res. Technol.* **2010**, *231*, 151–158. [CrossRef]
37. Chamari, M.; Djazayery, A.; Jalali, M.; Sadrzadeh yeganeh, H.; Hosseini, S.; Heshmat, R.; Behbahani Haeri, B. The effect of daily consumption of probiotic and conventional yogurt on some oxidative stress factors in plasma of young healthy women. *ARYA Atheroscler.* **2008**, *4*, 175–179.
38. Wang, Y.; Wu, Y.; Wang, Y.; Xu, H.; Mei, X.; Yu, D.; Wang, Y.; Li, W. Antioxidant properties of probiotic bacteria. *Nutrients* **2017**, *9*, 521. [CrossRef]
39. Kau, A.L.; Ahern, P.P.; Griffin, N.W.; Goodman, A.L.; Gordon, J.I. Human nutrition, the gut microbiome, and immune system: Envisioning the future. *Nature* **2011**, *474*, 327–336. [CrossRef]
40. Chiang, J.Y. Bile acid metabolism and signaling. *Compr Physiol.* **2013**, *3*, 1191–1212.
41. Di Ciaula, A.; Garruti, G.; Lunardi, B.R.; Molina-Molina, E.; Bonfrate, L.; Wang, D.Q.H. Bile Acid Physiology. *Ann. Hepatol.* **2017**, *16*, S4–S14. [CrossRef]
42. Bourgin, M.; Kriaa, A.; Mkaouar, H.; Mariaule, V.; Jablaoui, A.; Maguin, E.; Rhimi, M. Bile Salt Hydrolases: At the Crossroads of Microbiota and Human Health. *Microorganisms* **2021**, *9*, 1122. [CrossRef] [PubMed]
43. Luz, P.L.D.; Favarato, D.; Faria-Neto Junior, J.R.; Lemos, P.; Chagas, A.C.P. High ratio of triglycerides to HDL-cholesterol predicts extensive coronary disease. *Clinics* **2008**, *63*, 427–432. [CrossRef] [PubMed]

*Communication*

# The Potential Effect of Elevated Root Zone Temperature on the Concentration of Chlorogenic, Caffeic, and Ferulic acids and the Biological Activity of Some Pigmented *Solanum tuberosum* L. Cultivar Extracts

Hildegard Witbooi [1], Callistus Bvenura [1], Anna-Mari Reid [2], Namrita Lall [2,3], Oluwafemi Omoniyi Oguntibeju [4] and Learnmore Kambizi [1,*]

[1] Department of Horticulture, Faculty of Applied Sciences, Cape Peninsula University of Technology, Bellville 7535, South Africa; hildegardwitbooi@gmail.com (H.W.); bvenurac@cput.ac.za (C.B.)

[2] Department of Plant & Soil Sciences, Medicinal Plant Sciences, University of Pretoria, Pretoria 0002, South Africa; u29086354@tuks.co.za (A.-M.R.); namrita.lall@up.ac.za (N.L.)

[3] School of Natural Resources, University of Missouri, Columbia, MO 65211, USA

[4] Phytomedicine & Phytochemistry Group, Oxidative Stress Research Centre, Department of Biomedical Sciences, Faculty of Health and Wellness Sciences, Cape Peninsula University of Technology, Bellville 7530, South Africa; Oguntibejuo@cput.ac.za

* Correspondence: kambizil@cput.ac.za

**Citation:** Witbooi, H.; Bvenura, C.; Reid, A.-M.; Lall, N.; Oguntibeju, O.O.; Kambizi, L. The Potential Effect of Elevated Root Zone Temperature on the Concentration of Chlorogenic, Caffeic, and Ferulic acids and the Biological Activity of Some Pigmented *Solanum tuberosum* L. Cultivar Extracts. *Appl. Sci.* **2021**, *11*, 6971. https://doi.org/10.3390/app11156971

Academic Editors: Luca Mazzoni, Maria Teresa Ariza Fernández and Franco Capocasa

Received: 16 June 2021
Accepted: 15 July 2021
Published: 29 July 2021

**Publisher's Note:** MDPI stays neutral with regard to jurisdictional claims in published maps and institutional affiliations.

**Copyright:** © 2021 by the authors. Licensee MDPI, Basel, Switzerland. This article is an open access article distributed under the terms and conditions of the Creative Commons Attribution (CC BY) license (https://creativecommons.org/licenses/by/4.0/).

**Abstract:** Without a doubt, potatoes play a vital food and nutrition security role in the world as more than a billion people consume this vegetable. Furthermore, the polyphenolic constituents of pigmented potato cultivars and their associated health benefits have been reported. However, the antioxidant, anticancer, and antimycobacterial activity of pigmented cultivars are scanty. Therefore, the present study explores the phenolic acids and biological activities of cv. Salad Blue (SB) and non-pigmented control (BP1) extracts. The antiproliferative activity of *S. tuberosum* L. against human hepatocellular carcinoma (HepG2) was investigated, as well as the ability to inhibit *Mycobacterium smegmatis*. Chlorogenic acid was the most prominent phenolic acid in both treatments as well as cultivars. In the current trial, 24 °C significantly increased chlorogenic acid in cv. SB and BP1. Ethanolic extracts of all the samples showed no activity at the highest test concentration of 1000 µg/mL (ciprofloxacin MIC of 0.325 µg/mL) against *M. smegmatis*. The antiproliferative activity of the tuber samples against HepG2 liver cells had IC$_{50}$ values ranging between 267.7 ± 36.17 µg/mL and >400 µg/mL. Since the health benefits of these cultivars are highly valued, the present study provides useful information for future oncology studies, for human nutrition, as well as for how these underutilized cultivars can be fortified to improve their health benefits.

**Keywords:** *Solanum tuberosum*; antimycobacterial; antioxidant capacity; hepatocellular carcinoma; pigmented potatoes

## 1. Introduction

Of the crops that feed the world, potatoes are the third most consumed after rice and wheat, with more than a billion people relying on them for food and nutrition security [1]. This is because they are an important staple crop that is well endowed with complex carbohydrates and thus very high in energy, in addition to other nutritional benefits. This crop is easy to cultivate under a diverse range of climatic conditions except those found in Antarctica [2]. Latest statistics indicate that by 2019, over 17 million hectares were under potato cultivation. Global production also increased from 334.73 to 370.43 million metric tons (MMT) between 2009 and 2019 [2]. Current data [3] further show that global leaders in potato production by MMT are the following: China (91.92) > India (50.19) > Russia (22.07) > Ukraine (20.27) > USA (19.18) > Germany (10.60) > Bangladesh (9.65) > France (8.56) > Netherlands (6.96) > Poland (6.48).

In South Africa, potatoes also play a major food and nutrition security role, with about 677.46 km$^2$ under potato cultivation in the 2016/2017 agricultural season [4]. Furthermore, during the 2016/2017 agricultural season, about 2.5 MMT of potatoes were produced at a rate of approximately 3,617,200 kg/km$^2$ [4]. According to the same statistics, the average per capita consumption of potatoes in South Africa is 30 kg.

The usefulness of the potato peel, which is regarded as a waste product, has grown [5,6]. Due to its high antioxidant and antimicrobial effectiveness [7], it is used in food preservation. The potato peel has also been reported as a pharmaceutical ingredient in wound management [8,9], and glycoalkaloids in the leaves and other parts of the plant have been reported to offer natural protection against plant pests [10,11].

In the world today, the demand for polyphenolic and vitamin rich food sources is on the rise, and pigmented potatoes, among other crops, often possess these constituents [12].

Potatoes are rich in phenolic compounds; for example, about 49% to 90% of the total phenolics in this vegetable are in fact chlorogenic acid [13,14]. In addition, as shown in some epidemiological studies, a clear correlation has been established between consuming phenolic rich diets and better health [15–18]. Nevertheless, the antioxidant activity and some health promoting aspects of plant-derived phenolic compounds cannot be overemphasized. Anthocyanins in some pigmented potato cultivars have been reported to suppress stomach cancer in mice [19], prostate cancer cells [20], colon cancer cells [21], as well as liver cancer cells [22]. Furthermore, glycoalkaloids in these cultivars have been shown to inhibit the growth of colon, liver, stomach, and lymphoma cancer cells, among others [23,24]. However, the potential role of root zone temperature (RZT) on antioxidant, anticancer, as well as antimycobacterial activity of pigmented cultivar extracts has not been reported. We previously reported root zone temperature's role on physiological growth and polyphenolic contents in both pigmented and non-pigmented potato cultivars [25]. The results of this study showed the polyphenolic superiority of pigmented cultivars under a diverse range of root zone temperatures over the non-pigmented cultivar. Therefore, as an offshoot of this study, we tested the antiproliferative activity of the extracts of both the pigmented and non-pigmented cultivars against liver hepatocellular carcinoma (HepG2) cells and their antimycobacterial activity against *M. smegmatis*. In addition, chlorogenic and caffeic acid were significantly higher in SB under 24 °C in our previous study, and this informed our decision for the selected temperature and cultivar of the current study.

## 2. Materials and Methods

### 2.1. Plant Growth, Harvest, and Extraction

Potato tubers of cv. BP1 and SB were cultivated in a greenhouse, as described by [25], for 73 days to test the effect of controlled RZT (24 °C) and non-controlled RZT (ranging between 19 °C and 25 °C). Postharvest, the samples were separated into flesh and skin, and frozen at −80 °C in paper bags, after which they were freeze dried for 24 h in a VirTis genesis wizard 2.0 (UK). The samples were then powdered and sieved through a 40–60 mesh and stored at 4 °C until further use. About 200 g of the powdered and lyophilized tubers were extracted with 20 × the volume of 60% ethanol (EtOH) (Absolute; B&M Scientific) per mass (10 g > 200 mL) overnight and then ultra-sonicated for approximately 15 min at 40 °C. Samples were filtered using 0.22 µm syringe filters (25 mm) and concentrated using the Genevac miVac sample concentrator.

### 2.2. Phytochemical Analysis

A Dionex HPLC (Dionex Softron, Germering, Germany) was used to determine chlorogenic, caffeic, and ferulic acids in the samples. This HPLC has a Brucker ESI Q-TOF MS coupled autosampler and is equipped with a binary solvent manager. A reversed chromatography on a Thermo Fisher Scientific C18 column 5 µm, 4.6 × 150 mm (Bellefonte, PA, USA) was used to separate plant extract constituents via the use of a linear gradient of 0.1% formic acid in acetonitrile (solvent A) and water (solvent B) at 0.8 mL min$^{-1}$ flow rate, an electrospray voltage of +3500 V, an oven temperature of 30 °C, and a 10 µL injection volume. The negative mode was used to acquire the MS spectra. The nebulizer gas was set at 35 psi and the dry gas to 9 L min$^{-1}$ at 300 °C.

### 2.3. Antimycobacterial Activity of the Ethanolic Extract of Solanum tuberosum L.

*Mycobacterium smegmatis* (*M. smegmatis*) is a non-pathogenic and fast-growing species of mycobacterium. This model is most used in the physiology of mycobacteria, as it has relevance to the pathogenic species of *Mycobacterium tuberculosis*, the causative pathogen for tuberculosis. The minimum inhibitory concentration (MIC) values were determined according to [26], with slight modifications.

A stock solution of 20% DMSO (Sigma-Aldrich, Saint Louis, MO, USA) was used to dissolve all tuber ethanolic extracts in Sterile Middlebrook 7H9 media. Furthermore, using this sterile media, two-fold dilutions were made of each sample into a final assay yield volume of 200 μL. Ciprofloxacin (Sigma-Aldrich, Saint Louis, MO, USA) served as a positive drug control at a concentration range of 0.156–10 μg/mL. The solvent control (DMSO 2%) as well as the untreated bacterial control were carried out in triplicates. The plates were sealed using parafilm before incubation at 37 °C for 24 h. After incubation of 24 h, PrestoBlue (Thermofischer, South Africa) as a viability indicator was added to each well (20 μL). The minimum inhibitory concentration (MIC) values were defined as the concentration at which no color change was visible from blue to pink.

### 2.4. Antiproliferative Activity

The antiproliferative activity assay was carried out according to the method of [27]. In all, 100 μL of HepG2 cells with a cell density of 10,000 cells per well was seeded in 96 well plates after careful counting and left at 5% $CO_2$ and 37 °C overnight to incubate in order to allow for attachment. To prepare each sample, a stock solution of 2000 μg/mL was used. A final test concentration of 12.5 to 400 μg/mL in serial dilutions of the sample extracts in the cell-containing plates was made. The plates were then incubated for 72 h at 37 °C and 5% $CO_2$. Actinomycin D was used as the positive control (0.02 to 0.5 μg/mL) and DMSO at 2% as the solvent control. After incubation, PrestoBlue was added (20 μL) to each well, and the plates were left to incubate for a further 2–4 h. After incubation, the absorbance values were read (490 nm wavelength, including a reference wavelength of 690 nm) using a BIO-TEK Power Wave XS multi-well reader. The mean 50% inhibitory values (IC$_{50}$) were calculated, and statistical analysis was performed.

### 2.5. Statistical Analysis

Data were collected on 52 samples (13 plants per cultivar) per treatment. Statistically significant differences among treatment means were determined by two-way analysis of variance (ANOVA) at $p < 0.05$. Fisher's least significant difference (LSD) test was used to segregate means that were significantly different using a computer software program called STATISTICA (Palo Alto, California, USA). The mean IC$_{50}$ values (three replicates) were used to perform statistical analysis using GraphPad Prism (Version 7, San Diego, CA, USA) and two-way ANOVA. To identify significance in comparison to the control value, the Dunnett's MCT was performed. All experiments were conducted in triplicates.

## 3. Results

### 3.1. Caffeic, Chlorogenic, and Ferulic acid Content in Solanum tuberosum L. Exposed to Higher Root Zone Temperature

The results of the present study revealed the presence of chlorogenic, caffeic, and ferulic acids in the ethanolic extracts of the potato tuber cultivars, as shown in Table 1. The chromatographic peaks in the result profiles showed some variations in the mean concentrations among the two root zone temperatures and the cultivars. Chlorogenic acid was the most prominent phenolic acid in both treatments and cultivars. Cultivar BP1 flesh and skins increased chlorogenic acid by 13% and 26%, respectively, on exposure to an RZT of 24 °C. Similarly, cv. SB flesh and skins increased by 28% and 46%, respectively, on exposure to an RZT of 24 °C. Although the set RZT of 24 °C significantly lowered caffeic acid in cv. BP1 skins and flesh (0.29–0.03 µg/g), the same RZT significantly increased this specific phenolic acid in the cv. SB skins (0.39 µg/g) and the flesh (0.05 µg/g). Interestingly, the control temperature (no heat applied) significantly increased the concentration of caffeic acid in BP1 skins (0.367 µg/g) and decreased the concentration in the flesh (0.051 µg/g). In addition, as shown in Table 1, at 24 °C the concentration of chlorogenic acid in cv. BP1 skins was lowered to 0.530 µg/g and in the BP1 flesh to 0.531 µg/g; however, a set temperature of 24 °C had the ability to increase the concentration of chlorogenic acid in cv. SB skins (0.779 µg/g) and the flesh (0.707 µg/g). The control temperature significantly lowered the concentration of the chlorogenic acid in both cultivars. Ferulic acid was present, but in very low concentrations only in cv. SB. Using a two-way analysis of variance, a strong interaction was established between the specific cultivar and RZT on chlorogenic and caffeic acid contents in the present trial.

**Table 1.** The effect of root zone temperature on the phenolic acid content in *S. tuberosum* cv. BP1 and Salad blue.

| | Caffeic Acid (µg/g) | | Chlorogenic Acid (µg/g) | | Ferulic Acid (µg/g) | |
|---|---|---|---|---|---|---|
| | **Control** | **24 °C** | **Control** | **24 °C** | **Control** | **24 °C** |
| BP1 Skins | 0.37 ± 0.004 [aA] | 0.29 ± 0.003 [bB] | 0.39 ± 0.006 [bC] | 0.53 ± 0.002 [aC] | 0.00 ± 0.001 | 0.00 ± 0.001 |
| BP1 Flesh | 0.05 ± 0.001 [C] | 0.03 ± 0.000 [C] | 0.46 ± 0.050 [bB] | 0.53 ± 0.003 [aC] | 0.00 ± 0.001 | 0.00 ± 0.001 |
| SB Skins | 0.25 ± 0.003 [bB] | 0.39 ± 0.005 [aA] | 0.42 ± 0.005 [bBC] | 0.78 ± 0.014 [aA] | 0.01 ± 0.001 | 0.01 ± 0.001 |
| SB Flesh | 0.05 ± 0.002 [C] | 0.05 ± 0.001 [C] | 0.51 ± 0.007 [bA] | 0.71 ± 0.007 [aB] | 0.01 ± 0.001 | 0.01 ± 0.001 |

Values represent mean ± SD. Different small letters along the row per block represent significant differences at $p < 0.05$ and different capital letters down the column represent significant differences at $p < 0.05$. No heat was applied to the control. BP1 = Non-pigmented control; SB = Salad Blue.

### 3.2. Antimycobacterial Activity

The antimycobacterial activity of *Solanum tuberosum* (ethanol extracts) of both cultivars and treatments of BP1 and SB was investigated. The ethanolic extracts of all the tested samples did not show activity at the highest test concentration of 1000 µg/mL, as shown in Table 2. The positive drug control ciprofloxacin showed an MIC value of 0.325 µg/mL.

**Table 2.** Antimycobacterial activity against *M. smegmatis* (MIC µg/mL).

| | **Antimycobacterial Activity against *M. smegmatis* (MIC µg/mL)** |
|---|---|
| Control SB | NA |
| 24 °C SB | NA |
| Control BP1 | NA |
| 24 °C BP1 | NA |
| **Controls** | |
| Ciprofloxacin | 0.325 |

NA—Not Active at the highest test concentration of 1000 µg/mL.

### 3.3. Antiproliferative Assay

The antiproliferative ethanolic extract activity of *S. tuberosum* L. cultivars SB and BP1 subjected to two RZTs was tested against HepG2 liver cells. The IC$_{50}$ values of the samples ranged between 267.7 ± 36.17 µg/mL and >400 µg/mL, following 72 h of incubation as shown in Table 3. According to [28], after 72 h of incubation, plant extracts with IC$_{50}$ values greater than 100 µg/mL are non-cytotoxic to the particular cell line. However, there is an increase in activity when SB and BP1 varieties are compared.

**Table 3.** Antiproliferative activity against Hepatocellular carcinoma cells (HepG2) (IC50 µg/mL).

| | Antiproliferative Activity against Hepatocellular Carcinoma Cells (HepG2) (IC$_{50}$ µg/mL) |
|---|---|
| Control SB | 267.7 ± 36.17 |
| 24 °C SB | 290.8 ± 39.35 |
| Control BP1 | 393.0 ± 34.17 |
| 24 °C BP1 | NA |
| **Controls** | |
| Actinomycin | 0.49 ± 15.91 |

NA—Not Active at the highest test concentration of 1000 µg/mL.

## 4. Discussion

### 4.1. Caffeic, Chlorogenic, and Ferulic acid Content in Solanum tuberosum L. Exposed to Higher Root Zone Temperature

The results of the current study confirmed what was seen in the study conducted previously by [29], which showed that chlorogenic acid concentration was significantly lower in yellow-fleshed potatoes in comparison with the high values reported in colored potatoes. Furthermore, an elevated RZT showed minimal chlorogenic acid concentration increase. Interestingly, caffeic acid concentration increased when SB and BP1 were exposed to a higher RZT. The concentration of chlorogenic acid in red- or purple-fleshed cultivars has previously been reported to be 2.2 to 3.5 times higher than in yellow- and white-fleshed cultivars [30]. Similar results have been reported by other authors, including [31–35]. Phenolic compounds have a direct function in the type of response given by the plant when exposed to stress, such as from sun exposure or pathogen infection [36]. This is, therefore, a direct indication as to why the increase in RZT has a direct effect on the concentration of the phenolic, as seen in the current study.

### 4.2. Antimycobacterial Activity

Many literature studies have shown that the potato contains a variety of phenolic acids as a means of protection against microbes, viruses, and insects [37]. The mechanism of action of the antimicrobial potential of phenolic compounds has been proposed to be through the destabilization and permeation of the membrane of the microbe, which results in changes to the efflux activity and polarization; in addition, virulence factors, such as hydrophobicity, are directly affected [29]. A study conducted by [38] showed that the phenolic compound myricetin showed low antimycobacterial inhibition against *M. smegmatis* with an MIC value of 32 mg/L. Another study conducted by [39] showed that chlorogenic acid showed no inhibition against *M. smegmatis* with an MIC value of >2500 µg/mL. Moreover, during this study, a direct correlation of phenolic content to antimycobacterial activity could not be shown [39]. Due to the high levels of chlorogenic acid found in both cultivars, it could be concluded that this might be why no inhibitory activity was found against *M. smegmatis*. Further studies based on previous literature could focus on the activity of chlorogenic acid against other Gram-negative and Gram-positive bacteria and microbes.

### 4.3. Antiproliferative Activity

Several studies have concluded that phenolics are important sources of antioxidants and that a diet rich in antioxidants can have remarkable effects on the risk of developing cardiovascular and neurogenerative diseases including cancer and diabetes [40–43]. The anticancer activity of chlorogenic acid was investigated both in vitro and in vivo against the HepG2 cell line and HepG2 xenografts in nude mice. The study concluded that chlorogenic acid in greater concentrations had increased inhibition of HepG2 cells. The xenograft studies on nude mice achieved the same results and showed the suppression of the progression of the HepG2 xenograft [44]. Although there was no effective antiproliferative activity, as seen in the results against the HepG2 cell line by both the cultivars tested, it should be noted that the increase in chlorogenic acid content in the SB cultivar showed increased antiproliferative activity when compared with BP1. It should also be emphasized that although chlorogenic acid is present in high amounts in both cultivars, it is not the only phenolic compound, or compound in general, that is present, and the synergistic effects of all compounds in *Solanum tuberosum* L. should be noted when looking at antiproliferative activity.

## 5. Conclusions

The pigmented potato tubers' antioxidant capacity (through the presence of ferulic, chlorogenic, and caffeic acids), antiproliferative activity, and antimycobacterial activity are cultivar specific. In our study, increasing the RZT had a significant effect on caffeic and chlorogenic acid in the pigmented cultivar SB. The same effect was reported in the antiproliferative study. Our results may offer the opportunity to test the same and other cultivars of *Solanum tuberosum* against other cancer cell lines. These findings are of interest because they increase the availability of information on the experimental investigations of different cultivars found within Southern Africa.

**Author Contributions:** Conceptualization, L.K., N.L., O.O.O. and H.W.; methodology, L.K., O.O.O. and N.L.; software, validation, and formal analysis, A.-M.R.; investigation, H.W. and A.-M.R.; resources, L.K., O.O.O. and N.L.; data curation, A.-M.R. and H.W.; writing—original draft preparation, H.W.; writing—review and editing, C.B.; supervision, L.K., O.O.O., N.L. and C.B.; project administration, L.K.; funding acquisition, L.K., O.O.O. and N.L. All authors have read and agreed to the published version of the manuscript.

**Funding:** This research received no external funding.

**Institutional Review Board Statement:** Not applicable.

**Informed Consent Statement:** Not applicable.

**Data Availability Statement:** Not applicable.

**Conflicts of Interest:** The authors declare no conflict of interest.

## References

1. Birch, P.R.J.; Bryan, G.; Fenton, B.; Gilroy, E.M.; Hein, I.; Jones, J.T.; Prashar, A.; Taylor, M.A.; Torrance, L.; Troth, I.K. Crops that feed the world 8: Potato: Are the trends of increased global production sustainable? *Food Secur.* **2012**, *4*, 477–508. [CrossRef]
2. FAO. 2021a. Potato Production Worldwide from 2002 to 2019 (in Million Metric Tons). Statista Inc. Available online: https://www-statista-com.libproxy.cput.ac.za/statistics/382174/global-potato-production/ (accessed on 9 June 2021).
3. FAO. 2021b. Potato Production Worldwide in 2019, by Leading Country (in Million Metric Tons). Statista Inc. Available online: https://www-statista-com.libproxy.cput.ac.za/statistics/382192/global-potato-production-by-country/ (accessed on 9 June 2021).
4. POTATOPRO. 2021. South Africa Potato Statistics. Available online: https://www.potatopro.com/south-africa/potato-statistics (accessed on 9 June 2021).
5. Sonia, N.S.; Mini, C.; Geethalekshmi, P.R. Vegetable peels as natural antioxidants for processed foods—A review. *Agric. Rev.* **2016**, *37*, 35–41. [CrossRef]
6. Tiwari, B.K.; Valdramidis, V.; Donnell, C.P.O.; Muthukumarappan, K.; Bourke, P.; Cullen, P.J. Application of Natural Antimicrobials for Food Preservation. *J. Agric. Food Chem.* **2009**, *57*, 5987–6000. [CrossRef]
7. Pezeshk, S.; Ojagh, S.; Alishahi, A. Effect of plant antioxidant and antimicrobial compounds on the shelf-life of seafood—A review. *Czech J. Food Sci.* **2016**, *33*, 195–203. [CrossRef]
8. Orsted, H.; Keast, D.; Forest-Lalande, L.; Mégie, M.F. Basic principles of wound healing an understanding of the basic physiology of wound healing provides. *Wound Care Can.* **2004**, *9*, 1–5.
9. Dudek, N.; Maria, M.D.; Horta, P.; Hermes, D.; Sanches, C.C.; Modolo, L.V.; Lima, E.N. In vivo wound healing and antiulcer properties of white sweet potato (Ipomoea batatas). *J. Adv. Res.* **2013**, *4*, 411–415.
10. Ginzberg, I.; Tokuhisa, J.G.; Veilleux, R.E. Potato steroidal Glycoalkaloids: Biosynthesis and genetic manipulation. *LifeEarth Health Sci.* **2009**, *52*, 1–15. [CrossRef]
11. McCue, K.F. *Potato glycoalkaloids, Past Present and Future E-lycotetraose*; Harvard University: Chicago, IL, USA, 2009.
12. Witbooi, H.; Kambizi, L.; Oguntibeju, O. An alternative health crop for South Africa: Purple potato mini tuber production as affected by water and nutrient stress. *Afr. J. Food Agric. Nutr. Dev.* **2020**, *20*, 16818–16831. [CrossRef]

13. Friedman, M. Chemistry, Biochemistry, and Dietary Role of Potato Polyphenols. A Review. *J. Agric. Food Chem.* **1997**, *45*, 1523–1540. [CrossRef]

14. Riciputi, Y.; Diaz-de-Derio, E.; Akyol, H.; Capanoglu, E.; Cerretani, L.; Caboni, M.F.; Verardo, V. Establishment of ultrasound assisted extraction of phenolic compounds from industrial potato by-products using response surface methodology. *Food Chem.* **2018**, *269*, 258–263. [CrossRef]

15. Boker, L.K.; Van der Schouw, Y.T.; De Kleijn, M.J.; Jacques, P.F.; Grobbee, D.E.; Peters, P.H. Intake of dietary phytoestrogens by Dutch women. *J. Nutr.* **2002**, *132*, 1319–1328. [CrossRef] [PubMed]

16. Dragsted, L.O.; Strube, M.; Leth, T. Dietary levels of plant phenols and other non-nutritive components: Could they prevent cancer? *Eur. J. Cancer Prev.* **1997**, *6*, 522–528. [CrossRef]

17. Hertog, M.G.; Bueno-De-Mesquita, H.B.; Fehily, A.M.; Sweetnam, P.M.; Elwood, P.C.; Kromhout, D. Fruit and vegetable consumption and cancer mortality in the Caerphilly Study. *Cancer Epidemiol. Biomark. Prev.* **1996**, *5*, 673–677.

18. Knekt, P.; Kumpulainen, J.; Jarvinen, R.; Rissanen, H.; Heliovaara, M.; Reunanen, A.; Hakulinen, T.; Aromaa, A. Flavanoid intake and risk of chronic diseases. *Am. J. Clin. Nutr.* **2002**, *76*, 560–568. [CrossRef]

19. Hayashi, K.; Hibasami, H.; Murakami, T.; Terahara, N.; Mori, M.; Tsukui, A. Induction of Apoptosis in Cultured Human Stomach Cancer Cells by Potato Anthocyanins and Its Inhibitory Effects on Growth of Stomach Cancer in Mice. *Food Sci. Technol. Res.* **2006**, *12*, 22–26. [CrossRef]

20. Reddivari, L.; Hale, A.L.; Miller, J.C., Jr. Determination of phenolic content, composition and their contribution to antioxidant activity in specialty potato selections. *Am. J. Potato Res.* **2007**, *84*, 275–282. [CrossRef]

21. Madiwale, G.P.; Reddivari, L.; Holm, D.G.; Vanamala, J. Storage elevates phenolic content and antioxidant activity but suppresses antiproliferative and pro-apoptotic properties of colored-flesh potatoes against human colon cancer cell lines. *J. Agric. Food Chem.* **2011**, *59*, 8155–8166. [CrossRef]

22. Wang, Q.; Chen, Q.; He, M.; Mir, P.; Su, J.; Yang, Q. Inhibitory effect of antioxidant extracts from various potatoes on the proliferation of human colon and liver cancer cells. *Nutr. Cancer* **2011**, *63*, 1044–1052. [CrossRef]

23. Friedman, M.; Lee, K.-R.; Kim, H.-J.; Lee, I.-S.; Kozukue, N. Anticarcinogenic Effects of Glycoalkaloids from Potatoes against Human Cervical, Liver, Lymphoma, and Stomach Cancer Cells. *J. Agric. Food Chem.* **2005**, *53*, 6162–6169. [CrossRef]

24. Lee, K.-R.; Kozukue, N.; Han, J.-S.; Park, J.-H.; Chang, E.-Y.; Baek, E.-J.; Chang, J.-S.; Friedman, M. Glycoalkaloids and Metabolites Inhibit the Growth of Human Colon (HT29) and Liver (HepG2) Cancer Cells. *J. Agric. Food Chem.* **2004**, *52*, 2832–2839. [CrossRef]

25. Witbooi, H.; Bvenura, C.; Oguntibeju, O.O.; Kambizi, L. The role of root zone temperature on physiological and phytochemical compositions of some pigmented potato (*Solanum tuberosum* L.) cultivars. *Cogent Food Agric.* **2021**, *7*, 1905300. [CrossRef]

26. Lall, N.; Henley-Smith, C.J.; De Canha, M.N.; Oosthuizen, C.B.; Berrington, D. Viability Reagent, PrestoBlue, in Comparison with Other Available Reagents, Utilized in Cytotoxicity and Antimicrobial Assays. *Int. J. Microbiol.* **2013**, *2013*, 420601. [CrossRef]

27. Berrington, D.; Lall, N. Anticancer Activity of Certain Herbs and Spices on the Cervical Epithelial Carcinoma (HeLa) Cell Line. *Evid.-Based Complement. Altern. Med.* **2012**, *2012*, 564927. [CrossRef] [PubMed]

28. Lewis, C.E.; Walker, J.R.L.; Lancaster, J.E.; Sutton, K.H. Determination of anthocyanins, flavonoids and phenolic acids in potatoes. I: Coloured cultivars of *Solanum tuberosum* L. *J. Sci. Food Agric.* **1998**, *77*, 45–57. [CrossRef]

29. Akyol, H.; Riciputi, Y.; Capanoglu, E.; Caboni, M.F.; Verardo, V. Phenolic compounds in the potato and its by-products: An overview. *Int. J. Mol. Sci.* **2016**, *27*, 835. [CrossRef] [PubMed]

30. Ezekiel, R.; Singh, N.; Sharma, S.; Kaur, A. Beneficial phytochemicals in potato—A review. *Food Res. Int.* **2013**, *50*, 487–496. [CrossRef]

31. Furrer, A.; Cladis, D.; Kurilich, A.; Manoharan, R.; Ferruzzi, M.G. Changes in phenolic content of commercial potato varieties through industrial processing and fresh preparation. *Food Chem.* **2017**, *218*, 47–55. [CrossRef]

32. Külen, O.; Stushnoff, C.; Holm, D.G. Effect of cold storage on total phenolics content, antioxidant activity and vitamin C level of selected potato clones. *J. Sci. Food Agric.* **2013**, *93*, 2437–2444. [CrossRef]

33. Lachman, J.; Hamouz, K.; Musilová, J.; Hejtmánková, K.; Kotiková, Z.; Pazderů, K.; Domkárová, J.; Pivec, J.; Cimr, J. Effect of peeling and three cooking methods on the content of selected phytochemicals in potato tubers with various colour flesh. *Food Chem.* **2013**, *138*, 1189–1197. [CrossRef]

34. Stushnoff, C.; Holm, D.; Thompson, M.D.; Jiang, W.; Thompson, H.J.; Joyce, N.; Wilson, P. Antioxidant Properties of Cultivars and Selections from the Colorado Potato Breeding Program. *Am. J. Potato Res.* **2008**, *85*, 267–276. [CrossRef]

35. Gibbons, S. Anti-staphylococcal plant natural products. *Nat. Prod. Rep.* **2004**, *21*, 263–277. [CrossRef]

36. Geran, R.I.; Greenberg, N.H.; McDonald, M.M.; Schumacher, A.M.; Abbott, B.J. Protocols for screening chemical agents and natural products against animal tumour and other biological systems. *Cancer Chemother. Rep.* **1972**, *3*, 17–19.

37. Ngadze, E.; Coutinho, T.; Icishahayo, D.; van der Waals, J. Effect of calcium soil amendments on phenolic compounds and soft rot resistance in potato tubers. *Crop. Prot.* **2014**, *62*, 40–45. [CrossRef]

38. Lima, M.; de Sousa, C.P.; Fernandez-Prada, C.; Harel, J.; Dubreuil, J.; de Souza, E. A review of the current evidence of fruit phenolic compounds as potential antimicrobials against pathogenic bacteria. *Microb. Pathog.* **2019**, *130*, 259–270. [CrossRef] [PubMed]

39. Lechner, D.; Gibbons, S.; Bucar, F. Plant phenolic compounds as ethidium bromide efflux inhibitors in Myco-bacterium smegmatis. *J. Antimicrob. Chemother.* **2008**, *62*, 345–348. [CrossRef] [PubMed]

40. Da Cruz, R.C.; Agertt, V.; Boligon, A.A.; Janovik, V.; de Campos, M.M.A.; Guillaume, D.; Athayde, M.L. In vitro antimycobacterial activity and HPLC–DAD screening of phenolics from *Ficus benjamina* L. and *Ficus luschnathiana* (Miq.) Miq. Leaves. *Nat. Prod. Res.* **2012**, *26*, 2251–2254. [CrossRef] [PubMed]

41. Fernandez-Panchon, M.S.; Villano, D.; Troncoso, A.M.; Garcia-Parrilla, M.C. Antioxidant activity of phenolic compounds: From in vitro results to in vivo evidence. *Crit. Rev. Food Sci. Nutr.* **2008**, *48*, 649–671. [CrossRef]

42. Hollman, P.; Katan, M. Dietary Flavonoids: Intake, Health Effects and Bioavailability. *Food Chem. Toxicol.* **1999**, *37*, 937–942. [CrossRef]

43. Sato, M.; Ramarathnam, N.; Suzuki, Y.; Ohkubo, T.; Takeuchi, M.; Ochi, H. Varietal differences in the phenolic content and superoxide radical scavenging potential of red wines from different sources. *J. Agric. Food Chem.* **1996**, *44*, 37–41. [CrossRef]

44. Scalbert, A.; Manach, C.; Morand, C.; Rémésy, C.; Jiménez, L. Dietary Polyphenols and the Prevention of Diseases. *Crit. Rev. Food Sci. Nutr.* **2005**, *45*, 287–306. [CrossRef]

MDPI
St. Alban-Anlage 66
4052 Basel
Switzerland
Tel. +41 61 683 77 34
Fax +41 61 302 89 18
www.mdpi.com

*Applied Sciences* Editorial Office
E-mail: applsci@mdpi.com
www.mdpi.com/journal/applsci

www.ingramcontent.com/pod-product-compliance
Lightning Source LLC
LaVergne TN
LVHW071000170726
843515LV00006B/1034